Positional Astronomy

POSITIONAL ASTRONOMY

D. McNALLY
M.Sc., Ph.D., F.R.A.S.

A HALSTED PRESS BOOK

JOHN WILEY & SONS
NEW YORK

Published in the U.S.A. by Halsted Press, a Division of
John Wiley & Sons, Inc., New York

Library of Congress Catalog Card No. 74 4819

Printed and bound in Great Britain

ISBN: 0 470 589809

To my students in appreciation of their fortitude.

Contents

Preface

The field of astrometry is of great refinement both of concept and technique. It might therefore cause surprise that one who has made no contribution to astrometry should have the temerity to write a book on Positional Astronomy. The reason is that as I have been giving elementary lectures to first year students (studying for a B.Sc. in Astronomy) on this subject for some years, I appreciate their problems of assembling the necessary information from diverse sources. Much of the astrometric literature is at such a level and in such detail that a first year undergraduate is unable to find his way through it. Again, many modern astrometric methods presume easy access to computer controller measuring engines and a large computer system. An undergraduate student does not have these facilities.

I have therefore written this book at a level appropriate to that of an undergraduate in the first year of a British University. I have assumed that the student has not previously studied astronomy in depth. I also assume that at this stage the student is not wishing to study astrometry as a specialist science, but wishes to obtain a working knowledge of the principles and terminology of positional astronomy to the extent that he can use the data of the Astronomical Ephemeris with some appreciation of what underlies the formulae and numbers he is using. I have made no attempt to write an astrometric text—that is not a subject for me but for specialists in astrometry.

To write a text book for undergraduate use imposes certain constraints. Being at first year level, I have avoided extensive use of vector and matrix notation. Since both notations are being learnt at this stage and since I wished to keep the student in contact with what was happening on the sky, I have retained the notation and concepts of spherical trigonometry. This means that, apart from Taylor's Theorem, the basic mathematical background assumed is that acquired prior to university entrance. Since most students find positional concepts difficult I have kept the mathematical demands as small as possible. I have derived formulae in a way that makes the transition to either vector or matrix notation straightforward. The aim of the book is to give the student a feeling for the conceptual background which underlies the Astronomical Ephemeris, and I have derived formulae which can be used with data available from that source. This information is at its most useful in deriving first order corrections and I have restricted myself to derivations of formulae of that accuracy.

The chapters which proved the most difficult to write were those on orbital motion. There is an abundance of literature on orbit theory, but there is a gap between the orbit theory of university applied mathematics and the orbit theory of celestial mechanics. I have therefore attempted to fill that gap. While I have tried to give sufficient information on how to calculate geocentric positions of planets on the sky and a method of determining the elements of an orbit in the solar system, I have been extremely brief on the subject of binary stars. For this latter topic, there is an extensive literature. However, even in the case of visual and spectroscopic binary star orbits I have tried to give methods which can be regarded as quick laboratory methods for obtaining preliminary information.

My aim in writing the book, therefore, is to give an account of the principles underlying the definitions, methods and corrections in positional astronomy. The formulae should give first order accuracy for most everyday astronomical operations. I have not attempted to give an account with full astrometric precision. I would hope that a student could use this book to make a start on a preliminary calculation in any field covered by the book but any student specialising in a given part of the subject would, of course, work from the specialist literature and not this book.

I have found that in writing the book, it was best to approach the subject from a theoretical point of view. The derivation of formulae from the definitions is not obscured by observational detail. I have made no attempt to discuss practical observational matters. I have adopted the view of a user of astrometric information rather in parallel with the view of a user of a computer who wants to use a programming language and not the details of machine code and construction. This has its value since the majority of astronomers only want to know how to calculate a relevant parameter with some idea of the background.

I have resisted the temptation to put in examples. In my course at University College London, I set examples which extend the students' knowledge of positional astronomy and computer technique. I find that the effort of breaking the formulae into a suitable form for programming makes the students very much more aware of the nature of the problem than if they had only experimented with algebra. It is instructive to the student to see how actual values and numerological procedures affect a calculation. The solution of Kepler's Equation is a satisfactory way of introducing iterative techniques. Indeed there is a great deal of value in exposing first year students to the discipline of computation and numerology after a school career devoted almost exclusively to algebra. Other users of the book can supply examples best suited to their own courses.

I hope that this book will be of value to students in assisting them find their way into positional astronomy. For many younger astronomers positional astronomy is a chore. This is rather sad when astrometry can be a basic tool of great refinement for astronomical research. I hope many will come to recognise that positional astronomy can be a subject of great fascination and some may carry the subject to a more professional level. But if I have succeeded only in making life just a little easier for students of astronomy, the risk (undoubted) of arousing the wrath of my astrometric colleagues will have been worth it.

I owe a considerable debt to W. M. Smart's fine pedagogical text, *Text Book on Spherical Astronomy* (Cambridge University Press), for the direction it gives to the organisation of the material. I have tried to follow the spirit of that work in aiming to provide a treatment of the essentials of a variety of techniques which are founded in the astronomy of position. I have also profited from the admirable text *Spherical Astronomy* (Academic Press) by E. W. Woolard and G. M. Clemence. The Astronomical Ephemeris and the Explanatory Supplement to the Astronomical Ephemeris (Government Bookshop—Her Majesty's Stationary Office) have been of great value in suggesting the form in which formulae should be left. As far as is possible I have endeavoured to use a notation which is consistent with that used in the Astronomical Ephemeris, but this has not always been achieved. I have also made use of Lecture Notes prepared by R. Coutrez (Free University of Brussels) and *Principles of Astrometry* (Freeman) by P. van de Kamp. A useful introductory text on celestial mechanics is *Astrodynamics* (Macmillan) by A. E. Roy. For binary stars the review articles by W. H. van den Bos, R. M. Petrie and J. B. Irwin in *Stars and Stellar Systems Vol. II—Astronomical Techniques* (eds. G. P. Kuiper and B. M. Middlehurst, University of Chicago Press) have been very valuable. Both text and reviews lead into the previous literature of celestial mechanics. For recent literature the *Astronomical Journal* is a useful source to begin a survey. A useful compilation of astronomical data is *Astrophysical Quantities* by C. W. Allen (3rd Edition—Athlone Press).

While acknowledging my debt to the above authors, I must also acknowledge my indebtedness to my colleagues, Profs. C. W. Allen, A. H. Cook, R. H. Garstang, P. A. Sweet, Drs. T. Kiang, D. H. Sadler and Mr. C. A. Murray for clarification of many topics over a period of years. I am also grateful to Dr. E. L. G. Bowell for many discussions, in particular for the treatment of the phases of planets and for drawing my attention to the simplicity of Merton's treatment of the derivation of the elements of a planetary orbit. I wish to thank Mr. G. Cooke for his efforts in the preparation of diagrams and for reading much of the manuscript. He devised the necessary computer programs to draw representations of the celestial sphere. My thanks are also due to Mr. D. C. Underwood for his assistance to Mr. Cooke. I am indebted to the Controller of Her Majesty's Stationary Office for permission to reproduce Figs. 5.2, 8.15 and Tables 6.1, 6.2, 7.1.

I should also like to express my gratitude to Mrs. M. K. Beacham and Mrs. J. G. Cole who undertook with great grace the typing work involved in the preparation of the manuscript. Finally, but by no means least, I should like to thank my family for putting up with "the book" for so long, so cheerfully.

D. McNally

1974 June
University of London Observatory
Mill Hill Park
London NW7 2QS

Chapter I

Introduction

Astrometry is the science of measuring the angular separations of celestial objects.

The first attempts to measure the angular separations of stars were made about 500 BC in late Babylonian times. At the present time astrometry is the most precise section of astronomy. Astrometry is now concerned with the determination of fundamental frames of reference, the precise measurement of time for astronomical purposes, the refinement of the corrections required for precession and nutation, the determination of distance scales and motions within the Galaxy. By its nature astrometry works at the limit of observational technique. The development of computer controlled measuring machines of great precision has opened up a new range of astrometric possibilities which, like so many astrometric advances in the past, will be of immense importance in the furtherance and stimulation of astronomy.

Astrometry is one branch of fundamental astronomical research. However, the concepts of astrometry impinge in a direct manner on the everyday work of the astronomer engaged on other branches of astronomical research. An astronomer must be able to find his way on the sky and have an understanding of some of the phenomena which cause displacements of star position. A full understanding of astrometric techniques is not essential to developing an appreciation of what is involved in such diverse phenomena as refraction in the Earth's atmosphere, parallax, aberration, precession, proper motion, eclipse calculation, and motion under gravity. The aim of this book is to present a straightforward account of such matters leaving formulae at a stage where numerical answers sufficiently accurate for most practical purposes can be obtained. For greater accuracy, specialist discussions must be consulted. It has been assumed throughout that access is available to the Astronomical Ephemeris since that data is of basic importance to expeditious calculation.

The book has been written on the following plan. The formal geometrical definitions are given in Chapter II. In this chapter the necessary spherical trigonometry is developed to allow the handling of the parameters defining a spherical triangle. To complete the Chapter a derivation in terms of rectangular coordinates is given of the formulae necessary to calculate a new position of an object knowing the displacement and the undisplaced position. The form of this derivation allows an easy transformation to vector or matrix formalism. The entire mathematical formalism which is necessary is given in this chapter. The following Chapters, III and IV, are concerned with a formal definition of astronomical coordinate systems. Horizon and Equatorial coordinates are defined from the point of view of observationally realisable operations. Ecliptic and Galactic coordinates are defined in terms of equatorial coordinates. Formulae for making the transition between pairs of coordinate systems are given and the displacement formulae of Chapter II are expressed in terms of equatorial coordinates. Standard coordinates are discussed in view of their historical association with the Astrographic Catalogue. Although rarely used they have some value in a later chapter in deriving the method of dependences. Planetographic and Heliographic coordinates are also formally defined.

Chapter V is concerned with the definition of systems of time measurement. It is important to have a uniform measure of time. Unfortunately the period of the Earth's rotation which determines both sidereal and Universal time varies irregularly. Some of the variations can be predicted, but there remains unpredictable irregularities which can only be detected retrospectively by a comparison of the measures of ephemeris and Universal time. This is a continuing activity but one in which ephemeris time is being replaced by atomic time as a uniform measure of time. Since the aspect of the sky is determined by the rate of the Earth's rotation, astronomical measures of times must be retained and cannot be replaced by a uniform measure of time such as atomic time.

Chapters VI and VII are concerned with the nature of the displacements of star position and with the derivation of formulae whereby a first order correction for each phenomenon may be calculated. The phenomena of refraction in the Earth's atmosphere, parallax and aberration are considered together since these phenomena cause only displacements from the geometrical position of the celestial object. Precession, nutation and proper motion are different in character. Precession and nutation are a means of describing changes in the positions of the fundamental planes defining the coordinate systems. Proper Motion describes the changes of relative configuration caused by intrinsic motions of the stars. An alternative way of looking at these phenomena is the following. Refraction is a property of the structure of the Earth's atmosphere, parallax is a geometrical effect of the Earth's being in different places in space with respect to the Sun at different times of the year. Aberration is a kinematic effect of the Earth's being in motion with respect to the Sun. Precession and nutation are dynamical effects of the gravitational interaction of the Sun, Moon and planets with an Earth that (a) departs from perfect sphericity and (b) has a density distribution which suffers small time dependent variations from a mean. In view of the changes in density distribution within the Earth, the constants of precession and nutation show small variations which require the utmost refinement of astrometric technique to detect.

The methods of predicting solar and lunar eclipses, transits of Mercury and occultations of stars are given in Chapter VIII. While the information given in the Astronomical Ephemeris is usually adequate for most purposes, the derivation of the formulae indicates how the eclipse predictions are made and, in particular, allows the computation of preliminary times for conditions other than those specifically catered for in the Astronomical Ephemeris e.g. observation of an eclipse from rocket or satellite. An approximate method of calculating the times of transit in the case of Mercury is given though, in fact, transit calculations could equally well follow the same formulae as the calculation of an annular eclipse of the Sun.

The method of dependences is discussed in Chapter IX where it is shown that, while the method can be derived most easily in terms of standard coordinates, it may be used either with rectangular equatorial coordinates or equatorial coordinates. The method of dependences is not a highly regarded astrometric technique but nevertheless is a very useful laboratory approximate method. By means of simple and direct measurement of a photographic plate, the dependences may be calculated and an approximate position for an object obtained directly in either rectangular equatorial or equatorial coordinates.

Chapters X, XI, XII are concerned with orbital motion. These chapters are elementary in their approach in view of the vast literature on celestial mechanics and binary stars. The customary treatment of orbit theory from an applied mathematics viewpoint usually stops with the determination of the orbit in polar coordinates. This treatment is satisfactory when only a discussion of the dynamical situation is needed but is unsatisfactory if position in the orbit is to be translated into position on the sky. Therefore Kepler's Equation and the mean and eccentric anomalies are introduced in Chapter X. Kepler's Equation is easy to solve numerically and having determined the eccentric anomaly it is a simple matter to calculate position on the sky as set out in Chapter XI. In this chapter a method is given to derive approximate elements of an orbit from three observations. Geocentric planetary phenomena are also briefly discussed. Finally, in Chapter XII a brief discussion is given of binary star orbits. For visual binaries I have retained Kowalsky's method since it is a logical way to proceed in theory, and of course a logical theory is useful for gaining insight. However, Kowalsky's method is unsatisfactory in practice since it loses contact with the observations at the first stage of the analysis. A more practical laboratory method is that based on Thiele–Innes constants. Spectroscopic binary stars are treated to give a rapid means of determining a preliminary orbit and to give an idea of the terminology encountered in the literature. The section on eclipsing binaries is given for completeness. The simplest possible case only is treated since, in many cases, the discussion of eclipsing binaries is given as though the reader had considerable prior knowledge. There is a vast literature on eclipsing binary stars and further details should be sought there.

For the purpose of discussion, all celestial objects (excepting those when the radii of planetary and solar discs are to be determined) will be represented by geometrical points. Such a representation is useful since stars are sufficiently distant for their actual physical size to be irrelevant except in the case of observation using the most specialised interferometric equipment. Since stars are so distant, it is possible to generalise the visual impression of the sky so that the geometrical points representing the stars may be considered to be all at the same distance, distributed over the surface of a unit sphere. It is useful to think of the unit sphere as having infinite radius since problems concerned with the non-coincidence of the observer with the centre of the Earth disappear. The effects of the finite distance to an actual object can be treated through the phenomena of parallax and proper motion as displacements from this idealisation.

Diagrams involving the representation of curved surfaces on a plane surface always present a problem. When such diagrams occur in this book an attempt has been made to give a perspective view that best illustrates the problem under consideration. To this end, realistic values for defining parameters have been used where no conflict is caused with clarity. In some circumstances, such as the annual displacement caused by precession, it has been necessary to amplify parameters as an aid to clarity. The three dimensional celestial sphere is represented diagrammatically by projection onto a plane using orthographic oblique zenithal projection. Each projection was chosen to show the effect to be illustrated to best advantage. As a result of using this technique, the diagrams, produced under computer control, depart somewhat from the usual diagram; for example, the celestial pole is no longer placed on the boundary circle, but on a projection of the appropriate great circle. In the diagrams important great circles and the boundary circle are denoted by thick lines. Parts of circles and lines on the opposite side of the celestial sphere to that nearest the reader are denoted by broken lines and, where two lines or circles cross, the foremost remains unbroken. The result is a diagram which is easier to relate to the actual sky.

Chapter II

Mathematical Concepts, Definitions and Terminology

The fundamental positional measurements of astronomy are the relative angular separations of one star with respect to another. This chapter is concerned with the fundamental definitions of quantities relating to such measurements and the means of manipulating these quantities. The mathematical technique is that of spherical trigonometry. A cartesian method will also be discussed since, although three parameters are necessary to specify the position of a quantity in three dimensions, strictly only two parameters are needed to specify the position of an object on a defined surface. A full three-dimensional treatment therefore must contain a constraint. The justification for using a three-dimensional treatment is that it is useful in the development of formulae which can be used for numerical computation.

2.1 Definitions

It will be assumed that stars may be represented by geometrical points on the surface of a sphere, assumed to be infinitely distant but of unit radius. Three concepts can then be established immediately—the Great Circle, the Small Circle and the Pole of a Circle. In Fig. 2.1, C is the centre of a sphere of unit radius.

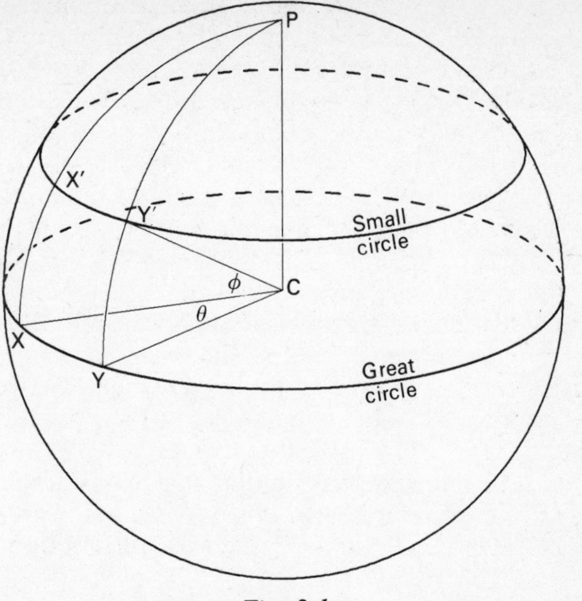

Fig. 2.1

(a) *Great circle.* A *great circle* is an arc on the surface of a unit sphere such that the plane defined by the arc *contains* the centre of the sphere. (The curve labelled Great Circle in Fig. 2.1 defines a plane containing the centre C.)

(b) *Small circle.* A *small circle* is an arc on the surface of a unit sphere such that the plane defined by the arc *does not contain* the centre of the sphere. (The curve labelled Small Circle in Fig. 2.1 defines a plane which does not contain the centre C.)

(c) *Pole of a circle.* The diameter perpendicular to the plane defined by any great circle (or any parallel small circle) cuts the unit sphere in two points called the poles of the circle. (In Fig. 2.1, P is an example of a pole.)

From these definitions it is clear that the great circles have a special significance. To every great circle there is an infinite number of parallel small circles. The length of the arc XY of the great circle is

$$XY = CX.\,\theta = CY.\,\theta = \theta, \tag{2.1}$$

where $\theta = X\hat{C}Y$ and $CX = CY = 1$ (since we have defined the sphere to be of unit radius). Therefore the lengths of arcs of great circles on a unit sphere may be expressed in terms of the angle subtended at the centre C of the sphere. Lengths of great circles only may be expressed in this way. (Lengths of arcs of small circles require the definition of an angle giving its elevation above the parallel great circle—clearly in Fig. 2.1 the length of the arc X′Y′ is $\theta \cos \phi$ where X′ lies on the great circle PX and Y′ lies on the great circle PY and ϕ is the angle $X'\hat{C}X = Y'\hat{C}Y$.) Therefore it is convenient to consider only points joined by arcs of great circles.

(d) *Spherical angle.* A spherical angle can only be defined between intersecting great circles. The *spherical angle* between two great circles is defined to be the angle between the respective tangents to the great circles at their point of intersection.

In Fig. 2.2 a unit sphere is centred at C. Consider the plane defined by the great circle AB. P is the pole of the great circle AB. The great circles PA and PB intersect at P. Let the tangents

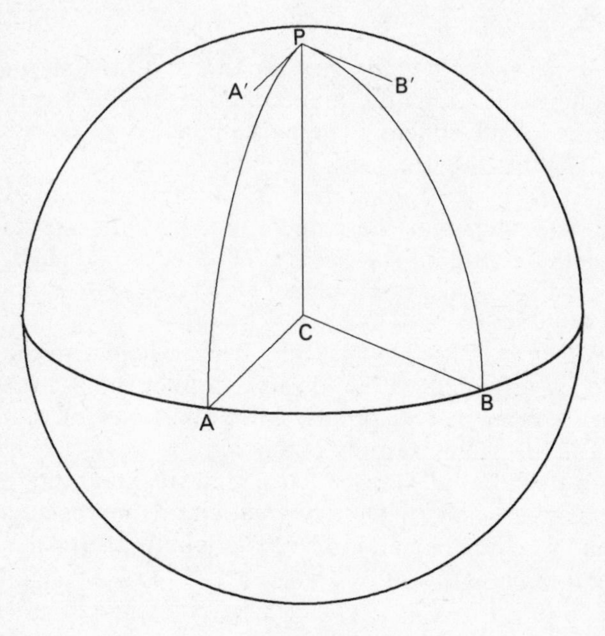

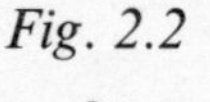
Fig. 2.2

at P to the great circles PA and PB be PA′ and PB′ respectively. Clearly PA′ is parallel to CA and PB′ is parallel to CB. The spherical angle between the great circles PA and PB at P is therefore $A'\hat{P}B' = A\hat{C}B$.

(e) *Spherical triangle.* Consider three points, A, B, C on the surface of a unit sphere (see Fig. 2.3). If the points are connected in pairs by arcs of great circle the figure so formed is called a *spherical triangle.* In Fig. 2.3 the spherical triangle is formed by the great circles AB, BC, CA. Since a sphere may be bisected so that any three points all lie on the same hemisphere, the length of a side of a spherical triangle on a unit sphere will not exceed 180°. A spherical triangle *cannot* be formed if any side is part of a small circle.

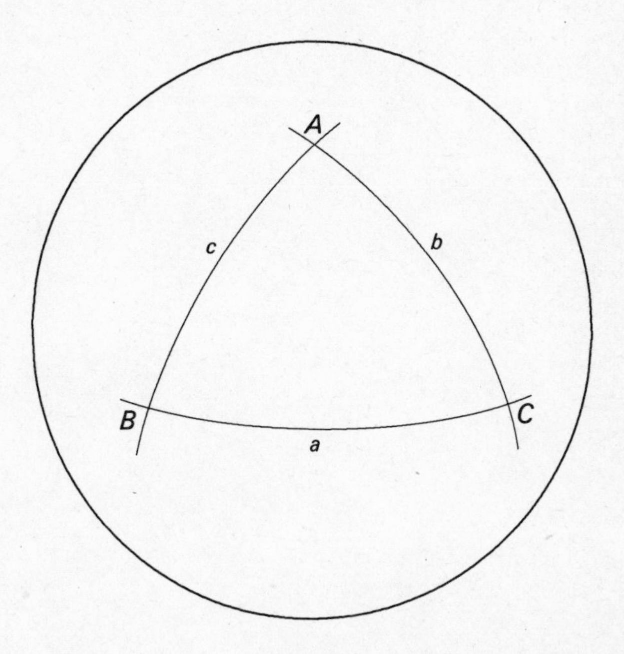

Fig. 2.3

It is conventional to denote the magnitudes of the angles of intersection between great circles and the corresponding vertices by upper case letters (e.g. A, B, C; see Fig. 2.3) and the lengths of the sides opposite the vertices by lower case letters (e.g. *a*, *b*, *c*; see Fig. 2.3). Therefore if the vertex formed by the intersection of the arcs AB, AC is denoted by A, the spherical angle between the great circles AB, AC is denoted by *A* while the opposite side (the arc of great circle BC) is denoted by *a*.

(f) *The polar triangle.* For any spherical triangle ABC it is possible to construct a polar triangle A′B′C′. Since AB (see Fig. 2.4) is a great circle, it will have two poles. Denote that pole of AB on the same side of AB as C by C′. A′ and B′ may be defined similarly. The spherical triangle A′B′C′ is the *polar triangle* of the spherical triangle ABC.

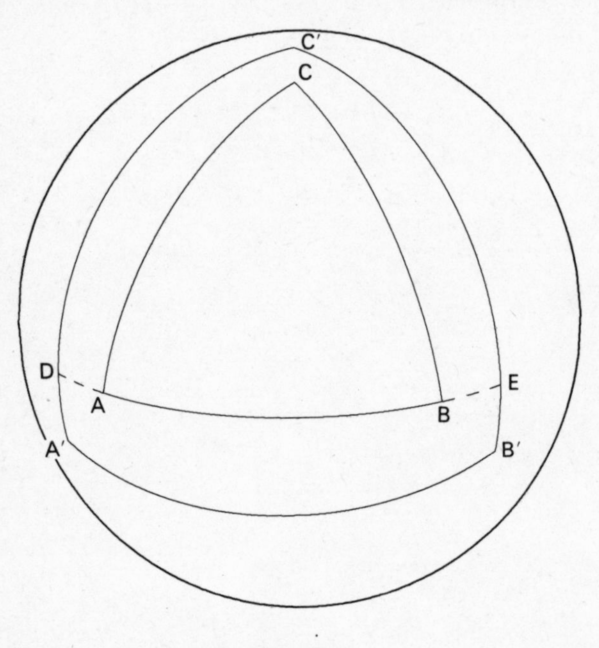

Fig. 2.4

Produce the great circle AB to cut the great circle C′A′ in D and C′B′ in E. Since A′ is the pole of the great circle CB any point on CB is 90° from A′. Similarly B′ is the pole of the great circle CA and is 90° from any point on CA. Hence A′ and B′ are 90° from C and so C is the pole of the great circle A′B′. Similarly A is the pole of the great circle B′C′ and B is the pole of the great circle A′C′. ABC is the polar triangle of the spherical triangle A′B′C′. Hence,

$$\mathrm{BD} = \mathrm{AE} = 90°, \qquad \mathrm{AB} = c,$$

and

$$\begin{aligned} \mathrm{DE} &= \mathrm{DA} + \mathrm{AB} + \mathrm{BE} \\ &= (90° - c) + c + (90° - c) \\ &= 180° - c = C', \end{aligned} \tag{2.2a}$$

since C′ is the pole of the great circle DABE. Similarly

$$A' = 180° - a, \tag{2.2b}$$

$$B' = 180° - b. \tag{2.2c}$$

Again,

$$\mathrm{A'\hat{C}B} = \mathrm{A\hat{C}B'} = 90°, \mathrm{A\hat{C}B} = C,$$

and

$$\begin{aligned} \mathrm{A'\hat{C}B} &= \mathrm{A'\hat{C}A} + \mathrm{A\hat{C}B} + \mathrm{B\hat{C}B'} \\ &= (90° - C) + C + (90° - C) \\ &= 180 - C = c', \end{aligned} \tag{2.3a}$$

since C is the pole of the great circle A′B′. Similarly

$$a' = 180° - A, \tag{2.3b}$$

$$b' = 180° - B. \tag{2.3c}$$

2.2 The Formulae of Spherical Trigonometry

Just as for plane triangles, trigonometric formulae may be established connecting the lengths of sides and spherical angles for spherical triangles. These formulae are as follows.

(i) The Cosine Rule:

$$\cos a = \cos b \cos c + \sin b \sin c \cos A. \tag{2.4}$$

(ii) The Transposed Cosine Rule:

$$\sin a \cos B = \cos b \sin c - \sin b \cos c \cos A. \tag{2.5}$$

(iii) The Sine Rule:

$$\frac{\sin A}{\sin a} = \frac{\sin B}{\sin b} = \frac{\sin C}{\sin c}. \tag{2.6}$$

In the cases of the Cosine Rule (eqn. (2.4)) or Transposed Cosine Rule (eqn. (2.6)) two further formulations of each rule may be obtained by changing the parameters cyclically. From these three fundamental equations a fourth may be derived—the Four Parts Rule.

(iv) The Four Parts Rule:

$$\cos a \cos C = \sin a \cot b - \cot B \sin C. \tag{2.7}$$

Two further formulations of the Four Parts Rule may be obtained by cyclic permutation of the parameters.

These four rules permit the manipulation of the parameters (namely A, B, C, a, b, c) defining a spherical triangle. The Cosine, Transposed Cosine and Sine Rules are the more generally useful in deriving formulae. The advantage of the Four Parts Rule lies in applications to situations where one spherical angle is a right angle.

For some purposes it is advantageous to use formulae which involve spherical angles rather than lengths of arc of great circle as primary parameters. Such formulae may be derived by replacing the parameters of eqns. (2.4)–(2.7) by their polar equivalents and evaluating using eqns. (2.2) and (2.3).

(v) Polar Cosine Rule:

$$\cos a' = \cos b' \cos c' + \sin b' \sin c' \cos A',$$

i.e.

$$\begin{aligned}\cos(180° - A) = {} & \cos(180° - B)\cos(180° - C) \\ & + \sin(180° - B)\sin(180° - C) \\ & \times \cos(180° - a),\end{aligned}$$

or

$$\cos A = -\cos B \cos C + \sin B \sin C \cos a. \tag{2.8}$$

(vi) Polar Transposed Cosine Rule:

$$\sin a' \cos B' = \cos b' \sin c' - \sin b' \cos c' \cos A',$$

or

$$\sin A \cos b = \cos B \sin C + \sin B \cos C \cos a. \qquad (2.9)$$

The Polar Sine and Four Parts Rules are identical with their non-polar counterparts. Polar formulae are frequently of use.

2.3 Derivation of the Cosine, Transposed Cosine, Sine and Four Parts Rules

The formulae of spherical trigonometry may be derived by considering a special spherical triangle with respect to two right-handed rectangular frames of reference, C(x, y, z), C(x', y', z') (see Fig. 2.5), where C is the centre of the unit sphere. The spherical triangle PQR is chosen so that its vertex P lies at the pole of the great circle defined by the plane containing the x- and y-axes, i.e. P lies on the z-axis. The

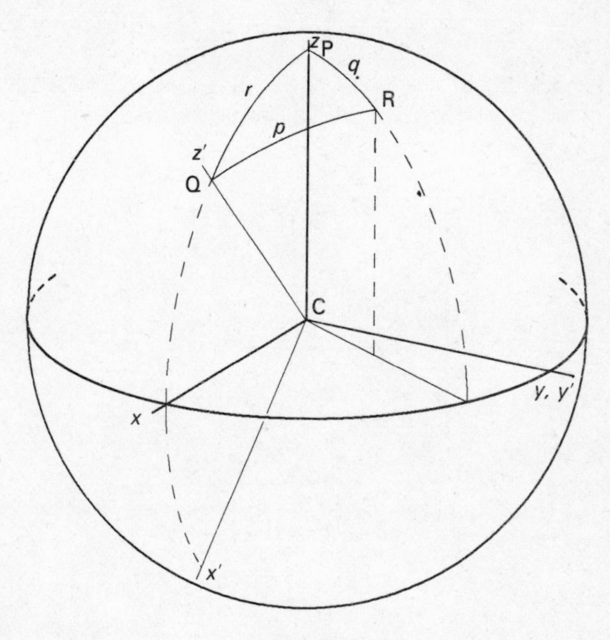

Fig. 2.5

vertex Q of the spherical triangle is chosen to lie in the plane defined by the x- and z-axes. The coordinate system C(x', y', z') is chosen such that the z'-axis passes through Q and the y'-axis is identical with the y-axis. The x'-axis therefore lies in the plane defined by the x-, z- and z'-axes. Let the coordinates of R be (x, y, z), (x', y', z') with respect to C(x, y, z), C(x', y', z') respectively. Then

$$\begin{aligned} x &= \text{CR} \cos(90° - q) \cos P = \sin q \cos P, \\ y &= \text{CR} \cos(90° - q) \sin P = \sin q \sin P, \\ z &= \text{CR} \sin(90° - q) = \cos q, \end{aligned} \qquad (2.10)$$

and

$$\begin{aligned} x' &= \text{CR} \cos(90° - p) \cos(180° - Q) \\ &= -\sin p \cos Q, \\ y' &= \text{CR} \cos(90° - p) \sin(180° - Q) \\ &= +\sin p \sin Q, \\ z' &= \text{CR} \sin(90° - p) = \cos p. \end{aligned} \qquad (2.11)$$

The relationship of the x-, z- and x'-, z'-axes can be seen from Fig. 2.6. The coordinates of D are respectively (x, z), (x', z').

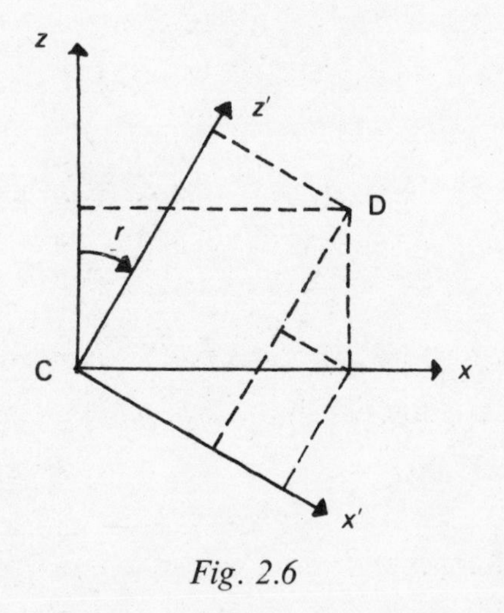

Fig. 2.6

2.3 *Derivation of the Formulae*

The inclination of the z' axis to the z axis is r. Then

$$\begin{aligned} x' &= x \cos r - z \sin r, \\ z' &= x \sin r + z \cos r. \end{aligned} \qquad (2.12)$$

Since $y' \equiv y$, evaluation of x', y', z', x, y, z using eqns. (2.10), (2.11) and (2.12) gives

$$\begin{aligned} -\sin p \cos Q &= \sin q \cos r \cos P - \cos q \sin r, \\ \sin q \sin P &= \sin p \sin Q, \\ \cos p &= \sin q \sin r \cos P + \cos q \cos r, \end{aligned} \qquad (2.13)$$

or, with rearrangement,

$$\sin p \cos Q = \cos q \sin r - \sin q \cos r \cos P$$

– Transposed Cosine Rule, (2.5)

$$\frac{\sin P}{\sin p} = \frac{\sin Q}{\sin q} \text{ – Sine Rule,} \qquad (2.6)$$

$$\cos p = \cos q \cos r + \sin q \sin r \cos P$$

– Cosine Rule. (2.4)

The Four Parts Rule may be derived using two applications of the Cosine Rule and the Sine Rule. From the Cosine Rule,

$$\cos q = \cos r \cos p + \sin r \sin p \cos Q,$$

$$\cos r = \cos p \cos q + \sin p \sin q \cos R.$$

So that substituting for $\cos r$ gives

$$\cos q = \cos^2 p \cos q + \sin p \cos p \sin q \cos R$$
$$+ \sin r \sin p \cos Q,$$

i.e.

$$\cos q \sin p = \cos p \sin q \cos R + \sin r \cos Q.$$

Division by $\sin q$ and use of the Sine Rule gives

$$\cot q \sin p = \cos p \cos R + \sin R \cot Q,$$

i.e.

$$\cos p \cos R = \sin p \cot q - \cot Q \sin R. \qquad (2.7)$$

The formulae in eqns. (2.4), (2.5), (2.6) and (2.7) will be used throughout this book. They will be used to evaluate expressions obtained for cartesian frames of references as well as more generally.

2.4 Rectangular Frames of Reference

To establish the positions of points on the surface of a sphere a frame of reference is necessary. A cartesian right-handed frame of reference will be used and may be established as follows. In Fig. 2.7. the great circle XY defines a plane

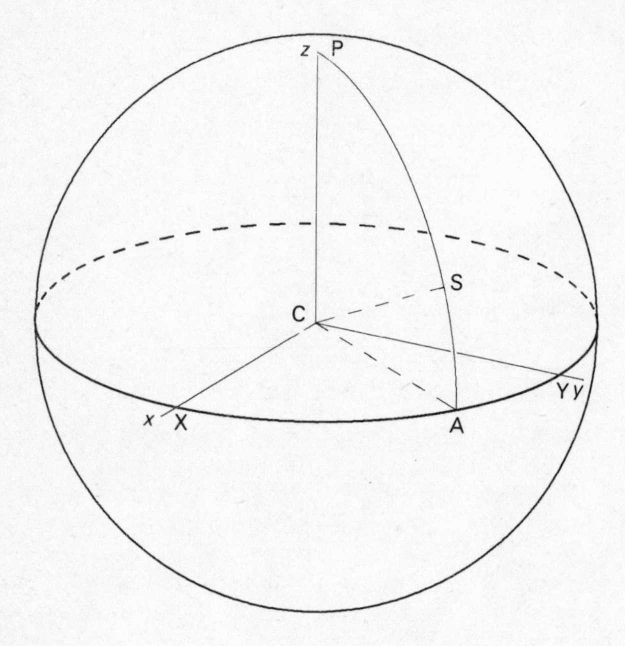

Fig. 2.7

containing the centre C of the unit sphere. The centre of the sphere C is chosen as the origin of coordinates. If P is the pole of the great circle XY then CP is perpendicular to the plane of the great circle XY. Denote by z the axis directed along CP. The x-axis may be chosen along the direction CX, where X is for the time being some arbitrary point in the plane of the great circle XY. Then, to complete the right handed set, the y-axis is directed along CY at right angles to CX in the plane of the great circle XY. (A right handed frame of reference is one such that if the plane of the great circle XY is observed from P an anticlockwise rotation of the x-axis will bring it into alignment with the y-axis.)

A point S on the surface of the sphere has coordinates (x, y, z) with respect to the coordinate axes established above. Clearly for a unit sphere the values of x, y, z are restricted to those satisfying

$$x^2 + y^2 + z^2 = 1. \tag{2.14}$$

It might seem that the use of a three dimensional treatment is unnecessary when all points involved are confined to a surface of a sphere. Indeed, it is clear that it is unnecessary to use three parameters to specify the position of S—two angular parameters only (namely $X\hat{C}A$ and $A\hat{C}S$—see Fig. 2.7) specify the position of S uniquely. However, there are certain advantages in using a three dimensional formulation in later developments of positional theory. These advantages occur when considering the displacement of star positions along great circles.

2.5 A General Displacement of a Star Image

Consider the displacement of a star at S_1 along the great circle S_1S_2 to S_2 (Fig. 2.8). The arc length S_1S_2 will be denoted by s. If the rectangular coordinates of S_1 and S_2 are (x_1, y_1, z_1) and (x_2, y_2, z_1) respectively with respect to $C(x, y, z)$ the problem is to determine (x_2, y_2, z_2) in terms of (x_1, y_1, z_1) and s.

To establish the connection let S_1 have position vector $\boldsymbol{r}_1$ and S_2 have position vector $\boldsymbol{r}_2$ where $\boldsymbol{r}_1, \boldsymbol{r}_2$ are vectors of unit length, and that

$$\boldsymbol{r}_1 = x_1\boldsymbol{i} + y_1\boldsymbol{j} + z_1\boldsymbol{k}, \tag{2.15}$$

$$\boldsymbol{r}_2 = x_2\boldsymbol{i} + y_2\boldsymbol{j} + z_2\boldsymbol{k}, \tag{2.16}$$

where $\boldsymbol{i}, \boldsymbol{j}, \boldsymbol{k}$ are the unit vectors in the directions of the respective axes. Thus

$$\boldsymbol{r}_1 \cdot \boldsymbol{r}_2 = \cos s = x_1x_2 + y_1y_2 + z_1z_2, \tag{2.17}$$

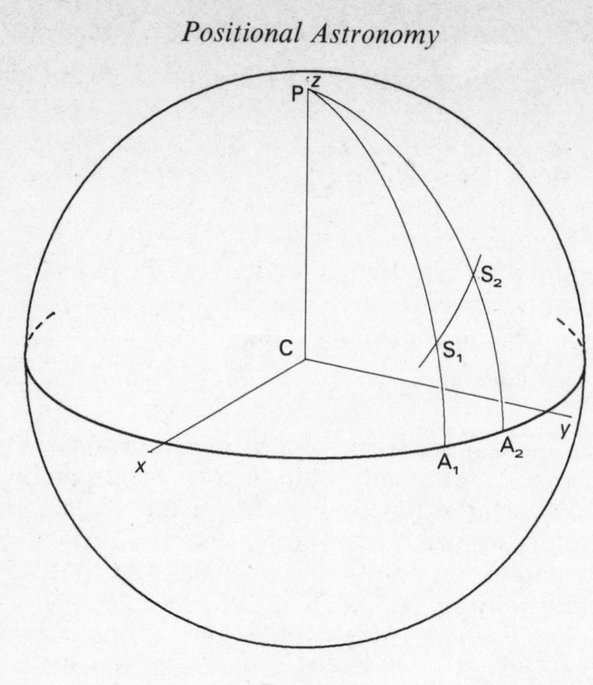

Fig. 2.8

and

$$r_1 \times r_2 = R \sin s = (y_1z_2 - y_2z_1)i + (z_1x_2 - z_2x_1)j + (x_1y_2 - x_2y_1)k, \tag{2.18}$$

where R is a unit vector perpendicular to the plane containing r_1, r_2. The vector R will intersect the surface of the sphere in a point O. Let the rectangular coordinates of O be (L, M, N). Then

$$y_1z_2 - y_2z_1 = L \sin s, \tag{2.19}$$

$$z_1x_2 - z_2x_1 = M \sin s, \tag{2.20}$$

$$x_1y_2 - x_2y_1 = N \sin s, \tag{2.21}$$

from eqn. (2.18). Since r_1 and R, r_2 and R are perpendicular,

$$x_1L + y_1M + z_1N = 0 = x_2L + y_2M + z_2N. \tag{2.22}$$

Eqns. (2.19) to (2.22) give four equations which may be solved for the three unknowns x_2, y_2, z_2. Eqns. (2.19) to (2.21) may be rewritten in the form

$$x_2 . 0 - y_2 . z_1 + z_2 . y_1 = L \sin s, \tag{2.19a}$$

$$x_2 . z_1 - y_2 . 0 - z_2 . x_1 = M \sin s, \tag{2.20a}$$

$$-x_2 . y_1 + y_2 . x_1 + z_2 . 0 = N \sin s. \tag{2.21a}$$

Subtract eqn. (2.21a) multiplied by y_1 from eqn. (2.20a) multiplied by z_1 obtaining

$$x_2(z_1^2 + y_1^2) - x_1(z_1 z_2 + y_1 y_2) = \sin s(z_1 M - y_1 N). \tag{2.23}$$

Therefore

$$x_2(x_1^2 + y_1^2 + z_1^2) - x_1(x_1 x_2 + y_1 y_2 + z_1 z_2)$$
$$= \sin s(z_1 M - y_1 N),$$

i.e.

$$x_2 = x_1 \cos s + \sin s(z_1 M - y_1 N), \tag{2.24}$$

making use of eqns. (2.14) and (2.17). Use of eqns. (2.19a) and (2.21a) gives

$$y_2 = y_1 \cos s + \sin s(x_1 N - z_1 L), \tag{2.25}$$

while use of eqns. (2.19a) and (2.20a) gives

$$z_2 = z_1 \cos s + \sin s(y_1 L - x_1 M). \tag{2.26}$$

Eqns. (2.24) to (2.26) give the required expressions for x_2, y_2, z_2 in terms of x_1, y_1, z_1 and s. However, the coordinates of O have been introduced. The method of evaluation of L, M, N will be deferred until Chapter III.

It is clear that if s is a small angle such that

$$\cos s \simeq 1,$$
$$\sin s \simeq s,$$

then

$$x_2 - x_1 = \mathrm{d}x = -x_1(1 - \cos s) + \sin s(z_1 M - y_1 N)$$
$$= s(z_1 M - y_1 N), \tag{2.27}$$

$$y_2 - y_1 = \mathrm{d}y = s(x_1 N - z_1 L), \tag{2.28}$$

$$z_2 - z_1 = \mathrm{d}z = s(y_1 L - x_1 M). \tag{2.29}$$

Use of eqns. (2.24) to (2.26) for finite displacements or eqns. (2.27) to (2.29) for infinitesimal displacements permit the determination of the rectangular coordinates for the displaced position. Before these equations can be further simplified, a system of coordinates will require definition. The arbitrary origin of the position of the x-axis must be fixed in some more rational manner. The establishment of a suitable coordinate system will form the subject of Chapter III.

Chapter III

Systems of Astronomical Coordinates I–Horizon, Equatorial and Ecliptic Coordinates

There are a variety of coordinate systems in use in astronomy. Each coordinate system has its uses in particular situations. While equatorial coordinates are very widely used, galactic coordinates (see Chapter IV) are of greater convenience when considering problems of galactic structure, and horizon coordinates are of more value in surveying—though their use in astronomy is reviving since many radio telescopes have alt-azimuth mountings. Indeed the Russian 6m telescope is an example of a recently built large optical telescope using an alt-azimuth mounting.

As was pointed out in the previous chapter, the positions of astronomical objects are determined by relative angular measurements. Since the measurements are relative any convenient zero point could be used. In practice, reference planes are used which are capable of observational realisation. This means that reference planes associated with the Earth will be used since the Earth is the only body in space for which detailed geometrical information exists. For the purpose of this chapter, the Earth will be regarded as spherical, concentric with the celestial sphere. Since the celestial sphere is regarded as infinitely distant no problem is involved by the fact that the observer is not at the centre of the Earth. (In some planetary and stellar problems this simplification cannot be made and special methods are used to deal with the situation. The effect of departures from sphericity in the figure of the Earth will also be considered at a later stage.)

3.1 Rectangular Equatorial Coordinates

The unit sphere on which star positions may be represented by geometrical points will now be called the *celestial sphere*. The projection of the Earth's equatorial plane onto the celestial sphere gives a fundamental great circle called the *celestial equator*. The poles of the celestial equator are called the *north/south celestial poles*: the north celestial pole is the projection of the terrestrial north pole and similarly for the south celestial pole. The projections of the small circles of latitude and the meridians (which are great circles) of longitude onto the celestial sphere would give a reference system whereby the positions of stars could be mapped. However, the system of terrestrial longitude rotates with the Earth and it would be desirable to establish some meridian plane relevant to astronomy to establish a non-rotating astronomical system of coordinates. A system of coordinates similar to terrestrial latitude and longitude can be established. In practice the astronomical system of coordinates does not correspond exactly with the terrestrial system of latitude and longitude. However, it is often useful to discount the small deviations of the two systems in order to consider the aspect of the sky as observed from different terrestrial latitudes.

For the present the small differences between terrestrial latitude and longitude and the astronomical system will be discounted. A fundamental plane is clearly the projection of the terrestrial equator to form the celestial equator. The zero of terrestrial longitude is the meridian which passes through Greenwich—the Greenwich Meridian. The Greenwich Meridian rotates with the Earth and is therefore unsuitable as a zero meridian for an astronomical system of coordinates. Some fixed point on the celestial equator must be determined which is defined on the basis of astronomical, and not terrestrial, observations. Such a fixed point is provided by the intersection with the equator of the plane defined by the annual motion of the Earth about the Sun. From the point of view of an observer on the Earth, the Earth's motion about the Sun is reflected by the apparent motion of the Sun, which defines a plane on the celestial sphere called the *plane of the ecliptic.* The intersection of the ecliptic with the celestial equator provides two fixed points on the celestial equator which could be used to define an origin of coordinates. The point chosen as origin is that point through which the Sun passes in Spring, and is known as the *Vernal Equinox.*

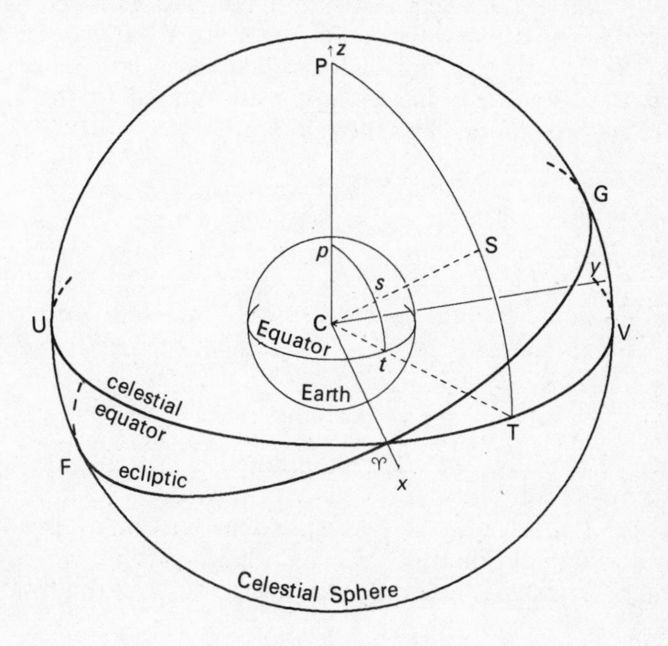

Fig. 3.1

In Fig. 3.1, C is the centre of the Earth and the centre of the celestial sphere. *p* is the north pole of the Earth and P is the north celestial pole. The plane of the Celestial Equator is U ♈ V and the ecliptic is the great circle F ♈ G. ♈ is the Vernal equinox. The position of any star S lying on the great circle or meridian PST through the north celestial pole P can clearly be specified by noting its angular height $\widehat{SCT} = ST$ above the celestial equator and the angular separation of the meridian PST from the meridian through P and ♈, i.e. $♈\hat{C}T = ♈T$. The meridian through the celestial poles and the Vernal Equinox is therefore the zero meridian of the astronomical reference system. ST is called the *Declination* (dec) of the star S and will be denoted by δ while ♈T is called the *Right Ascension* (R.A.) of the star S and will be denoted by α.

An equatorial system of rectangular coordinates $C(x, y, z)$ can be established. The origin of coordinates is at C, the centre of the Earth. The z-axis is defined by the direction of the north celestial pole, the x-axis by the direction of the Vernal Equinox, and the y-axis completes a right handed set (see Fig. 3.1). Clearly the coordinates of S are (x, y, z). The *rectangular equatorial coordinates* x, y, z are related to the *equatorial coordinates* (α, δ) by the relations

$$\begin{aligned} x &= \cos\alpha\cos\delta, \\ y &= \sin\alpha\cos\delta, \\ z &= \sin\delta, \end{aligned} \tag{3.1a}$$

and

$$\begin{aligned} \tan\alpha &= y/x, \\ \sin\alpha\cot\delta &= y/z. \end{aligned} \tag{3.1b}$$

Eqns. (3.1) allow the determination of x, y, z from α, δ or α, δ from x, y, z.

Fig. 3.1 illustrates the correspondence between terrestrial latitude and declination. The meridians of right ascension parallel terrestrial longitude but do not share in the rotation of the Earth.

The system of coordinates set out above, while giving a useful way of thinking about equatorial coordinates, is not exact. A more precise definition of the coordinates which indicates how such a system could be established in practice will be discussed in the next two sections of this chapter.

3.2 Horizon or Topocentric Coordinates

A primary direction must be established by some physically realisable means. For the Earth (assumed spherical) the direction of the gravitational field is radial. A plumb line can therefore be used to determine the direction of the Earth's centre. This primary direction, called the *astronomical vertical*, projected onto the celestial sphere cuts the celestial sphere in a point vertically over the observer. This point is called the *astronomical zenith* (conventionally denoted by the letter Z—see Fig. 3.2). The point on the celestial sphere diametrically

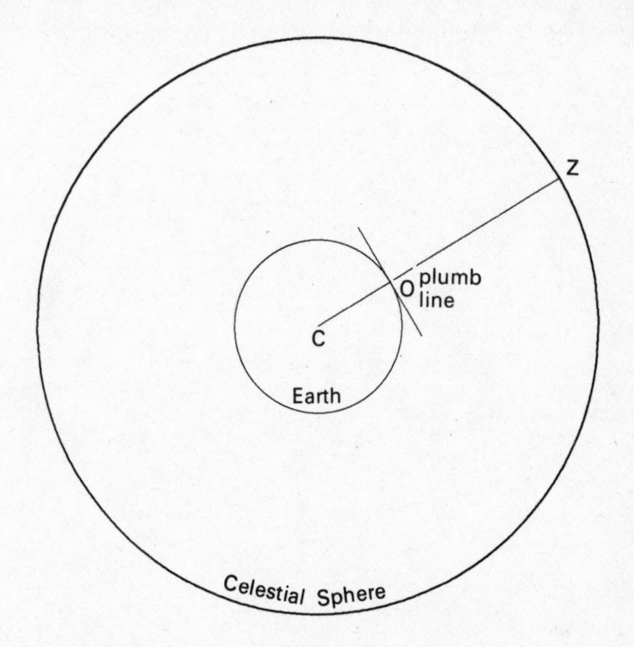

Fig. 3.2

opposite the zenith is called the *nadir*. The plane through the observer (denoted by the letter O) perpendicular to the astronomical vertical is called the *astronomical horizon*. The astronomical horizon does not necessarily coincide with the visible (or apparent) horizon because of variation of terrain. The approximation is usually close. From these definitions it follows that the zenith and nadir are the poles of the great circle of the horizon. Strictly the plane of the horizon on the sky should be defined by the intersection of a plane through C, parallel to the astronomical horizon, with the celestial sphere. However, for a celestial sphere at infinite distance, the distinction is of no practical significance.

The celestial sphere may now be drawn with the observer O at its centre. (The remarks above indicate that any observations made at the surface of a spherical Earth may be regarded as being made from the Earth's centre subject to certain restrictions to be discussed in a later chapter.) Having established the astronomical vertical and horizon, planes through O and Z (see Fig. 3.3) will cut the celestial sphere in great circles

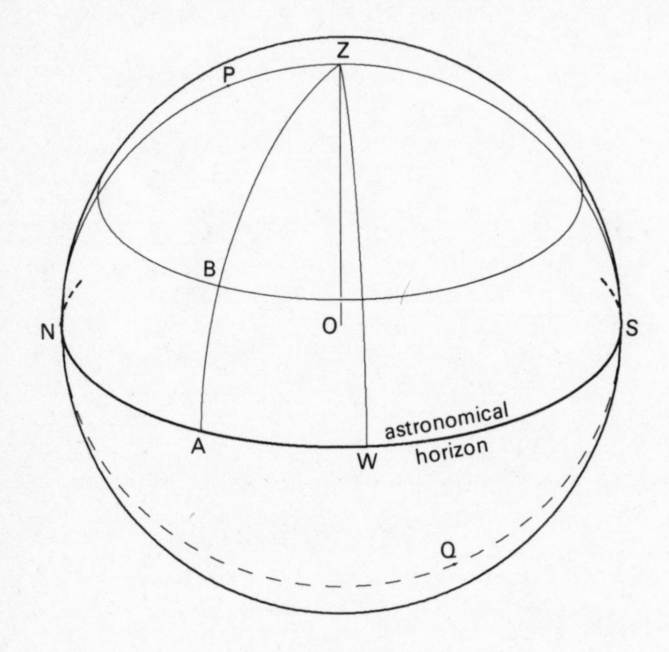

Fig. 3.3

called *vertical circles*. An example of a vertical circle is the great circle ZBA in Fig. 3.3. The plane perpendicular to OZ (i.e. the plane containing O and A in Fig. 3.3) is the astronomical horizon. Small circles may be drawn parallel to the astronomical horizon. Such small circles are called *parallels of equal altitude* since the objects on such a parallel are at the same elevation above the astronomical horizon. The altitude of the object at B is AB = $B\hat{O}A$. Altitude will be denoted by a. Of more value in astronomy is the distance of B from the zenith. BZ is the *zenith distance* of the object at B. z is used to denote zenith distances. Then

$$z = 90° - a. \tag{3.2}$$

Simple observation of the diurnal motion of the stars across the sky shows that the stars appear to move round two diametrically opposite points in the sky. These points are the *celestial poles*. Let P(Q) be the celestial pole in the northern (southern) hemisphere. The vertical circle ZPQ which contains the zenith and both celestial poles is called the *celestial* (*or observer's*) *meridian*. The celestial meridian determines a second fundamental plane. The line of intersection (NOS) of the celestial meridian with the plane of the horizon defines the *meridian line*. The points N, S in which the celestial meridian intersects the horizon are called the north point and the south point respectively. The vertical circle at right angles to the celestial meridian is called the *prime vertical*. The prime vertical intersects the horizon in two points E and W, the east point and west point respectively. The east point is on that side of the celestial sphere where celestial objects appear to rise and the west point of the horizon is on that side of the celestial sphere where celestial objects appear to set. N, S, E, W are called the *cardinal points*.

While the position of the object situated on the sky at B is partly defined by either its altitude a or its zenith distance z, its position on a given parallel of altitude is not defined. The vertical circle through B, ZB, cuts the horizon at A. The angular separation of A from the north point N (i.e. $N\hat{O}A$) unequivocally establishes the position of B. The arc length NA is called the *Azimuth* of B and is denoted by A. There are several conventions relating to the measurements of azimuth, e.g. $0° \leqslant A < 360°$ either measured eastwards or westwards from either the north or south points. For this book the convention will be adopted that $0 \leqslant A \leqslant 180°$ and the direction of measurement will be specified either eastwards or westwards from the north point. The horizon coordinates (a, A) of a given star change with time as the star moves about the celestial pole. The horizon coordinates of a given celestial object will also vary with the position of the observer on the Earth's surface. Consequently horizon coordinates are of no value in providing reference positions for celestial objects. However, because of the variation of (a, A) with time and with the position of the observer, these coordinates are of importance for navigation and surveying.

Horizon coordinates are also of importance astronomically since they give a method of realising fundamental directions and frames of reference. Before any other astronomical coordinate system can be established, horizon coordinates are established. A plumb line is used to give the astronomical vertical. The direction of the celestial pole is established by observation and so the celestial meridian is established.

However, a plumb line measures the local direction of gravity. Since the Earth is not spherical in shape, the direction defined by the plumb line will not pass through the centre of the Earth, except at locations on the equator and at the poles. Furthermore the direction defined by the plumb line will measure the local direction of gravity. The presence of mountain masses, for example, could deflect the plumb line from the true vertical. Therefore the horizon coordinates established by astronomical observation at any given site will depend on local conditions at that site.

3.3 Equatorial Coordinates

In section 3.1 a simple model for equatorial coordinates was set up. In this model, the celestial equator was the projection of the terrestrial equator on the sky and the poles were the projection of the geographic poles on the sky assuming a spherical Earth. In this context, right ascension corresponded to terrestrial longitude and declination to terrestrial latitude. While this correspondence is useful conceptually it leads to an improper astronomical definition of equatorial coordinates. These should be defined in a manner which follows the type of procedure described in section 2 above.

Horizon coordinates are a system of coordinates established purely by astronomical observation for any observer. The coordinate system so established is specific to the observer, the principal directions are fixed and the celestial objects move with respect to these fixed directions. The apparent rotation of the stars from east to west across the sky reflects the actual rotation of the Earth about its axis from west to east. If a system of coordinates could be established such that with respect to its principal planes, celestial objects had no motion, then the position of these objects on the sky could be described for all time by only two coordinates.

Having determined the *celestial pole* as described in section 2 above, the plane through the centre of the celestial sphere perpendicular to the line joining the observer O and the pole P may be defined. This plane intersects the celestial sphere in a great circle called the *celestial equator*. In Fig. 3.4, P is the north celestial pole and Q is the south celestial pole. P lies on the celestial meridian ZPNQS. Since W and E are the poles of the celestial meridian (since the prime vertical and celestial meridian are at right angles and W, E lie on the astronomical horizon), E or W are distant 90° from all points on the celestial meridian and in particular are 90°

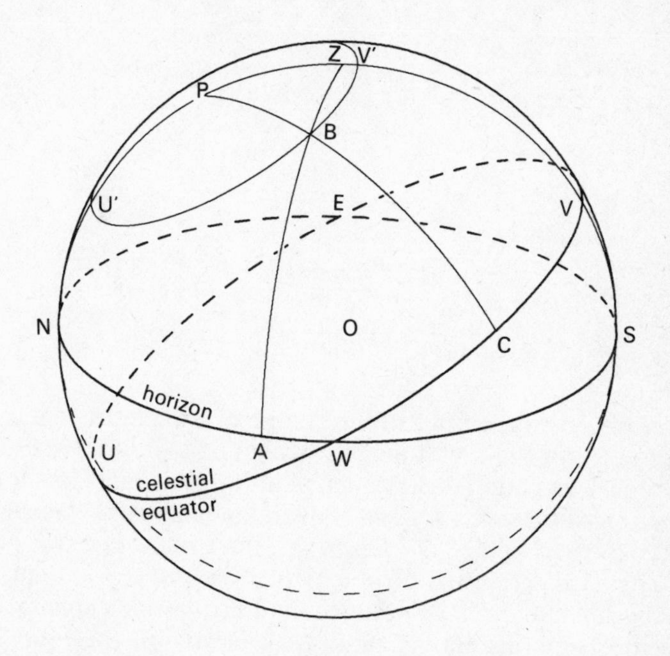

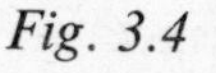
Fig. 3.4

from P and Q. Since P is the pole of the celestial equator it is clear that the celestial equator must pass through the points W and E. Great circles through P which intersect the celestial equator at right angles, such as PC, are called *meridians*.

Since the Earth has been assumed to be spherical with a radially directed gravitational field the terrestrial latitude ϕ of the observer is also the angular separation ZV (see Fig. 3.5)

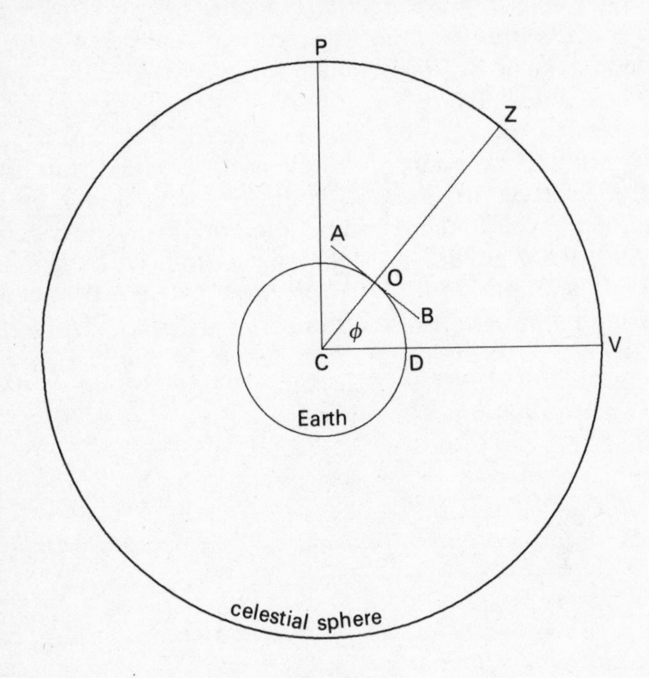

Fig. 3.5

of the observer's zenith from the plane of the celestial equator. The centre of the Earth is at C while the observer is at O in terrestrial latitude ϕ. Clearly a plumb line would define the direction of the line COZ. AOB is in the plane of the astronomical horizon while CDV lies in the plane of the equator both terrestrial and celestial. The terrestrial latitude ϕ is then the angular distance ZV. In practice the Earth is not a sphere and the direction of gravity has local variations. Hence ZV determined from astronomical measurements does not correspond exactly to terrestrial latitude. The angle ϕ will therefore be called the *astronomical latitude* of the observer since it is determined from astronomical measurements. The zenith distance of the north celestial pole P is therefore $90^\circ - \phi$. Once the zenith distance of the north celestial pole has been determined, the plane of the celestial equator can be determined with respect to the system of local horizon coordinates for the observer as illustrated in Fig. 3.4 and the astronomical latitude ϕ of the observer is then determined. The inclination of the plane of the celestial equator to the plane of the astronomical horizon for the observer O is also $90^\circ - \phi$.

Small circles (e.g. U′BV′, Fig. 3.4) parallel to the celestial equator may now be drawn. The apparent motion of celestial objects about the celestial poles will be along small circles parallel to the equator. The angular separation of the object situated at B (see Fig. 3.4) from the equator is BC and is called the *declination* of the object. Declinations are treated as algebraic quantities being positive north of the equator and negative south of the equator. Declination will be denoted by δ. Clearly any object on the small circle U′BV′ will have the same declination—such small circles are called *parallels of declination.* (As pointed out in section 3.1, celestial declination corresponds to terrestrial latitude; the declination of an object is, on the basis of the present definitions, very close to, but not identical with the latitude in which that object can be at the zenith.) A celestial object therefore has a single value of the declination for all time and so declination is suitable as one of the coordinates which fixes the star position. However, declination alone does not determine the position of an object uniquely. One way of fixing the position of the object is by using the angle between the celestial meridian (PNQSZ) and the meridian (PBC) through the star, i.e. the spherical angle $Z\hat{P}B$ (see Fig. 3.4). This angle is called the *Hour Angle* (H.A.) of the object and will be denoted by H. The hour angle is measured westwards from the celestial meridian and is expressed in units of sidereal time. The Earth makes one complete revolution on its axis in 24^h of sidereal time so that 24^h of sidereal time is equivalent to 360° in angular measure. Hence the hour angle H lies in the range $0^h \leqslant H < 24^h$.

The apparent rotation of celestial objects about the celestial pole is manifested for any individual object by its motion along a parallel of declination. Since the apparent motion is from east to west, in terms of Fig. 3.4, the motion is from V′ to B to U′ to V′. The point V′ at which the star has its minimum zenith distance is called *upper culmination* and the point U′ at which the star has its maximum zenith distance is called *lower culmination.* When an object is on the celestial meridian it is said to *transit.* If a star at lower culmination is above the horizon for an observer in a given latitude, the star is said to be *circumpolar* for that latitude. If, however, the parallel of declination on which the object is situated intersects the horizon then the star will set in the west and rise in the east by virtue of its apparent motion along the parallel of declination. Because of the apparent motion of objects along parallels of declination the hour angle of any particular object will increase with time within the range 0^h to 24^h. At the instant of transit $H = O^h$ and thereafter increases. The hour angle at setting is less than 12^h and, at the instant of rising, greater than 12^h.

The equatorial coordinates (H, δ) uniquely define the position of a celestial object at any instant. This coordinate system is used for practical observation with equatorially mounted telescopes. Such telescopes have their circles graduated in terms of declination and hour angle. However, for cataloguing purposes, the hour angle is an unsatisfactory coordinate to use because of its variation with time. It would therefore be an advantage if some natural origin on the celestial equator could be established. There is no property of the Earth which can give such an origin: however, advantage can be taken of the fact that the Earth moves in an orbit around the Sun. The plane of the Earth's orbit provides a further reference plane—the *plane of the ecliptic.* The motion of Earth is reflected in the apparent motion of the Sun. By determining the path of the Sun across the sky, the plane of the ecliptic can be established.

During the course of a year the declination of the Sun varies between the values $\pm\varepsilon$. The plane of the ecliptic is therefore inclined at an angle ε ($\approx 23^\circ{\cdot}5$) to the plane of the equator (see Fig. 3.6). ε is called the *obliquity of the ecliptic.*

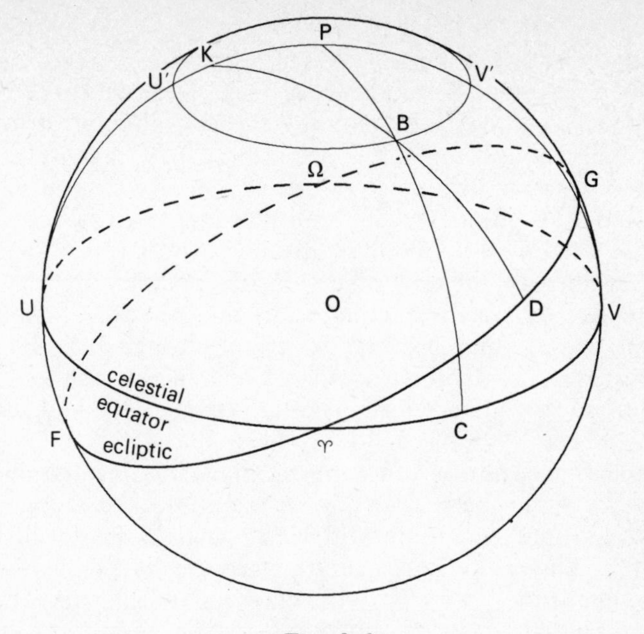

Fig. 3.6

The maximum declination of the Sun is GV = UF = G $\widehat{\Upsilon}$ V. The instant at which the Sun attains maximum declination is called the *solstice.* In the northern hemisphere the Sun attains maximum northerly declination ($+\varepsilon$) at the instant of *Summer Solstice* (G in Fig. 3.6) and its maximum southerly declination ($-\varepsilon$) at the instant of *Winter Solstice* (F in Fig. 3.6). (The names of the solstices are reversed in the southern hemisphere.)

However, on two occasions each year the Sun has zero declination and so lies on the equator. The point where the Sun's declination changes from negative to positive is called the *Vernal Equinox* (i.e., ♈ in Fig. 3.6) and the point where the Sun's declination changes from positive to negative is called the *Autumnal Equinox,* (i.e., Ω in Fig. 3.6). The equinoxes are therefore points which could be used to define an origin of coordinates. The rotation of the Earth causes the apparent movement of the Vernal Equinox across the sky just as it causes the apparent movement of the object at B. The position of the Vernal Equinox is independent of the Earth's daily (diurnal) rotation. Consequently B will always have the same angular separation from the Vernal Equinox.

The position of B with respect to the Vernal Equinox can be specified uniquely by defining the two angular distances ♈A, AB. AB is simply the declination of B as previously defined. A is called the *Right Ascension* (R.A.) of B and is denoted by α. Right Ascension is measured *eastwards* from the Vernal Equinox and $0^h \leqslant \alpha < 24^h$. The Right Ascension and declination of any object fix its position on the sky uniquely. The system of equatorial coordinates (α, δ) defined in this manner may be used to define rectangular equatorial coordinates (x, y, z) for a frame of reference which uses the astronomically determined celestial equator through equations (3.1).

Right Ascension is not a physically realisable parameter. Hour Angle has been shown to be realisable. The connection between Right Ascension and Hour Angle is made through Sidereal Time. The Earth rotates once on its axis every 24^h of sidereal time—it requires the elapse of 24^h of sidereal time for a celestial object to make a complete circuit about the celestial pole. In Fig. 3.7 the position of a star B is illustrated with respect to the astronomical horizon NWS and the celestial equator UWV. The Vernal Equinox is at ♈. The celestial (or observer's) meridian is NPZS. The Vernal Equinox, like any other point on the celestial sphere representing a celestial object, will transit the celestial meridian. The interval between successive upper culminations of the Vernal Equinox will be 24^h of sidereal time. Upper culmination of the Vernal Equinox on the celestial meridian may then be taken as the zero from which sidereal time can be measured. The sidereal time at any time subsequent to upper culmination is simply the Hour Angle of the Vernal Equinox. Let the Hour Angle of the Vernal Equinox be S_t then

$$\begin{aligned} S_t &= \mathrm{Z\hat{P}}\Upsilon = \mathrm{Z\hat{P}B} + \mathrm{B\hat{P}}\Upsilon \\ &= \mathrm{Z\hat{P}B} + \mathrm{C\hat{P}}\Upsilon \\ &= H + \alpha \end{aligned} \tag{3.3}$$

where H is the Hour Angle and α the Right Ascension of B. If the sidereal time of observation is known (and in practice

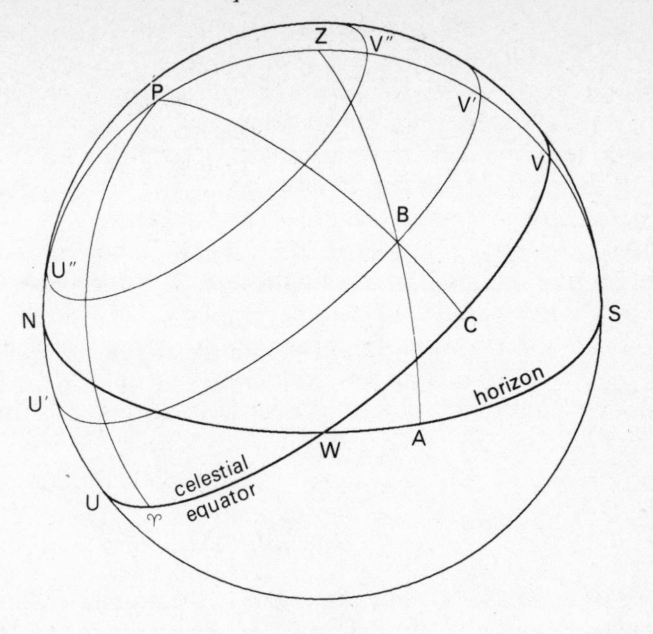

Fig. 3.7

this may be determined directly from a sidereal clock) for an object of known Right Ascension then the Hour Angle may be calculated from eqn. (3.3).

The system of equatorial coordinates is now complete. The position of stellar objects may be recorded in terms of Right Ascension and Declination. By introducing the sidereal time, Right Ascension may be converted to Hour Angle. The coordinates of Hour Angle and Declination are capable of realisation physically.

In terrestrial terms the apparent movements of the Sun may be understood as follows. In terms of section 3.1, when the declination of the Sun is zero, the Sun lies in the plane of the terrestrial equator. The Sun will be in the zenith at midday for observers on the equator. The Sun will rise at the east point of the horizon and set at the west point. The Sun will be above the horizon for as long as it is below the horizon —hence the length of the day will equal the length of the night —and so the term *equinox*. When the Sun has declination $+\varepsilon$

(i.e. at Summer Solstice for the northern hemisphere) the Sun will be in the zenith at midday in north latitude ε. The small circle of north latitude ε is called the *Tropic of Cancer*. Similarly when the Sun has declination $-\varepsilon$ (i.e. at Winter Solstice for the northern hemisphere) the Sun will be in the zenith at midday for south latitude ε. The small circle of south latitude ε is called the *Tropic of Capricorn*.

In Fig. 3.7 the parallel of declination U′BV′ cuts the horizon whereas the parallel of declination U″V″ does not. Any object on the parallel of declination U″V″ will never pass below the horizon whereas objects on the parallel of declination U′BV′ will rise and set. The condition that any object should be circumpolar for an observer in latitude ϕ is that

$$PU'' < PN$$

i.e.

$$90° - \delta < \phi \qquad (3.4)$$

For observers in latitudes $\phi > 90° - \varepsilon$ the Sun is, seasonally, circumpolar. In terrestrial terms, for observers north of the *Arctic circle* (the small circle of north latitude $90° - \varepsilon$) or south of the *Antarctic circle* (the small circle of south latitude $90° - \varepsilon$) the Sun will not rise above the horizon at one time (winter) of the year and at another time (summer) be above the horizon all day.

3.4 Ecliptic Coordinates

The introduction of the plane of the ecliptic to provide an origin of coordinates independent of the Earth's rotation, also gives the possibility of a further way of specifying the positions of celestial objects. The position of the object at B can be specified with respect to the pole P of the celestial equator U ♈ V and the celestial equator itself (see Fig. 3.6). It is also possible to specify the position of B with respect to the pole K and the ecliptic F ♈ G. The *ecliptic longitude* of B is ♈ D measured eastwards from ♈ and the *ecliptic latitude* of B is DB. Ecliptic longitude is denoted by λ and ecliptic latitude by β. In practice ecliptic coordinates are used mainly in problems related to planetary motions.

3.4 *Ecliptic Coordinates*

A system of rectangular coordinates, associated with equatorial coordinates, was defined in section 3.1 and the exact definition of the planes defining the equatorial system were discussed in section 3.3. In the same way a system of rectangular coordinates—ecliptic rectangular coordinates—can be defined in association with ecliptic coordinates.

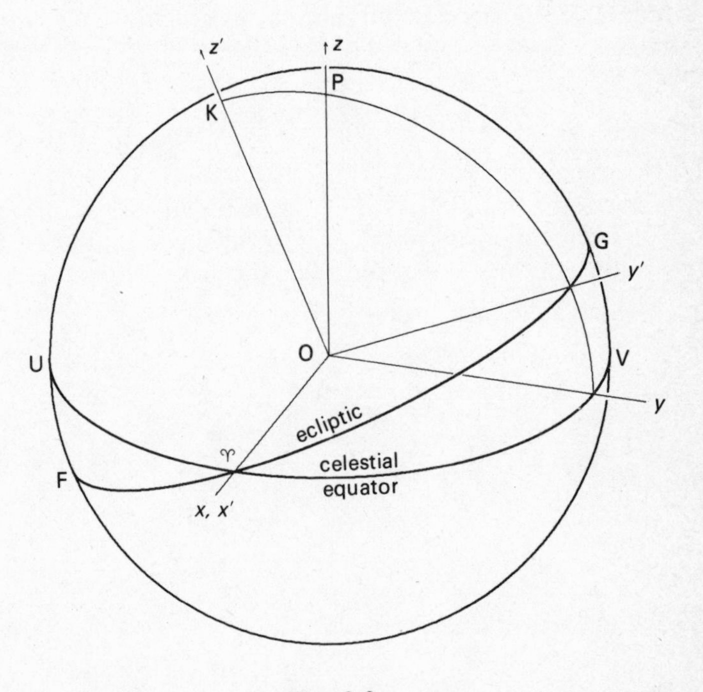

Fig. 3.8

In Fig. 3.8 the celestial sphere is centred at O, U ♈ V is the celestial equator, P is the north celestial pole, F ♈ G is the ecliptic, K is the pole of the ecliptic in the northern hemisphere. Equatorial rectangular coordinates are the system of axes $O(x, y, z)$ where the x-axis is in the direction O♈ and the z-axis is in the direction OP. The y-axis completes a right handed set. Ecliptic rectangular coordinates $O(x', y', z')$ may be established by taking the x'-axis in the direction O♈ (i.e., it is identical with the x-axis of equatorial rectangular coordinates); the z'-axis is in the direction OK and the y'-axis makes up the right-handed set. Then

$$\begin{aligned} x' &= \cos\lambda\cos\beta \\ y' &= \sin\lambda\cos\beta \\ z' &= \sin\beta \end{aligned} \qquad (3.5)$$

Rectangular equatorial and ecliptic coordinates are related by

$$\begin{aligned} x &= x' \\ y &= y'\cos\varepsilon - z'\sin\varepsilon \\ z &= y'\sin\varepsilon + z'\cos\varepsilon \end{aligned} \qquad (3.6)$$

where ε is the obliquity of the ecliptic. (*Note*: substitution for x, y, z from eqn. (3.1) and x', y', z' from eqn. (3.5) in eqn. (3.6) gives the last three equations of (3.8) below.)

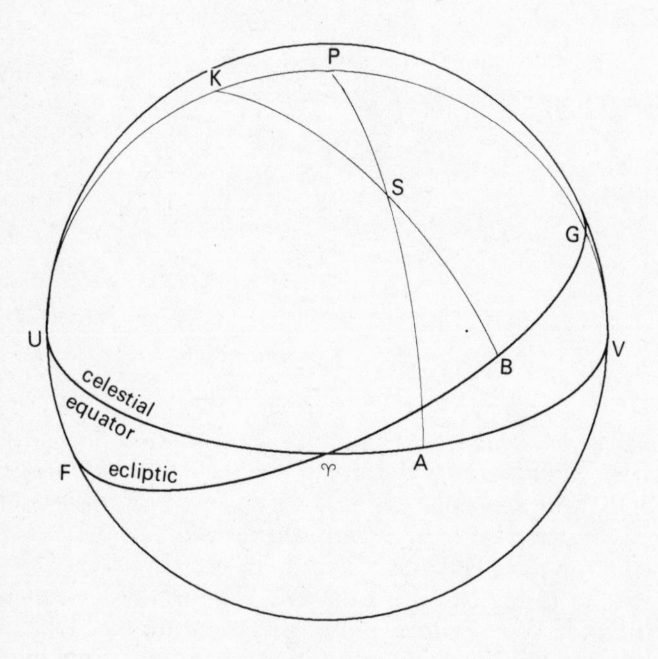

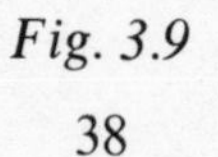

Fig. 3.9

3.5 Relationship Between Topocentric, Ecliptic and Equatorial Coordinates

(a) *Ecliptic and equatorial coordinates.* The apparent motion of celestial objects is irrelevant when either equatorial or ecliptic coordinates are considered. A celestial object may be represented by a fixed point on the celestial sphere for the purposes of these coordinate systems. Conversion between ecliptic (λ, β) and equatorial (α, δ) coordinates is readily established. With reference to Fig. 3.9 the equator is the great circle U ♈ V and has pole P while the ecliptic is the great circle F ♈ G and has pole K. The celestial object is at S, the point of intersection of the great circles PA and KB. The obliquity of the ecliptic is $\varepsilon = G\widehat{\Upsilon}V$. The ecliptic coordinates are (λ, β) and the equatorial coordinates are (α, δ) for S. The spherical triangle KPS has the following properties.

$$KP = \varepsilon, \qquad PS = 90^\circ - \delta, \qquad KS = 90^\circ - \beta,$$
$$P\hat{K}S = 90^\circ - \lambda, \qquad K\hat{P}S = 90^\circ + \alpha. \qquad (3.7)$$

Application of the Cosine, Transposed Cosine and Sine Rules to the spherical triangle KPS give

$$\sin\beta = \cos\varepsilon\sin\delta - \sin\varepsilon\cos\delta\sin\alpha$$

Cosine Rule,

$$\cos\beta\sin\lambda = \sin\varepsilon\sin\delta + \cos\varepsilon\cos\delta\sin\alpha,$$

Transposed Cosine Rule,

$$\cos\beta\cos\lambda = \cos\alpha\cos\delta, \qquad (3.8)$$

Sine Rule,

$$\sin\delta = \cos\varepsilon\sin\beta + \sin\varepsilon\cos\beta\sin\lambda,$$

Cosine Rule

$$\cos\delta\sin\alpha - \sin\varepsilon\sin\beta + \cos\varepsilon\cos\beta\sin\lambda,$$

Transposed Cosine Rule.

The first three equations of (3.8) may be used to convert equatorial to ecliptic coordinates while the last three may be used to convert ecliptic to equatorial coordinates.

(b) *Horizon and equatorial coordinates.* Because Horizon coordinates are time dependent, Right Ascension must be converted to Hour Angle using eqn. (3.3) to give the instantaneous value of H. In Fig. 3.10, B is the instantaneous position of a celestial object on the parallel of declination U′BV′. Z is

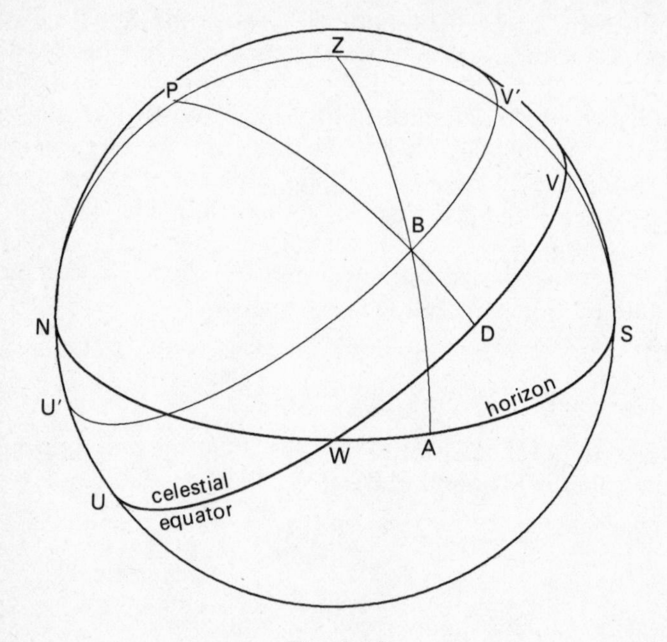

Fig. 3.10

the zenith and P is the north celestial pole. The great circles NWS and UWV are the astronomical horizon and celestial equator respectively. The celestial object lies at the point of intersection B of the great circles PD and ZA. The equatorial coordinates of B are (H, δ), where H is the hour angle, while the horizon coordinates are (a, A). The zenith distance z of B is $90° - a$. The spherical triangle PZB has the following properties.

$$\begin{aligned} &PZ = 90° - \phi,\ ZB = z,\ PB = 90° - \delta, \\ &Z\hat{P}B = H,\ P\hat{Z}B = A. \end{aligned} \tag{3.9}$$

Application of the Cosine, Transposed Cosine and Sine Rules

to the spherical triangle PZS give

$$\begin{aligned}
\cos z &= \sin\phi\sin\delta + \cos\phi\cos\delta\cos H, && \text{Cosine Rule,}\\
\sin z\cos A &= \cos\phi\sin\delta - \sin\phi\cos\delta\cos H, && \text{Transposed Cosine Rule,}\\
\sin z\sin A &= \cos\delta\sin H && \text{Sine Rule,}\\
\sin\delta &= \sin\phi\cos z + \cos\phi\sin z\cos A, && \text{Cosine Rule,}\\
\cos\delta\cos H &= \cos\phi\cos z - \sin\phi\sin z\cos A && \text{Transposed Cosine Rule.}
\end{aligned} \tag{3.10}$$

The first three equations of (3.10) may be used to convert from equatorial to horizon coordinates while the last three equations may be used to convert from horizon to equatorial coordinates.

3.6 General Displacement of a Star Image

In section 2.5 general formulae, eqns. (2.24, 25, 26), connecting the displaced position (x_2, y_2, z_2) with the true position (x_1, y_1, z_1) of a celestial object, were derived. The arbitrary rectangular coordinates (x, y, z) of Chapter II are now identified with equatorial rectangular coordinates. The position of a celestial object was displaced through an angle s along some great circle. The orientation of this great circle was determined through the direction cosines (L, M, N) of a vector $\boldsymbol{R}$ through its pole. Having defined equatorial coordinates in section 3.3 it is now possible to express L, M, N in terms of the equatorial coordinates (α, δ) or rectangular equatorial coordinates (x, y, z) of the star. In Fig. 3.11 UXAV is the celestial equator, P is the north celestial pole and S is the celestial object suffering an angular displacement s in the plane of the great circle XSB. Y is the pole of great circle XSB. ♈ is the Vernal Equinox and ♈A $= \alpha$, AS $= \delta$. A rotation in the plane of the great circle XSB will be deemed to be positive if it is anticlockwise when viewed from Y. Let S have a positive displacement s towards B. The spherical angle P$\hat{\text{S}}$B will be denoted by ψ and the spherical angle S$\hat{\text{X}}$A will be denoted by θ. Since X is both 90°

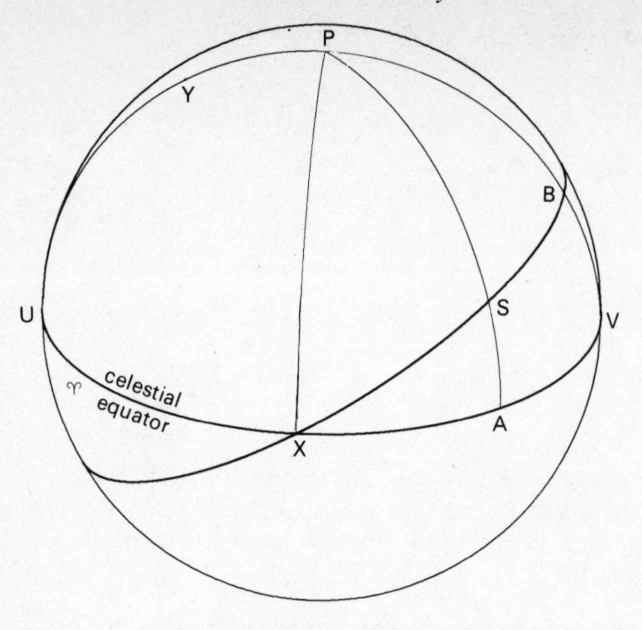

Fig. 3.11

from P and Y, X is 90° from any point on the great circle (PYUV) through P and Y. Then UX = 90° and ♈U = 90° $-\alpha_X$ where α_X is the right ascension of X. In the spherical triangle SXA

$$\begin{aligned} XA &= \alpha - \alpha_X, \qquad AS = \delta, \\ X\hat{A}S &= 90°, \qquad S\hat{X}A = \theta, \qquad X\hat{S}A = \psi \; (= P\hat{S}B). \end{aligned} \tag{3.11}$$

Application of the Sine Rule and Polar Cosine Rule to the spherical triangle SXA gives

$$\sin\delta \sin\psi = \sin\theta \sin(\alpha - \alpha_X), \quad \text{Sine Rule,} \tag{3.12a}$$

$$\cos\psi = \sin\theta \cos(\alpha - \alpha_X), \quad \text{Polar Cosine Rule,} \tag{3.12b}$$

$$\cos\theta = \sin\psi \cos\delta, \quad \text{Polar Cosine Rule.} \tag{3.12c}$$

Addition of eqn. (3.12a) multiplied by $\sin\alpha$ to eqn. (3.12b) multiplied by $\cos\alpha$ gives

$$\sin\theta \cos\alpha_X = \sin\alpha \sin\delta \sin\psi + \cos\alpha \cos\psi = -M, \tag{3.13a}$$

while subtraction of eqn. (3.12b) multiplied by $\cos\alpha$ from eqn. (3.12b) multiplied by $\sin\alpha$ gives

$$\sin\theta\sin\alpha_X = -\cos\alpha\sin\delta\sin\psi + \cos\psi\sin\alpha = L, \quad (3.13b)$$

and

$$\cos\theta = \cos\delta\sin\psi = N. \quad (3.13c)$$

The identification of L, M, N comes from eqn. (3.1) since the equatorial coordinates of Y are clearly $360° - (90° - \alpha_X)$, $90° - \theta$. Hence, evaluating (x, y, z) using eqn. (3.1),

$$\begin{aligned} Mz - Ny &= -(\sin\alpha\sin\psi + \cos\alpha\sin\delta\cos\psi), \\ Nx - Lz &= \cos\alpha\sin\psi - \sin\alpha\sin\delta\cos\psi, \\ Ly - Mx &= \cos\delta\cos\psi. \end{aligned} \quad (3.14)$$

Eqns. (3.14) can also be expressed in terms of x, y, z:

$$\begin{aligned} Mz - Ny &= -\frac{1}{\cos\delta}\{y\sin\psi + xz\cos\psi\}, \\ &= -(x^2 + y^2)^{-\frac{1}{2}}(y\sin\psi + xz\cos\psi), \\ Nx - Lz &= (x^2 + y^2)^{-\frac{1}{2}}(x\sin\psi - yz\cos\psi), \\ Ly - Mx &= (x^2 + y^2)^{\frac{1}{2}}\cos\psi. \end{aligned} \quad (3.15)$$

For any given displacement of a celestial object S, the angle ψ must be determined. ψ can only be defined when the nature of the displacement is specified. The general displacement of a celestial object S through an angle s along an arc of great circle may now be written in the form

$$\begin{aligned} x_2 - x_1 = dx &= -x_1(1 - \cos s) \\ &\quad - \sin s(y_1\sin\psi + x_1 z_1\cos\psi)/R, \\ y_2 - y_1 = dy &= -y_1(1 - \cos s) \\ &\quad + \sin s(x_1\sin\psi - y_1 z_1\cos\psi)/R, \\ z_2 - z_1 = dz &= z_1(1 - \cos s) \\ &\quad + \sin s \,.\, R\cos\psi, \end{aligned} \quad (3.16)$$

where $R = (x_1^2 + y_1^2)^{\frac{1}{2}} = (1 - z_1^2)^{\frac{1}{2}}$, (x_1, y_1, z_1) are the rectangular equatorial coordinates of the true position of the object and (x_2, y_2, z_2) are the rectangular equatorial coordinates of the displaced position. If s is small then

$$\begin{aligned} dx &= -s(\sin\alpha\sin\psi + \cos\alpha\sin\delta\cos\psi) \\ &= -s(y_1\sin\psi + x_1 z_1\cos\psi)/R, \\ dy &= s(\cos\alpha\sin\psi - \sin\alpha\sin\delta\cos\psi) \\ &= s(x_1\sin\psi - y_1 z_1\cos\psi)/R, \\ dz &= s(\cos\delta\cos\psi) = sR\cos\psi. \end{aligned} \tag{3.17}$$

Differentiation of eqns. (3.1a) gives

$$\begin{aligned} dx &= -\sin\alpha\cos\delta\, d\alpha - \cos\alpha\sin\delta\, d\delta \\ &= -y_1\, d\alpha - \frac{x_1 z_1}{R}\, d\delta, \\ dy &= \cos\alpha\cos\delta\, d\alpha - \sin\alpha\sin\delta\, d\delta \\ &= x_1\, d\alpha - \frac{y_1 z_1}{R}\, d\delta, \\ dz &= \cos\delta\, d\delta = R\, d\delta. \end{aligned} \tag{3.18}$$

From the first two eqns. of (3.18) it may be derived that

$$d\alpha = \frac{-y_1\, dx + x_1\, dy}{R^2}, \tag{3.19a}$$

while the third equation gives

$$d\delta = \frac{dz}{R}. \tag{3.19b}$$

Eqns. (3.19) may be used to find the displacements in the equatorial coordinates once the displacements in the rectangular equatorial coordinates have been determined.

3.7 The Determination of Equatorial Coordinates

To conclude this chapter on coordinate systems, a brief discussion of how equatorial coordinates are determined in practice will be given. Measurements of zenith distance are made of stellar objects using a meridian circle. A meridian circle is a refracting telescope mounted so that its optical axis lies in the plane of the celestial meridian and its axis of rotation is horizontal in the East-West direction. The meridian circle can therefore only move in the celestial meridian. The transit of a star across the optical axis (marked by a crosswire) can be timed and the measurement of zenith distance and the time of observation can be used to determine equatorial coordinates as follows.

In Fig. 3.10 a star B transits the celestial meridian SZPN at V′. The zenith distance of a star B is ZV′ = z at upper culmination. PZ = $90° - \phi$ and PB = $90° - \delta$ where δ is the declination of B. Then

$$\begin{aligned} z &= (90° - \delta) - (90° - \phi) \\ &= \phi - \delta. \end{aligned} \tag{3.20}$$

If the latitude of the observer is known, δ can be found. In practice observation of the same star at both upper and lower culmination gives both ϕ and δ. Suppose B is observed at lower culmination at V′ when its zenith distance is ZV′ = z'. Then

$$\begin{aligned} z' &= (90° - \delta) + (90° - \phi), \\ &= 180° - (\phi + \delta). \end{aligned} \tag{3.21}$$

Then

$$z' + z = 180° - 2\delta$$

and

$$z' - z = 180° - 2\phi, \tag{3.22}$$

or

$$\delta = 90° - \tfrac{1}{2}(z' \pm z)$$

and

$$\phi = 90° - \tfrac{1}{2}(z' \mp z). \tag{3.23}$$

The lower sign in the bracketed terms in eqns. (3.23) is used if upper and lower culminations are on the same side of Z. If similar observations are carried out for a large number of stars, clearly a very accurate value of ϕ can be determined. (Indeed accurate measures of ϕ indicate a small irregular variation of latitude.)

The determination of right ascension requires the sidereal time of transit. Suppose that at the instant of transit a perfectly regulated clock indicates time T. Since the hour angle of the star at V′ is zero then T is the right ascension of the star using eqn. (3.3). However, the clock requires calibration in order that the clock reading T should refer to the correct zero of right ascension. The determination of the calibration is made by observing the declination of the Sun. Let the Sun have equatorial coordinates $(\alpha_\odot, \delta_\odot)$. In Fig. 3.12, U ♈ V is the celestial equator and F ♈ G is ecliptic. If the Sun is at S then $\alpha_\odot = ♈\mathrm{A}$ and $\delta_\odot = \mathrm{AS}$. If $\mathrm{S}\widehat{♈}\mathrm{A} = \varepsilon$ is the obliquity of the ecliptic then in the spherical triangle S ♈ A

$$
\begin{aligned}
♈\mathrm{A} &= \alpha_\odot, & \mathrm{AS} &= \delta_\odot, \\
\mathrm{S}\widehat{♈}\mathrm{A} &= \varepsilon, & ♈\widehat{\mathrm{A}}\mathrm{S} &= 90°.
\end{aligned}
\tag{3.24}
$$

Application of the Four Parts Rule to the spherical triangle S ♈ A gives

$$
\sin \alpha_\odot = \tan \delta_\odot \cot \varepsilon. \tag{3.25}
$$

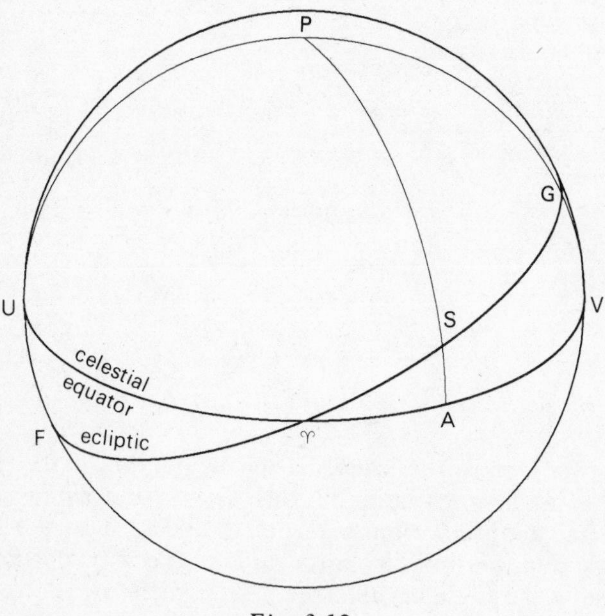

Fig. 3.12

Hence, if $\delta_{\odot}$ can be measured $\alpha_{\odot}$ is determined and so the clock can be calibrated. Unfortunately ε is not known *a priori* but may be determined by measuring the declination of the Sun at transit for several days before and after summer solstice. The maximum value of $\delta_{\odot}$ can be interpolated so giving

$$\varepsilon = (\delta_{\odot})_{\text{maximum}}. \tag{3.26}$$

Therefore, by observations of the Sun with a meridian circle, the obliquity of the ecliptic may be determined, and the solar right ascension determined directly by measurement of solar declination. By continued solar observation the regulation of a sidereal clock may be obtained (in practice the clock error is found) and consequently determination of the instant of transit of stars will give a direct determination of right ascension. In practice corrections must be applied for errors of timing, imperfections of the meridian circle and the small displacements of stellar position caused by the phenomena discussed in Chapters VI and VII. The determination of right ascension underlines the fundamental role played in astronomy by the measurement of time. The measurement of time will be discussed in Chapter V.

Chapter IV

Systems of Astronomical Coordinates II—Galactic, Standard, Planetographic, and Heliographic Coordinates

The coordinates discussed in this Chapter are used for special purposes. Galactic coordinates are used for problems which require exhibition of properties related to the structure of the Galaxy. Standard Coordinates are convenient to use where measurements on photographic plates are used. Planetographic and Heliographic coordinates are used to locate features observed on planetary or solar discs. Planetographic coordinates should not be confused with coordinates based on mapping of planetary surfaces. They are purely astronomical coordinate systems which specify relative angular separations.

4.1 Galactic Coordinates

For studies of Galactic structure it is useful to express positions of objects with respect to fundamental planes defined by the geometry of the Galaxy. The plane of the Galaxy (Milky Way) cuts the celestial sphere in a great circle giving a further fundamental reference plane. The plane of the Galaxy is inclined at angle I to the celestial equator. The points of intersection of the Galactic plane with the celestial equator are called the ascending and descending *nodes*. The ascending node is that node at which the Galactic Plane crosses the celestial equator from south to north when moving in the direction of increasing Right Ascension. Galactic coordinates are defined with respect to the Galactic plane and its poles. The relationship of the plane of the Galaxy and the celestial equator is illustrated in Fig. 4.1. In this figure, UN_AV is the great circle representing the celestial equator, JN_AL is the

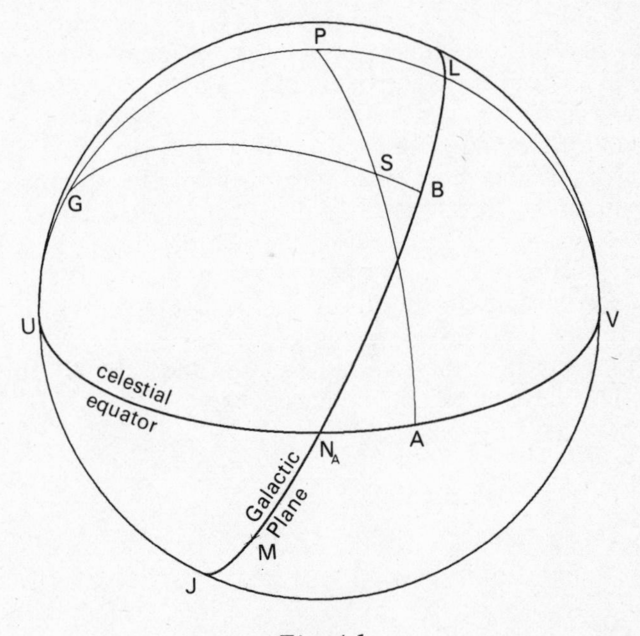

Fig. 4.1

great circle representing the Galactic plane. P is the north celestial pole and G is a pole of the Galactic plane. The celestial object S is situated at the intersection of the great circles PA and GB. The inclination of the Galactic plane to the celestial equator is $L\hat{N}_AV = I$. The Galactic coordinates of S are its Galactic longitude l = MB and Galactic latitude b = BS, where M at 33° from N_A is the origin of Galactic longitude. M is the adopted direction of the centre of the Galaxy (in the constellation of Sagittarius). The position of the Galactic Centre is defined so that the Galactic longitude of the north celestial pole is 123° (i.e. ML = 123° = $M\hat{G}P$).

In establishing a Galactic system of coordinates, a system was sought which would refer only to Galactic parameters. So far in this book it has been assumed that the fundamental reference systems can be established for all time. However, the effect of precession (see Chapter VII) is to move the position of the Vernal equinox. Galactic coordinates are independent of the effects of precession, and are referred to the equator and equinox of 1950·0, so that effects of precession subsequent or prior to 1950·0 must be removed from other systems of coordinates before conversion to Galactic Coordinates.

There are two systems of Galactic Coordinates in use. These are denoted by a superscript I or II (i.e. l^{I}, b^{I}; l^{II}, b^{II}). However, it has been recommended by the International Astronomical Union that the superscript II should be omitted and accordingly in this book the coordinates (l, b) will be understood to refer to system II. The use of the earlier system I is being progressively phased out, but the superscript I is used for distinguishing purposes.

With respect to the celestial equator and celestial pole for 1950·0 the equatorial coordinates of M are

$$\alpha_{\text{M}} = 17^{\text{h}}42^{\text{m}}\cdot4, \qquad \delta_{\text{M}} = -28°55'; \tag{4.1}$$

the equatorial coordinates of G are

$$\alpha_{\text{G}} = 12^{\text{h}}49^{\text{m}}\cdot0, \qquad \delta_{\text{G}} = +27°24' \; (I = 90° - \delta_{\text{G}} = 62°36'); \tag{4.2}$$

the equatorial coordinates of the ascending node N_A are

$$\begin{aligned} \alpha_{N_A} &= 18^{\text{h}}49^{\text{m}}\cdot0, \qquad \delta_{N_A} = 0. \\ &= 282°15'. \end{aligned} \tag{4.3}$$

The relationship between Galactic and equatorial coordinates may be established by using the spherical triangle GPS (see Fig. 4.1). For this triangle

$$\begin{aligned} &GP = I, \; PS = 90° - \delta, \qquad GS = 90° - b, \\ &S\hat{G}P = 90° - (l - 33°), \qquad G\hat{P}S = 90° + (\alpha - 282°\cdot25). \end{aligned} \tag{4.4}$$

Application of the Cosine, Transposed Cosine and Sine Rules to the spherical triangle GPS gives

$$\sin b = \sin \delta \cos I - \cos \delta \sin I \sin(\alpha - 282°{\cdot}25), \quad \text{Cosine Rule,}$$

$$\cos b \sin(l - 33°{\cdot}0) = \sin \delta \sin I + \cos \delta \cos I \sin(\alpha - 282°{\cdot}25), \quad \text{Transposed Cosine Rule,}$$

$$\cos b \cos(l - 33°{\cdot}0) = \cos \delta \cos(\alpha - 282°{\cdot}25), \quad \text{Sine Rule,} \qquad (4.5)$$

$$\sin \delta = \sin b \cos I + \cos b \sin I \sin(l - 33°{\cdot}0), \quad \text{Cosine Rule,}$$

$$\cos \delta \sin(\alpha - 282°{\cdot}25) = -\sin b \sin I + \cos b \cos I \sin(l - 33°{\cdot}0), \quad \text{Transposed Cosine Rule.}$$

The first three equations of (3.15) may be used to convert from equatorial to Galactic coordinates and the last three equations of (3.15) may be used to convert from Galactic to equatorial coordinates. In any of eqns. (3.15) the equatorial coordinates refer to the equator and equinox of 1950·0.

4.2 Standard Coordinates

The ubiquitous use of photographic plates in astronomy means that the positions of objects on the celestial sphere are recorded on plane surfaces (the photographic emulsion). There is a simple geometrical relationship between the disposition of objects on the sky and their images on a plane photographic plate. Advantage may be taken of this relationship to establish a system of plane coordinates called *Standard Coordinates* for use in photographic astrometry. Standard coordinates, introduced by H. H. Turner in 1893, give the positions of any object relative to the centre of the plate for an ideal telescope of standard focal length. Such a system of coordinates would therefore appear to be of limited usefulness. However, this is not the case; it will be shown in Chapter IX that the use of standard coordinates allows the ready elimination of instrumental errors and the effects of refraction and aberration. If specification of the position of the origin of coordinates is with respect to a given epoch, the axes of the standard coordinates are correctly orientated for that epoch. These properties make standard coordinates useful. The effects of parallax and proper motion cannot be automatically removed from the standard coordinates, thus making plate techniques useful in their determination.

Standard coordinates are defined with respect to a standard refractor of given focal length. The standard refractor was assumed free of all mechanical and optical imperfections. The centre of the celestial sphere is assumed to be at the centre of the object glass of the standard refractor. A plane—*the tangent plane*—is defined such that it is tangential to the celestial sphere at the point where the optical axis of the refractor produced, cuts the celestial sphere. The tangent plane is thus, by definition, parallel to the focal plane of the telescope and so parallel to the plane of the photographic plate being used to record star images. Star positions on the celestial sphere are therefore projected along the radii of the celestial sphere onto the tangent plane (note: projection is not parallel to the optical axis of the telescope). The geometrical situation is illustrated in Fig. 4.2. The centre of the celestial sphere coincides with the centre of the object glass at O. The tangent plane is tangential to the celestial sphere at A. The object at X on the celestial sphere projects along the radius OX to X′ on the tangent plane and is recorded at X″ in the focal plane.

Photographic plates are of finite size. Suppose that the plates are square (though the following argument will apply equally to rectangular or circular plates) and that the half width of a side is A″Y″. Thus on the tangent plane all stars projected within a square of side of length 2AY′ and centred at A will be recorded on the plate. The line AY′ corresponds to an arc of great circle AY on the celestial sphere. For a celestial sphere of unit radius the arc length AY gives the angular size of the field that can be observed for a plate whose side has half length A″Y″. If the angle $\hat{AOY}$ is denoted

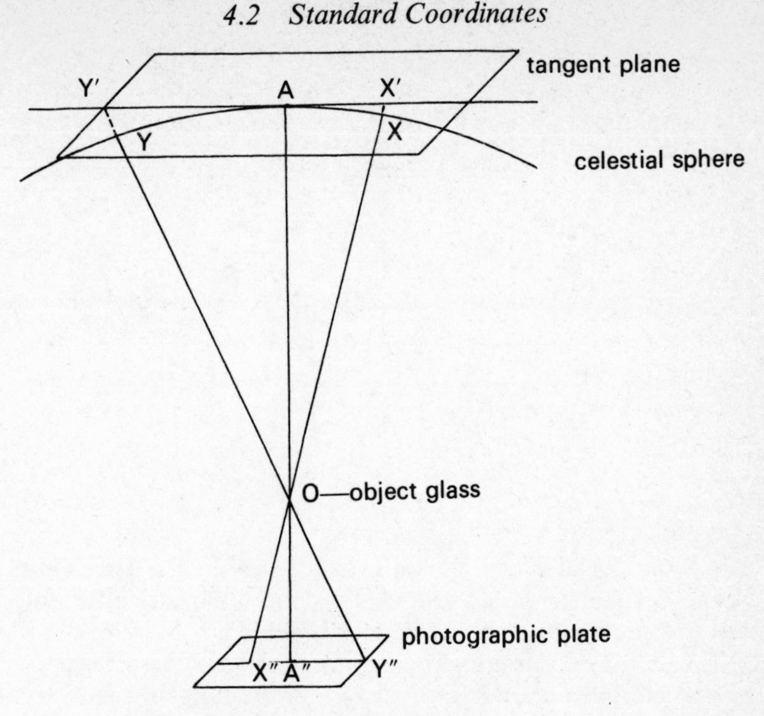

Fig. 4.2

by L and the plate has a side of length l then

$$l = 2\text{A}''\text{O} \tan L. \tag{4.6}$$

Eqn. (4.6) gives a quick conversion of linear plate scale to angular measurement on the sky since the focal length A″O is known for the telescope used.

In order to uniquely determine the position of stars on the tangent plane (and so photographic plate) an origin of coordinates and two axes must be chosen. The origin of coordinates is taken to be the point A at which the tangent plane touches the celestial sphere. The line on the tangent plane joining A to the projection of the north celestial pole defines one coordinate axis which is called the η' axis. The second coordinate axis is taken in an easterly direction at right angles to the η' axis at A and is called the ξ' axis. Axes on the photographic plate ξ, η, are taken at the point A″ and correspond respectively to the ξ', η' axes on the tangent plane.

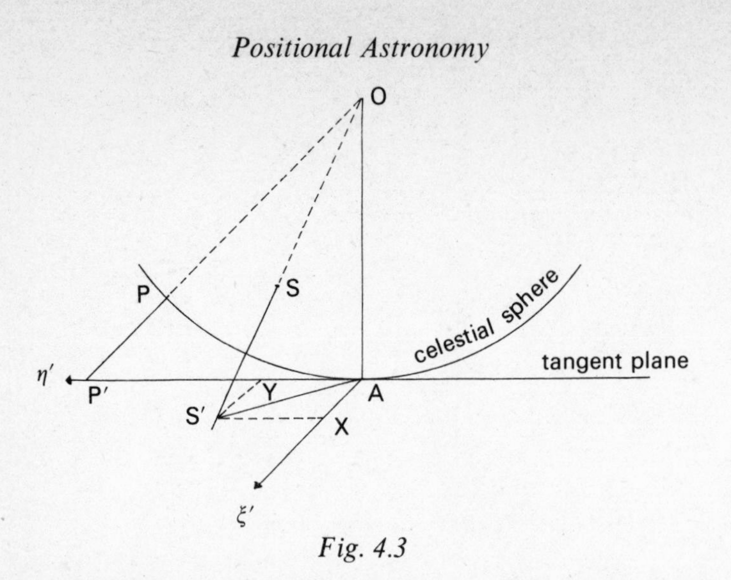

Fig. 4.3

The geometrical situation is illustrated in Fig. 4.3. The centre of the celestial sphere is at O. P is the north celestial pole and S is the position of some celestial object.

The standard refractor is pointing in the direction OA. The north celestial pole P projects to P′ on the tangent plane and S projects to S′. The line AP′ defines the direction of the η' axis while the ξ' axis is at right angles to AP′ in an easterly direction (i.e. ξ' increases in the direction of increasing right ascension).

Let S have equatorial coordinates (α, δ) and rectangular equatorial coordinates (x, y, z). Let A have equatorial coordinates (α_0, δ_0) and rectangular equatorial coordinates (x_0, y_0, z_0). (Strictly, equatorial coordinates should be defined with respect to a given epoch and this, while not stated explicitly, will be implicitly assumed in this section—see also chapter VII.) Set $\text{AS} = \phi$ and $\text{P}\widehat{\text{A}}\text{S} = \theta$. The coordinates of S′ in terms of the frame of reference $\text{A}(\xi', \eta')$ are

$$\begin{aligned} \xi' &= \text{AX} = \text{AS}' \sin\theta, \\ \eta' &= \text{AY} = \text{AS}' \cos\theta, \end{aligned} \tag{4.7}$$

since $\text{Y}\widehat{\text{A}}\text{S} = \text{P}\widehat{\text{A}}\text{S} = \theta$ by definition of spherical angles. Since $\text{SA} = \phi = \text{S}\widehat{\text{O}}\text{A}$

$$\text{AS}' = \text{AO} \tan\phi,$$

so that

$$\xi' = \text{AO} \tan \phi \sin \theta, \qquad \eta' = \text{AO} \tan \phi \cos \theta. \tag{4.8}$$

The relationship of the (ξ, η) axes to the (ξ', η') axes is illustrated in Fig. 4.4. From consideration of the similar triangles OA″Y″, OAY′ and OA″X″, OAX′ it is clear that

$$\frac{\xi'}{\text{OA}} = \frac{\xi}{\text{A}''\text{O}} \quad \text{and} \quad \frac{\eta'}{\text{OA}} = \frac{\eta}{\text{A}''\text{O}}. \tag{4.9}$$

Hence equations (4.8) may be written

$$\frac{\xi'}{\text{OA}} = \frac{\xi}{\text{A}''\text{O}} = \tan \phi \sin \theta, \qquad \frac{\eta'}{\text{OA}} = \frac{\eta}{\text{A}''\text{O}} = \tan \phi \cos \theta. \tag{4.10}$$

The focal length A″O of the standard telescope is taken as the unit of length so that

$$\xi = \tan \phi \sin \theta, \qquad \eta = \tan \phi \cos \theta. \tag{4.11}$$

ξ, η are known as the *standard coordinates.*

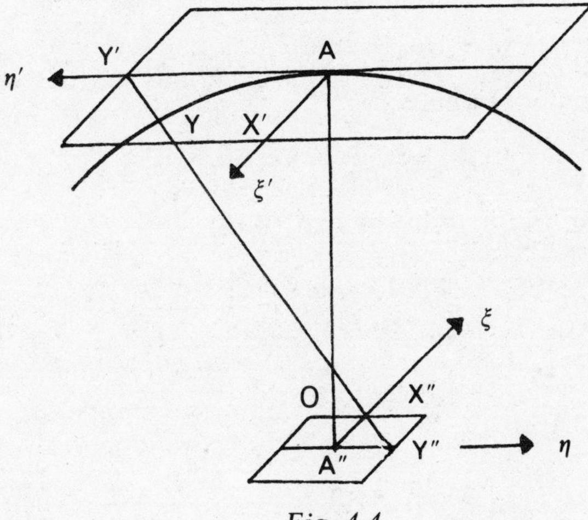

Fig. 4.4

It is clear that ξ, η may be expressed either in terms of the equatorial coordinates of the points S and A or in terms of their rectangular equatorial coordinates. For the spherical triangle PAS in Fig. 4.3,

$$\begin{aligned} &\mathrm{PA} = 90° - \delta_0, \qquad \mathrm{PS} = 90° - \delta, \qquad \mathrm{SA} = \phi, \\ &\mathrm{A\hat{P}S} = \alpha - \alpha_0, \qquad \mathrm{P\hat{A}S} = \theta. \end{aligned} \tag{4.12}$$

Application of the Cosine, Sine and Transposed Cosine Rules to the spherical triangle PAS gives

$$\cos\phi = \sin\delta\sin\delta_0 + \cos\delta\cos\delta_0\cos(\alpha - \alpha_0), \quad \text{Cosine Rule,} \tag{4.13}$$

$$\sin\phi\sin\theta = \cos\delta\sin(\alpha - \alpha_0), \quad \text{Sine Rule,} \tag{4.14}$$

$$\sin\phi\cos\theta = \sin\delta\cos\delta_0 - \cos\delta\sin\delta_0\cos(\alpha - \alpha_0), \quad \text{Transposed Cosine Rule.} \tag{4.15}$$

Hence dividing eqn. (4.14) by eqn. (4.13)

$$\xi = \frac{\cos\delta\sin(\alpha - \alpha_0)}{\sin\delta\sin\delta_0 + \cos\delta\cos\delta_0\cos(\alpha - \alpha_0)}, \tag{4.16}$$

$$= \frac{(-xy_0 + yx_0)(x_0^2 + y_0^2)^{-\frac{1}{2}}}{xx_0 + yy_0 + zz_0} = \frac{-xy_0 + yx_0}{RI}, \tag{4.17}$$

on using eqn. (3.1) where

$$R = (x_0^2 + y_0^2)^{\frac{1}{2}} = (1 - z_0^2)^{\frac{1}{2}}, \qquad I = xx_0 + yy_0 + zz_0.$$

Again dividing eqn. (4.15) by eqn. (4.13)

$$\eta = \frac{\sin\delta\cos\delta_0 - \cos\delta\sin\delta_0\cos(\alpha - \alpha_0)}{\sin\delta\sin\delta_0 + \cos\delta\cos\delta_0\cos(\alpha - \alpha_0)}, \tag{4.18}$$

$$= \frac{z}{(x_0^2 + y_0^2)^{\frac{1}{2}}(xx_0 + yy_0 + zz_0)} - \frac{z_0}{(x_0^2 + y_0^2)^{\frac{1}{2}}},$$

$$= \frac{1}{R}\left(\frac{z}{I} - z_0\right), \tag{4.19}$$

on using eqn. (3.1). Equations (4.16) and (4.18) may be rearranged to allow the determination of the equatorial coordinates (α, δ) of any object from its standard coordinates (ξ, η) given (α_0, δ_0) the equatorial coordinates of the plate centre. From eqn. (4.18)

$$\eta = \frac{\sin\delta\cos\delta_0\{1 - \cot\delta\tan\delta_0\cos(\alpha - \alpha_0)\}}{\sin\delta\cos\delta_0\{\tan\delta_0 + \cot\delta\cos(\alpha - \alpha_0)\}},$$

i.e.

$$\cot\delta\cos(\alpha - \alpha_0)\{\eta + \tan\delta_0\} = 1 - \eta\tan\delta_0,$$

whence,

$$\cot\delta\cos(\alpha - \alpha_0) = \frac{1 - \eta\tan\delta_0}{\eta + \tan\delta_0}. \tag{4.20}$$

Again eqn. (4.16) gives

$$\xi = \frac{\cot\delta\sin(\alpha - \alpha_0)}{\sin\delta_0 + \cot\delta\cos\delta_0\cos(\alpha - \alpha_0)},$$

i.e.

$$\cot\delta\sin(\alpha - \alpha_0) = \xi\sin\delta_0 + \xi\cos\delta_0 \cdot \frac{1 - \eta\tan\delta_0}{\eta + \tan\delta_0},$$

whence,

$$\cot\delta\sin(\alpha - \alpha_0) = \frac{\xi\sec\delta_0}{\eta + \tan\delta_0}. \tag{4.21}$$

Dividing eqn. (4.21) by eqn. (4.20) gives

$$\tan(\alpha - \alpha_0) = \frac{\xi\sec\delta_0}{1 - \eta\tan\delta_0}. \tag{4.22}$$

Therefore, if (α_0, δ_0) are known, eqn. (4.22) may be used to calculate α and either eqn. (4.20) or (4.21) used to determine δ.

Standard coordinates appear to be an idealised system. However, since the departures from ideal have terms which are linear in (ξ, η) allowing the easy determination of the error, standard coordinates have been useful in practical astrometry using photographic plates. The use of standard coordinates is discussed in Chapter IX. Standard coordinates are rarely used now since rectangular coordinates serve equally well.

Standard coordinates are the basic coordinate system of the Astrographic Catalogue.

4.3 Planetographic and Heliographic Coordinates

4.3.1 The general formulation of planetographic coordinates. The Sun and the planets are observed as discs. Markings such as sunspots or planetary features are seen projected onto the disc. It is often desirable to establish the angular separation of these features with respect to reference planes peculiar to the object studied. Such reference planes are usually chosen to be the equatorial plane of the Sun or planet (i.e. the plane through the centre of the object perpendicular to the axis of rotation,) and the central meridian (see below) at some epoch. These coordinates are purely astronomical and do not refer to topographical mapping which can now be carried out for Mars, Mercury and the Moon.

The determination of planetographic coordinates requires:

(i) a fundamental reference system for each planet,

(ii) the relationship between such a fundamental reference system and a practical system of measurement.

The fundamental reference system assumes that the Sun or planet is a sphere and that the planet is viewed geocentrically (i.e. the observer is at the centre of the Earth). The first fundamental direction which has to be established is the direction of the axis of rotation of the planet. This is defined by the equatorial coordinates (α_0, δ_0) of its point of intersection with the celestial sphere in the northern hemisphere of the ecliptic. In Fig. 4.5, P is the north celestial pole, K is the pole of the ecliptic. Any point lying on the celestial sphere above the ecliptic F♈G on the same side as K is said to be in the northern hemisphere of the ecliptic. To be strictly logical the *invariable plane* should be used in place of the ecliptic. (The planets do not lie in the ecliptic but lie close to it. If all the planets are considered a mean plane can be defined such that the actual departures from it are minimised and this is known as the invariable plane.) However, to maintain conformity with the Astronomical Ephemeris the ecliptic, not the invariable plane, will be used here. A is the point at which the axis of rotation of the planet intersects the celestial sphere.

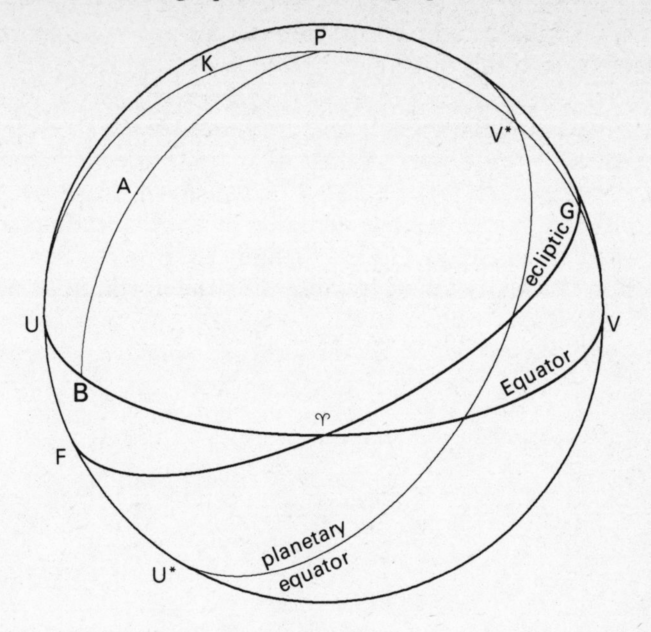

Fig. 4.5

Then,

$$♈\mathrm{B} = \alpha_0, \qquad \mathrm{BA} = \delta_0. \tag{4.23}$$

Having established the axis of rotation for the planet, its equatorial plane U*V* can be quickly established giving one fixed reference plane for planetographic coordinates. Planetographic latitude B gives the angular height of any feature above or below the planetary equator. However, the planet is rotating. Either some permanent feature on the planetary equator or some other means must be found to establish an origin of planetographic longitude L. Since permanent features on the planetary surface may not be well established, it is usual to assume that at an instant in time a defined planetographic meridian contained the geocentric line of sight. Then, knowing the period of rotation for the planet, the position of that plane at any subsequent (or earlier) period can be determined.

Before discussing the determination of (L, B) the practical system of measurement must be defined. The planet is regarded as a disc of finite radius. Surface features are assumed to be projected onto the disc in a direction parallel to the line of sight. To establish the position of a feature projected against the disc two angular measurements are made. In Fig. 4.6 C is the centre of the observed disc, F is the projected position of some planetary feature and CN is a fixed direction. The position of F is uniquely determined by the determination of the angular distance CF $= \rho^*$ and $N\hat{C}F = \theta$. The fixed direction is usually found by obtaining the north point of the

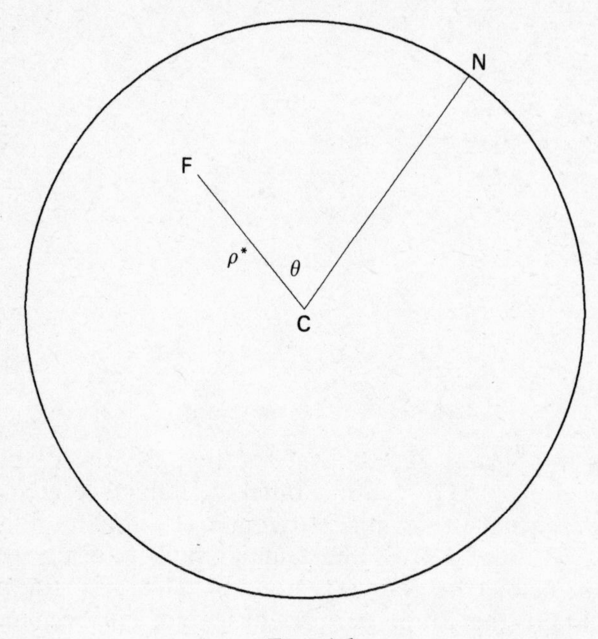

Fig. 4.6

disc. The north point N may be readily found since the motion of the planet across the sky is along a parallel of declination. The direction of drift of the planet in the telescope establishes the plane of the parallel of declination and the north direction is thus a line through the centre of the planet perpendicular to the parallel of declination. The separation ρ^* could be measured by micrometer eyepiece, for example. Having determined ρ^* and θ, the actual angular separation ρ of the surface feature, projected at F onto the planetary disc, from the line of sight must be determined. In Fig. 4.7 C is the centre of the planet and O is a geocentric observer. E is the point where the geocentric line of sight OC cuts the planetary surface. F′ is the position of a feature on the planetary surface corresponding to F (in Fig. 4.6) on the planetary disc. The angular separation of F′ from E, namely the arc F′E, is equal to ρ. The Angle $F'\hat{O}C$ is the measured separation ρ^* of F from C. Let the planet have linear radius R and angular semi-diameter S; let the distance OC be d. Then

$$\sin S = \frac{R}{d}, \tag{4.24}$$

and

$$\frac{R}{\sin \rho^*} = \frac{d}{\sin \{180° - (\rho + \rho^*)\}}, \tag{4.25}$$

using the sine rule for plane triangles applied to OF′C and since $E\hat{C}F' = \rho$. Then

$$\sin (\rho + \rho^*) = \frac{\sin \rho^*}{\sin S} \simeq \frac{\rho^*}{S}, \tag{4.26}$$

since ρ^* and S are small angles in most cases. ρ may be computed from eqn. (4.26).

The practical coordinates ρ, θ must be related to planetographic latitude and longitude. A distinction must be made between geocentric equatorial coordinates and planetocentric equatorial coordinates. Geocentric equatorial coordinates are defined with respect to the Earth's equator and orbit (the ecliptic): the origin of coordinates is the Vernal equinox. Correspondingly, planetographic equatorial coordinates may be defined with respect to the planetary equator and orbit. Planetocentric coordinates are related to planetographic latitude and longitude. In Fig. 4.8 the planetocentric celestial sphere is drawn about the centre of the planet C. The planetary equator is the great circle GHB, whose north pole is at A. The plane of the planetary orbit cuts the planetocentric sphere in the great circle FBE′. N_P is the north celestial pole. B is the descending node of

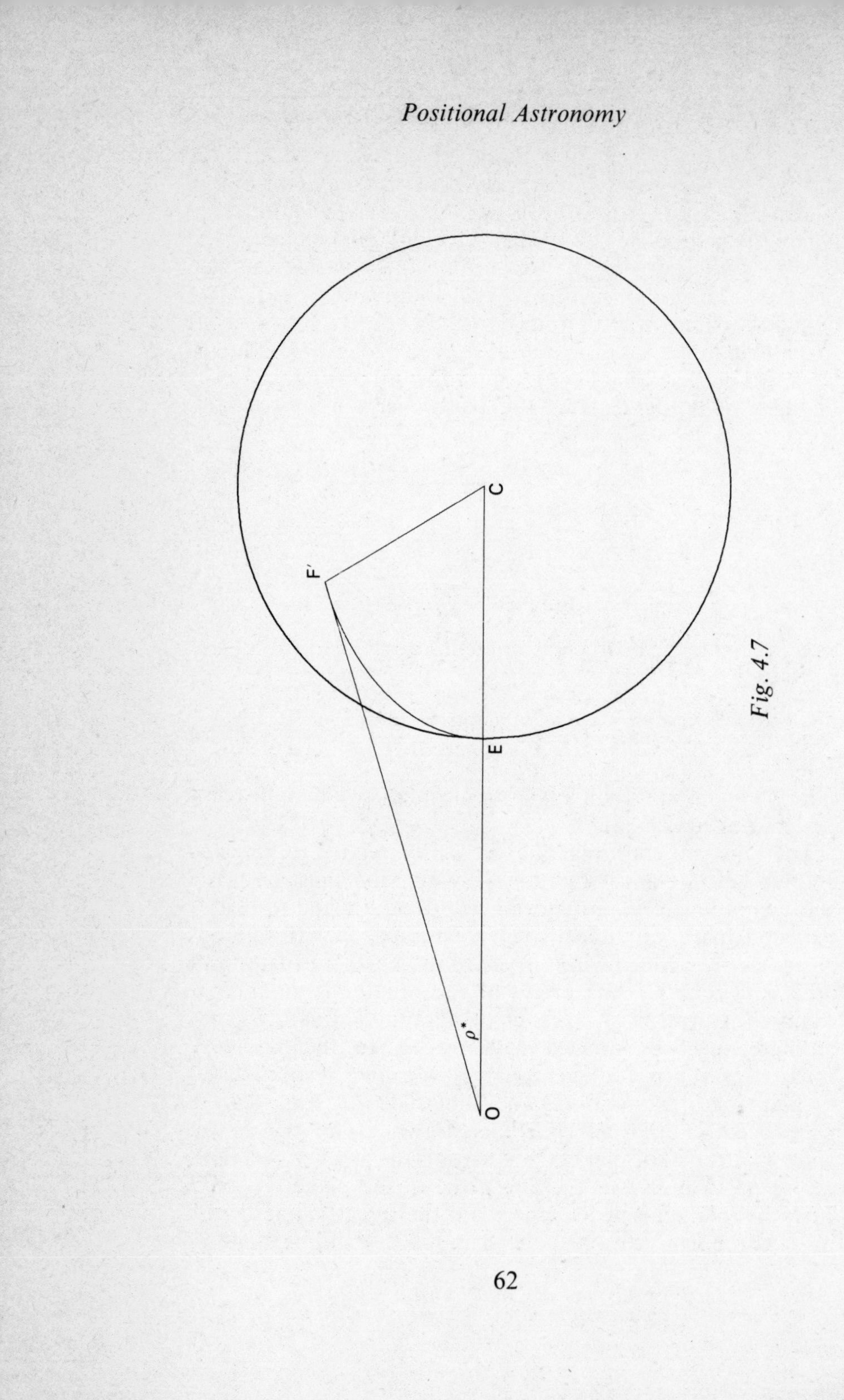

Fig. 4.7

the planetary orbit on the planetary equator and corresponds to the planetary autumnal equinox. Planes parallel to the Earth's equator and ecliptic cut the planetocentric sphere in the great circles ♈FG and ♈DH respectively. The inclination of the planetary orbit to the ecliptic is $B\widehat{D}H = i$. The inclination of the planetary orbit to the planetary equator is $M\widehat{B}E' = I$. The angle $D\widehat{F}G$ will be denoted by J. The arc lengths ♈F, ♈D, DB, FD and GB are denoted by N, ϖ, Ω, ω and Δ respectively.

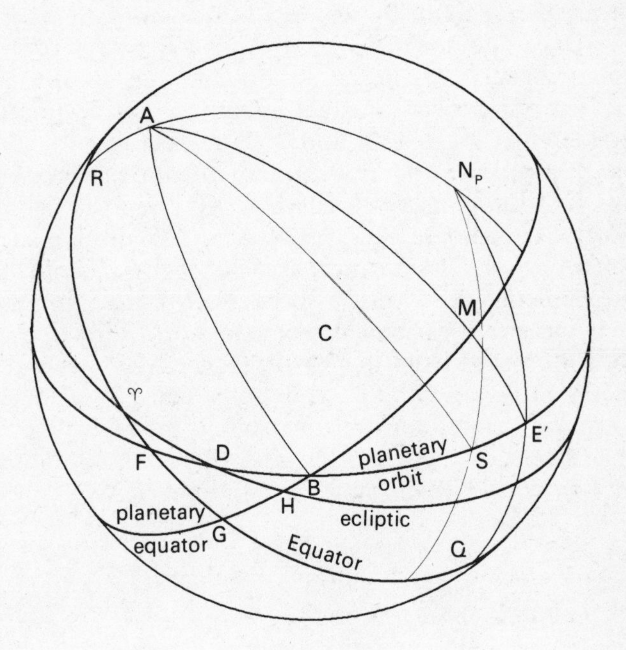

Fig. 4.8

If the angles $B\widehat{A}E'$ and ME′ could be established, a relationship with planetographic coordinates could be established. The geocentric equatorial coordinate of the planet are $\alpha =$ ♈Q, $\delta =$ QE′. E′, therefore, denotes the position of the planet in its orbit from a geocentric point of view. From a planetocentric point of view the line of sight to the Earth cuts the planetocentric sphere diametrically opposite the geocentric position of the planet. The planetocentric angular distance of the Earth from the planetary equator is therefore numerically equal, but of opposite sign, to the planetocentric angular distance of the planet, in its geocentric position E′, from the planetary equator. Let the angular separation in the plane of the planetary equator of the direction of the Earth from the planetary Vernal Equinox be A_E. Then $B\hat{A}E'$ is also A_E since B is the planetary Autumnal Equinox. If the planetocentric direction to the Earth cuts the planetocentric sphere at an angular height D_E above the planetary equator then $ME' = |D_E|$. The coordinates A_E, D_E will be referred to as the planetocentric right ascension and declination of the Earth. The planetocentric right ascension and declination of E′ are therefore $A_E + 180°$ and $-D_E$ respectively.

The quantities i and ϖ are known from the elements of the orbit of the planet (see Chapter XI*) and the geocentric equatorial coordinates (α_0, δ_0) of A and (α, δ) of E′ may be assumed known. The position angle P of the axis of rotation on the planetary disc must also be determined. The position angle is measured eastwards from the north direction as the planet is observed from the Earth. Hence $A\hat{E}'N_P$ is $360° - P$ (see Fig. 4.8). Again G is 90° from both A and N_P. Hence G is 90° from any point on the great circle through A and N_P. Then, since $G\hat{A}B = GB = \Delta$ and $B\hat{A}M = A_E$, $E'\hat{A}N_P = 90° - (A_E + \Delta)$. For the spherical triangle N_PAE',

$$N_PA = 90° - \delta_0, \qquad N_PE' = 90° - \delta,$$

$$AE' = 90° + D_E,$$

$$A\hat{N}_PE' = \alpha + 360° - \alpha_0, \qquad N_P\hat{A}E' = 90° - (A_E + \Delta),$$

$$= 360° + (\alpha - \alpha_0), \qquad N_P\hat{E}'A = 360° - P. \tag{4.27}$$

$$= \alpha - \alpha_0,$$

Application of the Cosine, Sine and Transposed Cosine Rules

*$\bar{\omega}$ is replaced by θ in Chapter XI.

to the spherical triangle N_PAE' gives

$$\cos D_E \sin P = -\cos\delta_0 \sin(\alpha - \alpha_0), \qquad \text{Sine Rule, (a)}$$

$$\cos D_E \cos P = \sin\delta_0 \cos\delta - \cos\delta_0 \sin\delta \cos(\alpha - \alpha_0), \qquad \text{Transposed Cosine Rule, (b)}$$

$$\sin D_E = -\sin\delta_0 \sin\delta - \cos\delta_0 \cos\delta \cos(\alpha - \alpha_0), \qquad \text{Cosine Rule, (c)} \quad (4.28)$$

$$\cos D_E \sin(A_E + \Delta) = \sin\delta \cos\delta_0 - \cos\delta \sin\delta_0 \cos(\alpha - \alpha_0), \qquad \text{Transposed Cosine Rule, (d)}$$

$$\cos D_E \cos(A_E + \Delta) = \cos\delta \sin(\alpha - \alpha_0), \qquad \text{Sine Rule. (e)}$$

Clearly division of eqn. (4.28a) by (4.28b) gives $\tan P$, division of eqn. (4.28d) by (4.28e) gives $\tan(A_E + \Delta)$ and eqn. (4.28c) gives $\sin D_E$. Then P, $A_E + \Delta$, D_E are known. Before A_E can be found, Δ must be determined.

Δ may be determined by using the spherical triangle FGB. G is 90° from any point on the great circle through A and N_P. Therefore RG = 90°. Hence FG = $90° - N - (360° - \alpha_0) = (\alpha_0 - N) - 270°$, since ♈F $= N$ and ♈R $= 360° - \alpha_0$. In the spherical triangle FGB

$$FG = (\alpha_0 - N) - 270°, \qquad GB = \Delta,$$

$$BF = BD + DF = \Omega + \omega, = 90° - (N - \alpha_0), \quad (4.29)$$

$$F\hat{B}G = I, \qquad B\hat{F}G = J, \qquad F\hat{G}B = 90° + \delta_0.$$

Application of the Polar Cosine, Sine and Polar Transposed Cosine Rules to the spherical triangle FGB give

$$\sin I \sin \Delta = \sin J \cos (N - \alpha_0),$$

Sine Rule, (a)

$$\sin I \cos \Delta = \cos \delta_0 \cos J - \sin \delta_0 \sin J \sin (N - \alpha_0),$$

Polar Transposed Cosine Rule, (b)

$$\cos I = \cos J \sin \delta_0 + \sin J \cos \delta_0 \sin (N - \alpha_0),$$

Polar Cosine Rule, (c) (4.30)

$$\sin I \cos (\Omega + \omega) = - \sin \delta_0 \sin J + \cos \delta_0 \cos J \sin (N - \alpha_0),$$

Polar Transposed Cosine Rule, (d)

$$\sin I \sin (\Omega + \omega) = \cos \delta_0 \cos (N - \alpha_0),$$

Sine Rule. (e)

Clearly division of eqn. (4.30a) by (4.30b) gives $\tan \Delta$, division of eqn. (4.30c) by (4.30d) gives $\tan (\Omega + \omega)$ and eqn. (4.30c) gives $\cos I$. Hence Δ, $\Omega + \omega$ and I may be found though at the expense of introducing the quantities N and J. N, J and ω may be determined from the elements i, ϖ of the planetary orbit by making use of the spherical triangle ♈FD where

$$♈\text{D} = \varpi, \qquad ♈\text{F} = N, \qquad \text{FD} = \omega,$$
$$\text{D}\widehat{♈}\text{F} = \varepsilon, \qquad ♈\hat{\text{F}}\text{D} = 180° - J, \qquad ♈\hat{\text{D}}\text{F} = i. \qquad (4.31)$$

Again application of the Polar Cosine, Sine and Polar Transposed Cosine Rules to the spherical triangle ♈FD gives:

$$\sin J \sin N = \sin i \sin \varpi, \qquad \text{Sine Rule,} \qquad \text{(a)}$$

$$\sin J \cos N = \cos i \sin \varepsilon + \sin i \cos \varepsilon \cos \varpi,$$
$$\text{Polar Transposed Cosine Rule,} \qquad \text{(b)}$$

$$\cos J = \cos i \cos \varepsilon - \sin i \sin \varepsilon \cos \varpi,$$
$$\text{Polar Cosine Rule,} \qquad \text{(c)} \qquad (4.32)$$

$$\sin J \cos \omega = \sin i \cos \varepsilon + \cos i \sin \varepsilon \cos \varpi,$$
$$\text{Polar Transposed Cosine Rule,} \qquad \text{(d)}$$

$$\sin J \sin \omega = \sin \varepsilon \sin \varpi, \qquad \text{Sine Rule.} \qquad \text{(e)}$$

Division of eqn. (4.32a) by (4.32b) gives $\tan N$, division of eqn. (4.32e) by (4.32d) gives $\tan \omega$ and eqn. (4.32c) gives $\cos J$. Therefore having found N, J and ω the problem is determined. From the elements of the planetary orbit i and ϖ and the obliquity of the ecliptic, the auxiliary variables N, J, ω can be found using eqns. (4.32) and then the further auxiliary variables I, Δ and Ω may be found from eqns. (4.30). Finally P, A_E, and D_E are found from eqns. (4.28). (α_0, δ_0) must be referred to the true equinox and equator of date. Allowance must therefore be made for the precession and nutation (see Chapter VII) of the Earth's axis of rotation and for precession of the planetary axis of rotation. The precessional variation to be used is the sum of the general precession of the Vernal Equinox and the precession of the Vernal Equinox of the Planet. Values of α_0, δ_0, ω, Δ, Ω, $\sin I$, $\cos I$ are listed in the Explanatory Supplement to the Astronomical Ephemeris for the beginning of each year from 1950·0 to 1975·0 for Mars and Jupiter. The corrections to α_0, δ_0, and Δ to convert to the values at the true equinox of date are also given.

The formulae given so far do not determine the planetographic coordinates of any feature on the surface of a planet. However, once the planetocentric direction of the Earth has been established the coordinates of any other point may be determined. The longitude of the central meridian at the instant of observation has to be established. A convention for the direction of measurement of longitude has also to be established. For the planets, longitude is measured from 0° to 360° eastwards on a *geocentric* celestial sphere. Since, with the exception of Uranus and Venus, the planets rotate in the same sense as the Earth (from west to east), longitude is measured in the opposite direction to the sense of planetary rotation (as viewed from the Earth). However, heliographic coordinates adopt the opposite convention—that is, heliographic longitude is measured westwards in the same direction as the observed solar rotation. The expression giving the longitude of the central meridian is the same for both planetographic and heliographic coordinates.

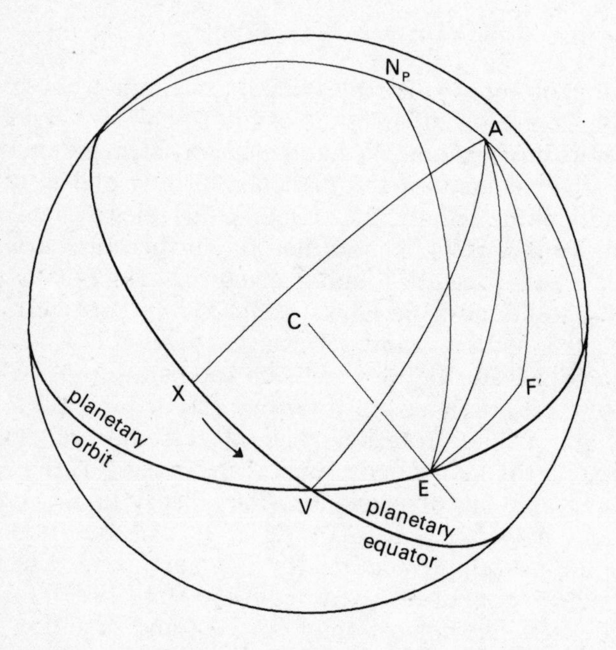

Fig. 4.9

In Fig. 4.9 the planetocentric sphere is drawn showing the direction CE of the line of sight from the point of view of the planet. V is the ascending node of the planetary orbit on the planetary equator and hence corresponds to the planetary vernal equinox. E is the direction in which the Earth is observed and therefore has planetocentric equatorial coordinates (A_E, D_E). The observed central meridian is therefore AE where A is the pole of the planetary equator. Let F′ be the position of an observed feature on the surface of the planet. EF′ is therefore the quantity ρ which is determined from observation. Since the planet is being observed from the direction EC, its observed direction of rotation will be from X to V in the direction of the arrow. Planetary longitudes are therefore measured in the direction VX, i.e. towards the east. Hence

$$\mathrm{A\hat{E}F'} = \mathrm{N_P\hat{E}F'} - \mathrm{N_P\hat{E}A},$$
$$= P - \theta. \tag{4.33}$$

Let X be the zero of planetary longitude and let L_0, B_0 be the planetographic coordinates of E. Clearly $B_0 = D_E$. Suppose at time $t = t_0$ the meridian through A and V has a longitude $\mathrm{V}(t_0)$. At any subsequent time t the longitude of the meridian through V will be $V(t)$:

$$V(t) = V(t_0) + \frac{360°}{T}(t - t_0), \tag{4.34}$$

where T is the period of planetary rotation and T, $t - t_0$ are measured in the same units. (It should be noted that the meridian AX is assumed to share in the rotation of the planet while V, being the intersection of the planetary equator with the plane of the planetary orbit, does not.) In the case of heliographic coordinates where longitude is measured in the same direction as rotation the plus sign is replaced by a minus namely

$$V(t) = V(t_0) - \frac{360°}{T}(t - t_0). \tag{4.35}$$

The longitude of the central meridian is therefore

$$L_0 = V(t) - A_E, \tag{4.36}$$

in the case of planetographic coordinates and

$$L_0 = V(t) + A_E, \tag{4.37}$$

in the case of heliographic coordinates. Finally the planetographic coordinates of F′ may be found by making use of the spherical triangle AEF′ which has properties

$$\begin{aligned} \mathrm{AE} &= 90^\circ - B_0(= 90^\circ - D_E), \\ \mathrm{AF'} &= 90^\circ - \mathrm{B}, \qquad \mathrm{EF'} = \rho, \\ \mathrm{E\hat{A}F'} &= L_0 - L(\text{planetographic}), \qquad \mathrm{A\hat{E}F'} = P - \theta. \\ &= L - L_0(\text{heliographic}). \end{aligned} \tag{4.38}$$

Then, using the Cosine and Sine Rules of spherical trigonometry,

$$\sin \mathrm{B} = \sin B_0 \cos \rho + \cos B_0 \sin \rho \cos (P - \theta), \qquad \text{Cosine Rule,} \tag{4.39}$$

$$\sin (L_0 - L) = \sin \rho \sin (P - \theta) \sec B, \qquad \text{Sine Rule} \tag{4.40}$$

B may be determined from eqn. (4.39) and L from eqn. (4.40). In the case of heliographic coordinates eqn. (4.40) is replaced by

$$\sin (L - L_0) = \sin \rho \sin (P - \theta) \sec B. \tag{4.41}$$

In this way planetographic or heliographic coordinates may be established.

4.3.2 Heliographic coordinates. In the special case of heliographic coordinates certain simplifications of the above treatment are possible since the orbit of the Earth defines the ecliptic. Again it is customary to dispense with (α_0, δ_0) and replace them by Ω. For heliographic coordinates $i = \varpi = \omega = N = 0$, $J = \varepsilon$. Under these conditions eqns. (4.32) give no useful information and eqns. (4.30) simplify to give expressions for $\tan \Delta$ and $\tan \Omega$, namely:

$$\tan \Delta = \frac{\sin \varepsilon \cos \alpha_0}{\cos \delta_0 \cos \varepsilon + \sin \delta_0 \sin \varepsilon \sin \alpha_0}, \tag{4.42}$$

$$\tan \Omega = -\frac{\cos \delta_0 \cos \alpha_0}{\sin \delta_0 \sin \varepsilon + \cos \delta_0 \cos \varepsilon \sin \alpha_0}. \tag{4.43}$$

In the case of the Sun, Ω, I are given and (α_0, δ_0) are not used. Therefore expressions must be found in which Ω and I are used.

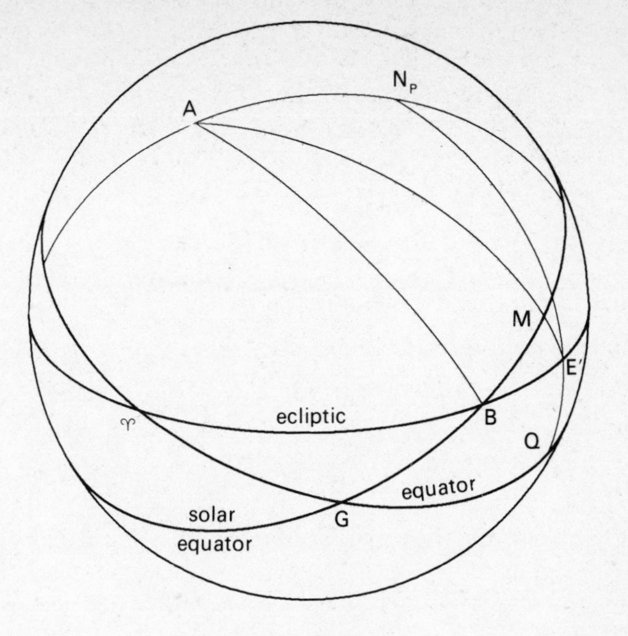

Fig. 4.10

Simplification of eqn. (4.28) may be used to find A_E. However, it is simpler to use Fig. 4.10 where the notation is the same as that used in Fig. 4.8. $\mathrm{B\hat{A}E'} = A_E$, $\mathrm{ME'} = -D_E$, $\mathrm{G\hat{♈}B} = \varepsilon$, $\mathrm{G\hat{B}♈} = I$, $♈\mathrm{B} = \Omega$ and $♈\mathrm{E'} = \odot$ where $\odot$ is the geocentric longitude of the Sun. In the spherical triangle MBE′

$$\mathrm{MB} = A_E, \qquad \mathrm{ME'} = -D_E, \qquad \mathrm{BE'} = \odot - \Omega,$$
$$\mathrm{M\hat{B}E'} = I, \qquad \mathrm{B\hat{M}E} = 90°.$$

Then, using the Transposed Cosine Rule,

$$0 = \cos(\odot - \Omega)\sin A_E - \sin(\odot - \Omega)\cos A_E \cos I, \qquad (4.44)$$

or

$$\tan A_E = \tan(L_0 - V) = \tan(\odot - \Omega)\cos I, \qquad (4.45)$$

using eqn. (4.37) for heliographic coordinates. Again using the Sine Rule

$$-\sin D_E = \sin B_0 = \sin(\odot - \Omega)\sin I, \qquad (4.46)$$

where L_0, B_0 are the coordinates of the centre of the disc from the geocentric point of view. Eqns. (4.45) and (4.46) are the equations normally used for determining heliographic coordinates. The determination of P, i.e. the determination of the position angle of the solar axis, is most readily obtained from eqn. (4.28) since

$$\tan P = -\frac{\cos\delta_0 \sin(\alpha - \alpha_0)}{\sin\delta_0 \cos\delta - \cos\delta_0 \sin\delta \cos(\alpha - \alpha_0)}. \tag{4.47}$$

In the spherical triangle ♈E′Q

$$\begin{aligned} &♈\mathrm{Q} = \alpha, \qquad \mathrm{QE}' = \delta, \qquad ♈\mathrm{E}' = \odot, \\ &\mathrm{Q}\widehat{♈}\mathrm{E}' = \varepsilon, \qquad ♈\widehat{\mathrm{Q}}\mathrm{E}' = 90°. \end{aligned} \tag{4.48}$$

Use of the Cosine, Sine and Transposed Cosine Rules gives

$$\begin{aligned} \cos\odot &= \cos\delta\cos\alpha, && \text{Cosine Rule,} \\ \sin\odot\sin\varepsilon &= \sin\delta, && \text{Sine Rule,} \\ \sin\odot\cos\varepsilon &= \cos\delta\sin\alpha, && \text{Transposed Cosine Rule} \end{aligned} \tag{4.49}$$

The point G is 90° from both A and N_P and so is 90° from any point on the great circle through A and N_P. Hence for the Spherical triangle ♈GB

$$\begin{aligned} ♈\mathrm{B} &= \Omega, ♈\mathrm{G} = 90° - (360° - a_0), \mathrm{GB} = \Delta, \\ &= \alpha_0 - 270°, \\ \mathrm{B}\widehat{♈}\mathrm{G} &= \varepsilon, ♈\widehat{\mathrm{B}}\mathrm{G} = I, \\ ♈\widehat{\mathrm{G}}\mathrm{B} &= 180° - (90° - \delta_0), \\ &= 90 + \delta_0. \end{aligned} \tag{4.50}$$

Use of the Polar Cosine, Sine, Polar Transposed Cosine Rules and 4 Parts Rule gives

$$\sin \delta_0 = \cos I \cos \varepsilon - \sin I \cos \Omega \sin \varepsilon, \quad \text{Polar Cosine Rule,} \quad \text{(a)}$$

$$\cos \alpha_0 \cos \delta_0 = \sin I \sin \Omega, \quad \text{Sine Rule,} \quad \text{(b)}$$

$$\sin \alpha_0 \cos \delta_0 = -(\cos I \sin \varepsilon + \sin I \cos \varepsilon \cos \Omega), \quad \text{Polar Transposed Cosine Rule,} \quad \text{(c)} \qquad (4.51)$$

$$\tan \Delta = \frac{\sin \Omega \sin \varepsilon}{\sin I \cos \varepsilon + \cos I \cos \Omega \sin \varepsilon}, \quad \text{4 parts Rule.} \quad \text{(d)}$$

Eqn. (4.51d) is the more useful equivalent of eqn. (4.42). Substitution of eqns. (4.49) and eqns. (4.51a,b,c) into eqn. (4.47) gives

$$\tan P = \frac{-\cos \odot \tan \varepsilon - \cos(\odot - \Omega) \tan I}{1 - \cos \odot \tan \varepsilon \,.\, \cos(\odot - \Omega) \tan I}. \qquad (4.52)$$

Clearly $\tan P$ may be written in the form

$$\tan P = \tan(P_1 + P_2), \qquad (4.53)$$

where

$$\tan P_1 = -\cos \odot \tan \varepsilon, \quad \tan P_2 = -\cos(\odot - \Omega) \tan I. \qquad (4.54)$$

P_1 is the position angle of the line joining the centre of the solar disc to the pole of the ecliptic and P_2 is the position angle of the solar axis of rotation (projected on the solar disc) with respect to that line. Equations (4.45), (4.46) and (4.52) serve to determine the parameters required to define heliographic coordinates. The adopted values for Ω, I and the period of solar rotation are

Period of Rotation	25·38 mean solar days	
I	$7°15'$	(4.55)
Ω	$73°40' + 50''{\cdot}25t$	

where t is the time in years from 1850·0. In determining Ω only the integral number of years elapsed is used. $\odot$ however must be corrected for precession, nutation and aberration to date. In the case of the Sun, V is denoted by M in the Astronomical Ephemeris. The determination of M is made assuming that at Greenwich Mean Noon on 1854 Jan. 1: (JD 2 398 220·0) the zero meridian of solar longitude passed through the ascending node of the Solar equator on the ecliptic. In terms of Fig. 4.10 the arc AB had longitude zero on this date. At subsequent times the longitude of the ascending node B is given by

$$M(\equiv V) = 360^\circ - \frac{360^\circ}{25{\cdot}38}(\text{J.D.} - 2\,398\,220{\cdot}0). \qquad (4.56)$$

An alternative form for M is

$$M - 180^\circ = 112^\circ{\cdot}766 + (2\,430\,000{\cdot}5 - \text{J.D.}) \times 14^\circ{\cdot}184\,397\,16, \qquad (4.57)$$

where $14^\circ{\cdot}184\,397\,16$ is the daily rotation of the Sun. Eqn. (4.57) is the more convenient form.

4.3.3 Planetary data relevant to planetographic coordinates. The general theory given above and the details for planets below refer to the geometric, and not the illuminated, disc.

(*i*) *Mars.* Planetographic coordinates for Mars are determined from the following constants.

$$\alpha_0 = 21^{\text{h}}11^{\text{m}}10^{\text{s}}{\cdot}42 + 1^{\text{s}}{\cdot}565(t - 1950{\cdot}0)$$
$$\delta_0 = 54^\circ 39' 27'' + 12''{\cdot}60(t - 1950{\cdot}0)$$
at the beginning of year t.

$$\text{Period of rotation} = 24^{\text{h}}37^{\text{m}}22^{\text{s}}{\cdot}6689 \text{ (ephemeris time)}$$
$$= \text{a daily motion of } 350^\circ{\cdot}891962. \qquad (4.58)$$

Longitude of central meridian on 1909 Jan. 15·5 U.T.

(= J.D. 2 418 322·0) = $344^\circ{\cdot}41$.

The corrections to be applied to (α_0, δ_0), namely $\Delta\alpha_0, \Delta\delta_0$, to allow for terrestrial and planetary precession and nutation from the beginning of the year are given by

$$\Delta\alpha_0 = 0^\circ{\cdot}000\,376A + 0^\circ{\cdot}000\,291B - 0^\circ{\cdot}001\,013\tau,$$
$$\Delta\delta_0 = 0^\circ{\cdot}000\,206A + 0^\circ{\cdot}000\,186B - 0^\circ{\cdot}000631\tau, \qquad (4.59)$$

where A, B are Besselian Day Numbers (see Chapter VII) and τ is the fraction of the year. Tables of $\alpha_0, \delta_0, \varpi, i, \Delta, \Omega$, $\sin I$, $\cos I$, are given in the Explanatory Supplement to the Astronomical Ephemeris (p. 336) for the beginning of each year from 1950 to 1975 and these must be corrected by the appropriate amount given by eqn. (4.59) to give the value of (α_0, δ_0) at the instant of observation.

The longitude of the central meridian does not of itself determine V. Let t_0 be the epoch at which the zero of longitude is fixed (in the case of Mars 1909 Jan. 15·5 UT). However, at the initial instant

$$V(t_0) = L_0 + A_E(t_0). \tag{4.36a}$$

$A_E(t_0)$ can be obtained in the manner outlined in section 4.3.1 and so $V(t_0)$ can be computed. The Explanatory Supplement to the Astronomical Ephemeris gives $V(t)$ in the form

$$V(t) + 180° = 325°{\cdot}845 + 350°{\cdot}891962\,(\text{J.D.} - 2\,418\,322{\cdot}0), \tag{4.60}$$

from which $V(t)$ at any time t may be determined.

(*ii*) *Jupiter*. Planetographic coordinates for Jupiter are determined from the following constant.

$$\alpha_0 = 17^h52^m0^s{\cdot}84 + 0^s{\cdot}247\,(t - 1910{\cdot}0), \quad \text{at the beginning of the year } t.$$

$$\delta_0 = +64°33'34''{\cdot}6 - 0''{\cdot}60\,(t - 1910{\cdot}0), \tag{4.61}$$

Period of rotation

$$9^h50^m30^s{\cdot}003 \quad 877°{\cdot}90 \quad \text{System I,}$$
$$= \text{daily rotation}$$
$$9^h55^m40^s{\cdot}632 \quad 870°{\cdot}27 \quad \text{System II.}$$

Longitude of central meridian at 1897 July 14·5 U.T. (J.D. 2414120·0)

$$= 47°{\cdot}31 \quad \text{System I,}$$

$$= 96°{\cdot}58 \quad \text{System II.}$$

A table giving α_0, δ_0, Δ, ϖ, i, Ω, $\sin I$, $\cos I$ (all slowly or very slowly varying quantities) is given in the Explanatory Supplement to the Astronomical Ephemeris (p. 339) for the beginning of each year from 1960 to 1975. Corrections for precession and nutation within a year can be effected by simple interpolation if considered essential. As for Mars $A_E(t_0)$ has to be determined in order to establish $V(t_0)$. The expressions for V are

$$V + 180^\circ = 100^\circ{\cdot}978 + 877^\circ{\cdot}90\,(\text{J.D.} - 2\,414\,120{\cdot}0) \quad \text{System I,}$$

$$= 149^\circ{\cdot}976 + 870^\circ{\cdot}27\,(\text{J.D.} - 2\,414\,120{\cdot}0) \quad \text{System II.} \tag{4.62}$$

(*iii*) *Venus.* A convention for Venus has been recommended by the International Astronomical Union (IAU) (Transactions XIVB p. 128) as follows.

$$\left.\begin{aligned} \alpha_0 &= 18^{\text{h}}12^{\text{m}}{\cdot}0 = 273^\circ{\cdot}0 \\ \delta_0 &= +\,66^\circ{\cdot}0 \end{aligned}\right\} \text{for epoch } 1950{\cdot}0.$$

$$\text{Period of rotation} = 243^{\text{d}}{\cdot}0 \text{ (retrograde).} \tag{4.63}$$

Longitude of the central meridian =

320°·0 on 1964 June 20·0 U.T. (J.D. 2 438 566·5).

It should be noted that Venus rotates in a retrograde manner. The IAU has suggested that the convention be adopted that the planetographic longitude of the central meridian should be measured from 0° to 360° in such a way that it increases with time. This convention has been adopted in the case of Mars and Jupiter. Longitudes on Venus therefore increase towards the west.

(*iv*) *Mercury.* A convention for Mercury has been recommended by the IAU (Transactions XIVB p. 128) as follows.

Rotational and orbital axes of rotation are coincident.

Period of rotation = $58^{\text{d}}{\cdot}6462$.

Zero of longitude contained the subsolar point at first perihelion passage of Mercury in 1950 (J.D. 2 433 292·63).

In using eqns. (4.28), (4.30), (4.32) the point E′ on Fig. 4.8 must be regarded as the heliocentric position of Mercury and heliocentric coordinates used in place of geocentric coordinates.

(*v*) *Moon.* In principle the determination of selenographic coordinates is simpler. On the other hand the Moon, being closer to Earth, exhibits motions called librations which must be allowed for in the determination of selenographic coordinates. The determination of the librations is not a subject for this book. Selenographic coordinates in the sense of this section must not be confused with reference systems used in the preparation of charts of the topography of the lunar surface.

Chapter V

Time

As was foreshadowed in Chapter III, accurate measurement of time is required by astronomy. While there have been many philosophical reflections on the nature of time, all that is required in astronomical practice is some realisable system of time measurement. It is sufficient that man possesses a sense which allows the ordering of events in an earlier-later sequence and it is the rationalisation and refinement of crude sense which leads to a precise formulation of time measurement.

It might also be argued that astronomical measurements of time have an accuracy considerably less than that which can be attained with an atomic clock. Atomic clocks have a stability to 1 part in 10^{13} whereas astronomical measurements are accurate to about 1 part in 10^3 (in terms of a unit of 1 second). However, astronomy is concerned with the accurate location of position on the sky and there is no direct connection between atomic clock time (A.T.) and the systems of time keeping used in astronomy. A correction, transforming from A.T. to an astronomical time system, can only be made retrospectively, as explained below. Clocks, no matter how accurate, give an integrated time interval: that is, they add on units of a fundamental time interval. On the other hand the astronomically observed time scales give the time at an instant.

Astronomical time is not uniform. The reason for this is that the Earth's period of rotation undergoes both systematic and irregular variations. The aspect of the sky which is seen by an observer depends upon the Earth's rotation. To maintain celestial coordinate systems in a meaningful way astronomical time must be kept with due regard to variation in the period of the Earth's rotation. Although atomic time can be maintained with precision and is not considered to undergo long term secular changes (though this must be thoroughly investigated), it is not suitable for astronomical purposes. Since irregular variations in the Earth's period of rotation are not known until after they occur, clearly the differences between astronomical and atomic clock time can only be established retrospectively. Therefore, the need for astronomical measures of time are in no way superseded by refinements in the technique of time keeping.

The rotation of the Earth is reflected in the apparent diurnal motion of the stars across the sky. The period of the Earth's rotation therefore determines the rate at which the stars cross the sky. If, as was originally believed, the period of rotation for the Earth on its axis was uniform and invariable, then the stars would move across the sky at a uniform rate. Suppose that a sudden small increase in period took place and thereafter rotation at the slower rate was maintained. With respect to the original, shorter, period of rotation the stars would appear to transit late. This would be equivalent to an increase in stellar right ascension. However, if the system of time keeping was adjusted to allow for the change in period then the increase in right ascension would not be required. It might seem odd that time should be adjusted on the basis of an erratic time-keeper—it is contrary to normal practice. However, since time is measured through observation of stars, in ignorance of the variation of the period of the Earth's rotation, the measured time determined astronomically makes this allowance automatically. Therefore astronomical measures of time are not uniform.

It is essential that a uniform measure of time be established. For example, Newton's first law—that bodies moving under the influence of no forces will maintain their uniform motion in a straight line—may be interpreted to show that such bodies will always require the elapse of the same time interval to cover any given unit of distance. The concept of uniform measure of time can therefore be associated with uniform motion. Indeed the time used in describing dynamical motion could be used as a fundamental measure of time. In particular, the time parameter which describes the motion of the Moon could be used in this way as a measure of uniform time. This is in fact the operational description of *Ephemeris Time* (E.T.) There is no reason to suppose that ephemeris time has any connection with any astronomical measure of time. Indeed there is no reason why ephemeris time should be used as a uniform measure of time—atomic clock time would be just as suitable since there is no connection between either system of uniform time determination and time determined astronomically.

Therefore astronomy faces two problems—that of establishing a uniform measure of time and relating this to a system of time measurement which is practical for astronomical observations. Since the aspect of the sky seen by an observer at any instant depends on the rotation of the Earth, *Sidereal Time* will be considered first.

5.1 Sidereal Time

A *Sidereal Day* may be defined as the interval between successive upper culminations of any star on the observer's meridian. This simple definition of a sidereal day assumes that the Earth rotates at a constant rate. It also assumes that the equator and ecliptic are fixed planes. As will be shown in Chapter VII this is not the case because of the effects of precession and nutation. The Vernal Equinox moves westwards because of precession. The Vernal Equinox may be regarded as a point on the celestial sphere and, just as for any star, the interval between successive transits of the Vernal Equinox across the observers meridian is 24 hours of sidereal time. However, because of the westwards motion of the Vernal Equinox the interval between two successive transits of the moving equinox will be $0^{s}{\cdot}008$ less than the interval between two successive transits of a fixed equinox. The interval between successive transit of the moving equinox is defined to be a sidereal day of 24^{h}. The instant of transit of the equinox is sidereal noon. The interval between successive transits of the fixed equinox is the period of the Earth's rotation.

5.1 *Sidereal Time*

It is not the period of the Earth's rotation but the sidereal day which is of importance in astronomy. The *Sidereal Time* at any instant is thus the hour angle of the Vernal Equinox. A distinction must be made between *Apparent Sideral Time* and *Mean Sidereal Time.* Apparent sidereal time refers to the true equinox of date—that is, the true equinox of date found by making allowance for both precession and nutation. Mean sidereal time refers to the mean equinox of date—that is the mean equinox of date found by making allowance only for precession. The difference between apparent and mean sidereal time is called the *equation of the equinoxes* (called in the past, nutation in right ascension). Observation of stars determines apparent sidereal time. Apparent sidereal time is corrected by allowing for nutation, so giving mean sidereal time. Sidereal clocks are adjusted to keep mean Sidereal Time.

In Chapter III eqn. (3.3) a relation between sidereal time S_t, Hour Angle H and right ascension α was given, namely

$$S_t = H + \alpha. \tag{3.3}$$

In this equation, S_t is the apparent sidereal time and H, α are referred to the true equator and equinox of the instant of observation. The time S_t^* shown by a sidereal clock is then given by

$$S_t = S_t^* + E_E, \tag{5.1}$$

where E_E is the equation of the equinoxes. However, for practical purposes not concerned with long term astrometric determinations, E_E is so small ($\sim 0^s{\cdot}0005$/day) that S_t and S_t^* may be taken to be the same.

The Greenwich Sidereal Date (G.S.D.) and Greenwich Sidereal Day Number have been introduced to give a measure of the time elasped since J.D. = 0·0 (see later in this chapter) in sidereal time units. The integral part of G.S.D. is the Greenwich Sidereal Day number. The non-integral part is the Greenwich Sidereal Time. For 1973 Jan. 0·73 the Greenwich Sidereal Day Number is 2 448 369.

Sidereal time refers to times of transit over a meridian. Each meridian could in principle determine its own sidereal time but such a situation is unnecessary since the sidereal times on different meridians can be readily calculated. Therefore sidereal times for one standard meridian are sufficient—the standard meridian being that of Greenwich.

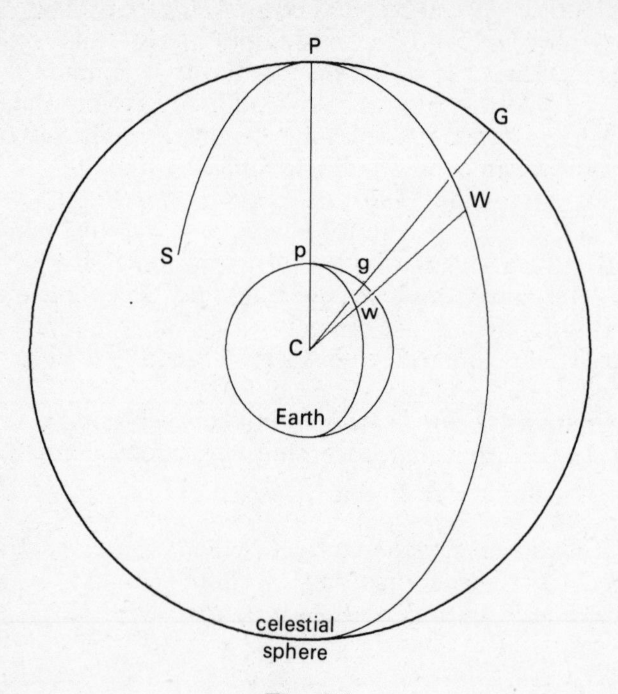

Fig. 5.1

In Fig. 5.1 the celestial sphere is drawn concentric with the Earth centred at C. Suppose the star S is observed from Greenwich (g) and Washington (w). The zenith at Greenwich is G and the zenith at Washington is W. P is the north celestial pole and p is the north pole of the Earth. Let H_g be the Greenwich Hour angle of S and let H_w be the Washington Hour angle. Then

$$\begin{aligned} H_g = \mathrm{G\hat{P}S} &= \mathrm{G\hat{P}W} + \mathrm{W\hat{P}S}, \\ &= H_w + \widehat{\mathrm{gpw}}, \\ &= H_w + l_w, \qquad (5.2) \end{aligned}$$

where l_w is the longitude *west* of Greenwich of Washington.

Then, since

$$S_g = H_g + \alpha,$$

and

$$S_w = H_w + \alpha,$$
$$S_w = S_g - l_w, \tag{5.3}$$

where S_g, S_w are the sidereal times at Greenwich and Washington respectively. In eqn. (5.3) the positive sign is used for longitudes *east* of Greenwich.

5.2 Universal Time

Sidereal Time is unsuitable for civil use. Civil Time is regulated by the hours of daylight and therefore depends on the annual and diurnal motion of the Sun. It is, however, clear that the interval between sunrise and sunset cannot be used to define a day since this interval is much shorter in December than in June (in the northern hemisphere). Furthermore the Earth moves in an elliptical orbit about the Sun. This means that the apparent motion of the Sun against the stars varies with the season and so the interval between successive transits of the Sun across the observer's meridian will vary. The interval between successive transits of the Sun across the observer's meridian is called a *solar day*. The solar day averaged over a year is called the *mean solar day* and is defined to be 24^h of mean solar time. Formerly it was convenient to measure civil time in terms of a fictitious body called the Mean Sun. The Mean Sun was assumed to move in the plane of the celestial equator at a uniform rate. Therefore the interval between successive transits of the Mean Sun across the observer's meridian would be a mean solar day. Just as for any other celestial object the sum of the Hour Angle (H_{MS}) and right ascension (α_{MS}) of the Mean Sun would give the local sidereal time S_l, i.e.

$$\begin{aligned} S_l &= H_{MS} + \alpha_{MS}, \\ &= H_{\odot} + \alpha_{\odot}, \\ &= H_* + \alpha_*, \end{aligned} \tag{5.4}$$

where a subscript $\odot$ denotes the Sun and a subscript * denotes any other star. Clearly

$$\alpha_{\text{MS}} - \alpha_{\odot} = H_{\odot} - H_{\text{MS}} = E_T, \tag{5.5}$$

where E_T is the *equation of time*. For astronomical purposes the equation of time is not now used though it is useful for navigational and surveying purposes. On the basis of this simple interpretation Mean Solar time and Universal time are identical.

The instant of transit of the Mean Sun over the observer's meridian defines *local mean noon*. Similarly transit of the Mean Sun over the Greenwich meridian defines *Greenwich mean noon*. For many astronomical purposes it is convenient to start the day at noon since no change of date is required during the hours of darkness. On the other hand it would be inconvenient if the date were to change at mid-day and consequently for civil purposes the day runs from midnight to midnight. Time measured from noon to noon is called *Greenwich Mean Astronomical Time* (G.M.A.T.) and that measured from midnight to midnight is *Greenwich Mean Time* (G.M.T.). Clearly

$$\text{GMT} = \text{GMAT} + 12^{\text{h}}. \tag{5.6}$$

Since transit of the Mean Sun defines mean noon for each meridian it would be wholly inconvenient if the reckoning of Civil Time varied from village to village and town to town. The Earth is therefore divided into *time zones* which adopt the time of some central meridian. The addition of each $15°(= 360°/24)$ of longitude corresponds to a time difference of 1 hour. The central meridians of time zones, in principle, should therefore be spaced at 15° intervals starting from the zero of longitude (the meridian through Greenwich) and the time zone extend $7\frac{1}{2}°$ on either side of the central meridian. In practice this can be inconvenient for the normal pattern of living and so time zones are adjusted so that such patterns are disrupted as little as possible.*

* The recent controversy in the United Kingdom following the adoption of Central European Time is an interesting case. Many of the countries in Western Europe should keep the time of the Greenwich Meridian—for example most of Spain and all of Portugal lie to the west of the Greenwich meridian. However, because of a historical controversy and other reasons,

Zone time is the Greenwich time plus or minus the longitude l_{CM} (in degrees) divided by 15 of the central meridian of the zone concerned, i.e.

$$ZMT = GMT \pm (l_{CM}/15) \tag{5.7}$$

The plus sign refers to longitudes *east* of Greenwich and the minus sign to longitudes *west* of Greenwich. The true mean time for any meridian is given by an equation similar to (5.7) where l_{CM} is replaced by the longitude of the meridian.

It should not be thought that Greenwich Mean Time is determined solely on the basis of the apparent motion of the Sun. The definition of a mean solar day involves both the rotation of the Earth on its axis and the apparent motion of the Sun. The right ascension of a star α_* does not change during a year (apart from minor changes from parallax, aberration and precession which do not affect the essence of the argument) whereas the right ascension of the Mean Sun α_{MS} does change during the year. The uniform variation of α_{MS} is adjusted so that its period of rotation is precisely that of the true Sun. The annual motion of the Mean Sun, like the true Sun, is such that its right ascension is increasing. The interval between transits of the mean Sun over the observer's meridian is therefore longer than a sidereal day. This follows because the motion of celestial objects is from east to west on the sky and right ascension increases eastwards.

Suppose that the sidereal time is S_1 when the Hour Angle of the Mean Sun is H_1 and its right ascension is α_1. Let S_2 be

---

these countries adopted Central European Time. It was subsequently considered more convenient for Britain to do so too. However, the effect of adopting the time of a meridian 15° to the east of Greenwich was very marked in the more western parts of the British Isles lying up to 10° west of Greenwich and so 25° west of the central meridian for Central European Time. The view that the minority living in Western districts and those who require daylight as early as possible in the working day should not be inconvenienced prevailed and the country returned to GMT. Summer Time is simply a device to maximise the amount of daylight during the working day. A real objection to the experiment of adopting Central European Time was that it was called British Standard Time. Such misuse of the appelation "standard" could lead to confusion, the more particularly when two names CET and BST applied to the same central meridian and both had claims to being "standard".

the sidereal time exactly one mean solar day later. From the definition of a mean solar day the Hour Angle of the Mean Sun will have increased by 24^h. This means that 24 hours of mean solar time have elapsed. Then,

$$S_1 = H_1 + \alpha_1 \tag{5.8}$$

and

$$S_2 = (H_1 + 24) + \alpha_2. \tag{5.9}$$

On subtraction of eqn. (5.8) from (5.9),

$$S_2 - S_1 = 24^h + (\alpha_2 - \alpha_1) = 24^h + \frac{360^\circ}{365^d{\cdot}2422},$$

$$= 24^h + 3^m 56^s{\cdot}556, \tag{5.10}$$

where 365·2422 mean solar days is the length of the Tropical Year (see later). A solar day is therefore $24^h 3^m 56^s{\cdot}556$ in the measure of sidereal time. Conversely, a sidereal day is $23^h 56^m 04^s{\cdot}090$ in the measure of mean solar time. Clearly, therefore, mean solar time can be related to sidereal time by means of an algebraic expression. The need for mean time is a civil rather than an astronomical requirement; the rotation of the Earth is still fundamental to the establishment of civil time and the variation of rotation period which affects sidereal time will also affect civil time.

With this digression on mean time as background, it is desirable to define *Universal Time* (U.T.) more formally. Universal Time is defined to be the Greenwich hour angle of a specified fiducial point on the celestial equator plus 12^h. The addition of 12^h is to ensure that the Civil Day begins at midnight. U.T. conforms closely with the mean diurnal motion of the Sun. The definition of the fiducial point is such that it is at a position near that of the Sun at the beginning of the year. Suppose that with respect to some epoch (see Chapter VII) the right ascension of the fiducial point is α_0. The point is given a uniform motion θ in right ascension. Allowance must be made for the movement of the origin of the right ascension resulting from general precession of amount $(a_1 t + a_2 t^2)$ where t is the time elapsed since the defined epoch. At time t the right ascension of the fiducial point is therefore

$$\alpha = \alpha_0 + \theta t + a_1 t + a_2 t^2. \tag{5.11}$$

The value of α_0 is fixed so that the fiducial point is nearly in the same position as the Sun at the beginning of the year. Newcomb evaluated eqn. (5.11) obtaining

$$\alpha_U = \alpha_0 + at + a_2t^2 = 18^h38^m45^s{\cdot}836 + 8\,640\,184^s{\cdot}542T_U + 0^s{\cdot}0929T_U^2, \qquad (5.12)$$

where T_U is measured in Julian centuries (of 36525 days in the measure of U.T.) from Mean Noon on 1900 Jan 0, and $a = \theta + a_1$.

Since, from its definition, U.T. has the character of an Hour Angle, an equation of type (3.3) can be written down in terms of Greenwich Mean Sidereal Time (GMST). Let H_U be the Greenwich hour angle of the fiducial point, so that

$$\text{U.T.} = 12^h + H_U. \qquad (5.13)$$

Then

$$\text{GMST} = H_U + \alpha_U = 24^h + H_U + \alpha_U, \\ = \text{U.T.} + 12^h + \alpha_U. \qquad (5.14)$$

Since α_U is known at any time from eqn. (5.12) and the Greenwich Mean Sidereal Time can be established by observation, the Universal Time may be calculated. The observationally established Universal Time is denoted by U.T.0. Correction of U.T.0 for polar motion gives U.T.1 while correction of U.T.1 for the extrapolated seasonal variation of the Earth's rotation gives U.T.2. Variability of the length of the Mean Solar Day is not eliminated but U.T.2 is almost free of periodic variation. U.T.C. (Universal Coordinated Time) is the time which is disseminated by the time services, and is a smoothed form of U.T.2. U.T.C. should not depart from U.T.1. by more than $0^s{\cdot}7$, but in practice this tolerance is extended to 1^s. The irregular variations in the Earth's period of rotation carry through into the determination of U.T. irrespective of whether it is U.T.0, U.T.1 or U.T.2. Hence U.T. and sidereal time are compatible and may be used together for practical observational purposes. In the Astronomical Ephemeris a correction ΔU.T.1 = (U.T.1 − U.T.C.) is listed. This is a retrospective correction.

5.3 Ephemeris Time

Because the Earth is not a perfect time-keeper it is desirable to have a uniform measure of time. Any measure of time which had uniformity* would be appropriate. Clearly, since such a system could be arbitrarily constructed, it need have no connection with astronomical measures of time and the difference between the uniform measure and astronomical time would require to be determined in arrear. Had an atomic clock of sufficient precision been available when the irregular variability of sidereal time was detected, it is likely that atomic time would have been taken as the standard uniform measure. However, one was not available, and it was decided that the uniform measure of time should be the time argument of the dynamical equations governing the motions of the solar system. In practice it is the motion of the Moon which can be most readily determined with accuracy and *ephemeris time* refers to the dynamical theory of the motion of the Moon. The determination of ephemeris time is therefore only as good as the theory of the lunar orbit. At present the theory of the lunar orbit is adequate, given the limits of precision with which observations can be carried out.

The measure of ephemeris time was chosen so that it closely agreed with universal time during the nineteenth century. At 1901·5 the difference was $-2^s{\cdot}54$, at 1950·5 the difference was $+29^s{\cdot}42$ and at 1973·5 the difference was $+43^s{\cdot}6$ (extrapolated). A variation of about 1^m in the century is to be expected. Since ephemeris and universal times are almost equal it is necessary to avoid confusion by stating which system of time measurement is being used. In this chapter we shall use a subscript E to denote ephemeris time. (In many texts ephemeris time is denoted by T.)

The fundamental epoch from which ephemeris time is measured is 1900 Jan. 0, 12^h E.T. The geometric mean longitude of the Sun referred to the mean equinox of date at that instant was taken to be $279°41'48''{\cdot}04$. Although the definition is perfectly clear the instant has to be determined observa-

* Uniformity here should be interpreted to mean the closest approach to uniformity that is currently available, since departures from uniformity can only be detected if a *more* uniform measure is known.

tionally. In reducing the observations a value must be assumed for the constant of aberration (see Chapter VI). This may change as better fundamental constants become available and, if changed, will result in a new determination of the instant of the fundamental epoch. It is to be noted that the fundamental epoch for ephemeris time is not identical with the fundamental epoch for Universal Time. The fundamental epoch for ephemeris time is about 4^s later than that for Universal Time.

The primary unit of ephemeris time is the tropical year at the fundamental epoch. The tropical year is the interval during which the Sun's mean longitude increases by 360°. Newcomb's expression for the mean longitude L of the Sun is

$$L = 279°41'48''\cdot04 + 129\,602\,768''\cdot13 T_E + 1''\cdot089 T_E^2. \tag{5.15}$$

Then

$$\begin{aligned} \frac{dL}{dT_E} &= 129\,602\,768''\cdot13 + 2''\cdot178 T_E, \\ &= 129\,602\,768''\cdot13, \text{ for } T_E = 0. \end{aligned} \tag{5.16}$$

In the above equation T_E is the ephemeris time measured in units of centuries of 32 625 ephemeris days. The length of the tropical year at 1900 Jan. 0 at 12^h E.T. will therefore be

$$\begin{aligned} \frac{2\pi}{(dL/dT_E)/100} &= \frac{360 \times 60 \times 60}{1\,296\,027\cdot681\,3} \times 365\cdot25 \times 86\,400, \\ &= 31\,556\,925\cdot974\,7 \text{ ephemeris seconds.} \end{aligned} \tag{5.17}$$

The above definition of ephemeris time leads to the definition of the length of an ephemeris second.

In practice ephemeris time is obtained by observation of the position of the Moon (or the Sun and other planets—it so happens that the geocentric motion of the Moon is greater than the motion of the other bodies and gives greater accuracy of measurement). The Universal Time of observation can be determined very accurately and with little delay. The difference ΔT where

$$\Delta T = \text{E.T.} - \text{U.T.}, \tag{5.18a}$$

can then be found. U.T. is the universal time of observation of the Moon (or planet) and E.T. is obtained from the dynamical theory (ephemeris) by interpolation to find its value when the planet is in its observed position. That is, the fundamental parameter is the observed position of the body—U.T. is the universal time of that observation and E.T. is the theoretically predicted time at which the body should be in its observed position. Clearly the correction ΔT can only be made in arrear when an accurate determination of the lunar position has been completed. Although values of ΔT can be extrapolated (so that ephemeris time can be converted to Universal Time for practical observation), its accurate determination is not available until some time after the observations have been made.

Ephemeris time is derived from the equation of motion of the Moon. However, the equation of motion undergoes refinement and the assessment of Ephemeris time changes. The International Lunar Ephemeris gives an ephemeris time denoted by E.T.0 while modifications of this Ephemeris give E.T.1 and E.T.2.

The advent of atomic clock time has given a physically realisable uniform measure of time as far as can at present be ascertained. Consequently the need for ephemeris time, other than for astronomical problems, has disappeared. Prior to 1972 atomic time was disseminated in a form denoted by A.1. As from 1972 January 1 atomic time is disseminated as International Atomic Time (I.A.T). For most practical purposes E.T. is related to I.A.T. by the relation

$$\text{E.T.} = \text{I.A.T.} + 32^{\text{s}}{\cdot}18. \qquad (5.18\text{b})$$

A first approximation to ΔT namely $\Delta T(A)$ is tabulated in the Astronomical Ephemeris. $\Delta T(A)$ is defined in terms of practical measures of time such as I.A.T. and U.T.1 to be

$$\Delta T(\text{A}) = \text{I.A.T.} + 32^{\text{s}}{\cdot}18 - \text{U.T.1} \qquad (5.18\text{c})$$

Prior to 1972, $\Delta T(\text{A})$ was defined by

$$\Delta T(\text{A}) = 32^{\text{s}}{\cdot}15 + \text{A.1} - \text{U.T.C.} \qquad (5.18\text{d})$$

The form for $\Delta T(\text{A})$ given in eqn. (5.18c) is now used in the Astronomical Ephemeris where it is tabulated (to 4 significant figures) retrospectively from 1956 to 1972 Jan. 1 and extrapolated values given (to three significant figures) until 1975 Jan. 1 (1974 Astronomical Ephemeris). This degree of accuracy is sufficient for most practical purposes.

In section 5.2 on Universal Time it was shown that a mean sidereal day of 24^h (=86 400 mean sidereal seconds) was equivalent to $23^h56^m04^s{\cdot}090$ (=86 164·09054 mean solar seconds) of mean solar time. The ratio of the mean solar and mean sidereal days is, therefore, 1·002 737 909 265 $+0{\cdot}589\,T_U\,.\,10^{-10}$. The correction ΔT is determined only to the second place of decimals and so it is clear that the same factor may be used with ephemeris time as with universal time.

Ephemeris time is a system of time measurement which is independent of the period of the Earth's rotation. Clearly practical calculation of Hour angle cannot be carried out in terms of ephemeris time. However, it is useful to consider E.T. in the same terms as U.T. A fundamental meridian called the *ephemeris meridian* may be defined. This may be considered to be the equivalent of the Greenwich Meridian for a uniformly rotating earth. At 12^h U.T. the E.T. is $12^h + \Delta T$. The ephemeris meridian, if ΔT is positive, therefore lies to the East of the Greenwich Meridian by an angular amount 1·002 738 ΔT (the factor 1·002 738 gives the conversion from ephemeris to sidereal time). *Ephemeris Sidereal Time* (E.S.T.) is then the hour angle of the equinox measured from the ephemeris meridian. The *Ephemeris Hour Angle* (E.H.A.) of any object is its hour angle measured from the ephemeris meridian. Then

$$\text{E.S.T.} = \text{E.H.A.} + \alpha_E. \tag{5.19}$$

where α_E is the right ascension of the object defined with respect to the ephemeris meridian.

Ephemeris longitude is the longitude of any station measured from the ephemeris meridian. Clearly the precise position of the ephemeris meridian is not known until ΔT has been found, but all calculations involving U.T. now have a parallel in E.T. In particular if

$$\begin{aligned}\text{U.T.} + 12^h &= \text{Greenwich Hour angle of a point on the celestial equator whose right ascension is } \alpha_U\\ &= \text{G.S.T.} - \alpha_U,\end{aligned}$$

then

$$\text{E.T.} + 12^{\text{h}} = \text{Ephemeris Hour angle of a point on the celestial equator whose right ascension is } \alpha_{\text{E}}$$
$$= \text{E.S.T.} - \alpha_{\text{E}},$$

where

$$\alpha_{\text{U}} = 18^{\text{h}}38^{\text{m}}45^{\text{s}}{\cdot}836 + 8\,640\,184^{\text{s}}{\cdot}542T_{\text{U}} + 0^{\text{s}}{\cdot}092\,9T_{\text{U}}^2, \text{ (time in terms of U.T. epoch 1900 Jan 0, } 12^{\text{h}}\text{U.T.)} \tag{5.20}$$

and

$$\alpha_{\text{E}} = 18^{\text{h}}38^{\text{m}}45^{\text{s}}{\cdot}836 + 8\,640\,184^{\text{s}}{\cdot}542T_{\text{E}} + 0^{\text{s}}{\cdot}092\,9T_{E}^2, \text{ (time in terms of E.T. epoch 1900 Jan 0, } 12^{\text{h}}\text{E.T.)}$$

Hence from eqn. (5.20)

$$\begin{aligned} 0 &= \text{U.T.} + 12^{\text{h}} - \text{GST} + \alpha_{\text{U}}, \\ 0 &= \text{E.T.} + 12^{\text{h}} - \text{EST} + \alpha_{\text{E}}. \end{aligned} \tag{5.21}$$

So that on subtraction of the above eqns. (5.21)

$$\begin{aligned} 0 &= (\text{E.T.} - \text{U.T.}) + \{12^{\text{h}}(\text{E.T.}) - 12^{\text{h}}(\text{U.T.})\} \\ &\quad - (\text{E.S.T.} - \text{G.S.T.}) + (\alpha_{\text{E}} - \alpha_{\text{U}}), \\ &= \Delta T + 0 - 1{\cdot}002\,738\Delta T + (\alpha_{\text{E}} - \alpha_{\text{U}}), \end{aligned}$$

i.e.,

$$\alpha_{\text{E}} = \alpha_{\text{U}} + 0{\cdot}002\,738\Delta T, \tag{5.22}$$

since the difference between 12^{h}(E.T.) and 12^{h}(U.T.) is effectively zero. The relationship between ephemeris and universal time is illustrated in Fig. 5.2.

From the illustration it is clear that E.T. and U.T. cannot be algebraically related since the value of ΔT can only be determined by observation.

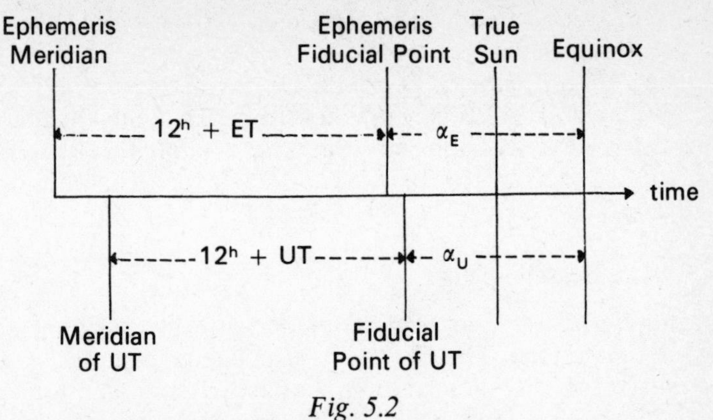

Fig. 5.2
derived from Fig. 3.1, p. 78, Explanatory Supplement to the Astronomical Ephemeris

5.4 Julian Date

It is frequently necessary to determine dates either in the future or in the past and it is useful to have some way of denoting astronomical days. To be useful the sequence should begin at some instant preceding recorded astronomical history. *Julian Day Number* 0 is defined to start at Greenwich Mean Noon on Jan 1 4713 BC. The Julian Day Number therefore denotes the number of days which have elapsed since that origin and Greenwich Mean Noon on any subsequent day. Any instant can therefore be defined by the fraction of the day from Greenwich mean noon until the instant and the Julian Day Number. The *Julian Date* of any event therefore breaks down into an integral part giving the Julian Day Number and a fractional part giving the fraction of the day elapsed since Greenwich Mean Noon. The Julian date (J.D.) for 1973 June 24 at 18^h U.T. would be J.D.2 441 858·25. Because Julian Date runs from mid-day to mid-day, no change of Day Number occurs at night. For this reason Julian Date is used as the time parameter in double and variable star observations. In the Explanatory Supplement to the Astronomical Ephemeris the Julian Day numbers are tabulated at the beginning of each month for each year from 1900 to 1999. The date Jan 0 or March 0 is found. This simply means that the Julian date may be found on Jan 24 by adding 24 to the value for Jan 0. Jan 0 of one year is identical with December 31 of the previous year.

The *Julian Ephemeris Date* (J.E.D.) began day 0 at 12^h E.T. on Jan 1 4713 BC. J.E.D. will differ from J.D. at any instant by the value of ΔT at that instant. The *Modified Julian Date* (M.J.D.) is defined to be J.D. − 2 400 000·5. The origin of M.J.D. is 1858 Nov 17 at *0^h U.T.* (*not* 12^h U.T.).

5.5 The Besselian Year

The Besselian Year is used in deriving precessional and other corrections. The *Besselian Year* begins when the right ascension of the fiducial point whose right ascension α_E, determined by equation (5.20) duly altered by the correction for aberration, attains the value 18^h40^m. This value is attained near the beginning of the year. The beginning of the Besselian year is denoted by attaching "·0" to the year in question. The Besselian year 1973·0 begins on 1973 Jan 0·494 E.T. The beginning of the Besselian Year is tabulated from 1900–1999 in the Explanatory Supplement to the Astronomical Ephemeris on pages 434, 435. It should be noted that while the length of the Besselian Year is shorter than the tropical year by an amount $0^s{\cdot}1418\,T_E$ (T_E being the ephemeris time in centuries after 1900) it is treated as being identical for practical purposes.

5.6 The Calendar

The calendar is a means whereby the passage of time may be measured in some agreeable manner for the purposes of civil life in all its aspects. The standard intervals of time measurement such as the year, the month and the day are based on astronomical periods, namely the periods of the Earth's motion about the Sun, the Moon's motion about the Earth and the daily rotation of the Earth. The week may be an artificial construction deriving from both the Jewish and Roman calendars and passing into the Christian era.

The complexity of calendar compilation derives from the fact that no astronomical body has a period which can be measured in integral units. The *synodic month* is 29·5306 days, the tropical year is 365·2422 days, the number of lunations in a tropical year is 12·3683—and so on. Calendars based on lunar months were faced with the problem of intercalating a month when the seasons and calendar became out of step. With the passage of time more refined methods of intercalating were adopted. However, it is the Roman Calendar which most affects the calendar in use today. Like other ancient calendars the Roman calendar was lunar in basis, intercalation being fixed by the authorities. Prior to 46 BC the continually necessary intercalation became neglected with the result that the calendar and seasons became two months out of step. Julius Caesar, on the advice of the Alexandrian astronomer Sosigones, revised the calendar, adopting a value of 365·25 days for the length of the year. Intercalation was standardised to the insertion of one intercalary day every fourth year—that is, a cycle of 3 years of 365 days followed by a year of 366 days. This calendar seems to have been finally established by AD 8.

The Julian calendar in many respects was a different way of implementing the cycle of intercalation introduced by the Greek astronomer Callipus. His cycle had a period of 76 years (=940 lunations = 27759 days). The Julian calendar rationalised intercalation of months to an intercalary day every fourth year. The length of a year based on the Callipic cycle of 27759 days is close to 365·25 days as in the Julian Calendar.

However, by the fifteenth century AD it was becoming clear that calendar and seasons were again out of step. The Gregorian calendar was introduced in 1582 and is based on a 400 year cycle. The essential difference between the Julian Calendar and the Gregorian calendar is the omission of 3 leap years in the 400 year cycle. The centennial year may only be a leap year if the number of the year is divisible by 400. In the Gregorian calendar 1600 and 2000 are leap years but not 1700, 1800 or 1900, whereas in the Julian calendar these latter years would have been leap years. On the basis of the Julian calendar each interval of 400 years would have 146 100 days whereas, obviously, on the Gregorian system the same interval has 146 097 days. The length of the Gregorian year is 365·2425 days and is therefore only 0·0003 days longer than a tropical year. The Gregorian calendar will be only 3 days in error after 10^4 years and in consequence there is little point in complicating the Gregorian calendar for current civil use. In establishing the Gregorian calendar, 10 days were omitted in 1582 so that the day following Oct 4 was Oct 15, restoring the date of the Vernal Equinox to March 21. The Gregorian calendar was not accepted in England until 1752. (In that year the beginning of the year was transferred to Jan 1 from March 21 and the 10 days were omitted so that September 14 followed September 3.)

The system of chronology using the beginning of the Christian era begins with AD 1 as the first year of that era. The previous year was 1 BC. For astronomical purposes the year preceeding AD 1 is denoted by 0 and is a leap year. The year -1 corresponds to 2 BC. It should be noted that with this system of chronology the century should end on the 31st December of the centennial year i.e. 1900 Dec 31 or 2000 Dec 31 and the new century begin with next succeeding year i.e. 1901 Jan 1 or 2001 Jan 1. Conventional usage now begins the century on Jan 1 of the centennial year and ends on Dec 31 of the year preceeding the centennial year, i.e. 1900 Jan 1 to 1999 Dec 31.

5.7 The Seasons

The Sun appears to move across the celestial sphere in the plane of the ecliptic. In Fig. 5.3 the celestial equator is the great circle U♈V while the ecliptic is the great circle F♈G. The pole of the equator is P and the pole of the ecliptic is K. The Sun crosses the celestial equator at the equinoxes, i.e. the Sun is instantaneously at ♈ (or Ω) on or about March 21 (or September 21). The Sun attains its greatest northerly declination on or about June 21 (i.e., is at G) and its most southerly declination on or about December 21. The equinoxes and solstices therefore allow the division of the year into four parts called *the seasons*. The interval between Vernal Equinox and Summer Solstice is called *Spring*; the interval between Summer Solstice and Autumnal Equinox is called *Summer*; the interval between Autumnal Equinox and Winter Solstice is called *Autumn* and the interval between Winter Solstice and Vernal Equinox is called *Winter**. It is clear that the seasons so defined do not correspond exactly with the common usage of the terms. Because of the non-uniform

* In the Northern hemisphere.

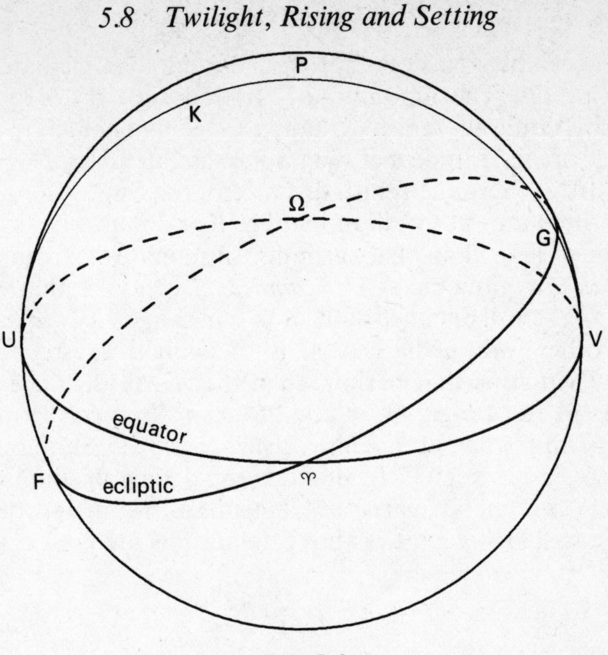

Fig. 5.3

motion of the Earth in its elliptical orbit the lengths of the seasons are not equal. The mean durations of Spring, Summer, Autumn and Winter in the Northern Hemisphere are, respectively, $92^d.84$, $93^d.60$, $89^d.78$ and $89^d.02$.

5.8 Twilight, Rising and Setting

Although these phenomena are not strictly related to the determination of time, they are related to the observation of objects and must be taken into account in practice. Times for twilight, rising and setting do not require high precision since such quantities are only required for planning observational programmes and are not required for determination of parameters of precision.

5.8.1 Twilight. The Earth's atmosphere contains particulate matter which scatters sunlight. (This may be deduced from (a) the red appearance of the Sun at Sunset or Sunrise or from (b) the blue colour of the sky since it is known that light of this colour is most easily scattered by small particles.)

Therefore the sky does not become dark immediately the Sun has set. The amount of scattered sunlight, depending critically on the zenith distance of Sun, changes by nine orders of magnitude between solar zenith distances of 90° and 110°. Where the zenith distance of the Sun is in excess of 108° the amount of illumination from scattered sunlight becomes less than background illumination from other sources (air glow etc.). *Astronomical Twilight* is that period when the zenith distance of the Sun is in range $90° < z < 108°$. Two other intermediate periods of twilight are recognised. *Civil Twilight* is that period when the zenith distance of the Sun is in the range $90° < z < 96°$ and *Nautical Twilight* is that period when the zenith distance of the Sun is in the range $90° < z < 102°$. In practice the degree of illumination during twilight is variable depending on meteorological conditions. However, the above definitions are convenient.

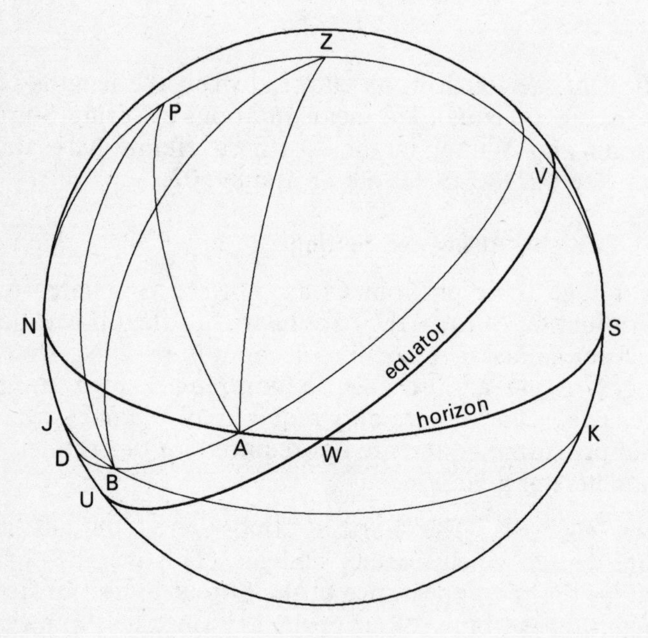

Fig. 5.4

5.8 *Twilight, Rising and Setting*

In Fig. 5.4 the great circle NWS is the horizon, UWV is the celestial equator, NPZS is the observer's meridian, P is the north celestial pole and Z is the observer's zenith. Suppose the Sun has declination $+\delta$ on the day it sets at A. When the Sun reaches the point B, its zenith distance is 108° and astronomical twilight is ending. In the spherical triangle PZB

$$\text{PZ} = 90° - \phi,\ \text{PB} = 90° - \delta,\ \text{ZB} = z_t = 108°,$$
$$\text{Z}\hat{\text{P}}\text{B} = H_t, \tag{5.23}$$

where ϕ is the latitude of the observer. Similarly in the spherical triangle PZA

$$\text{PZ} = 90° - \phi,\ \text{PA} = 90° - \delta,$$
$$\text{ZA} = z_s = 90°,\ \text{Z}\hat{\text{P}}\text{A} = H_s. \tag{5.24}$$

Application of the Cosine Rule to each triangle gives

$$\cos z_t = \cos 108° = \sin\phi \sin\delta + \cos\phi\cos\delta\cos H_t \tag{5.25}$$

and

$$\cos z_s = 0 = \sin\phi\sin\delta + \cos\phi\cos\delta\cos H_s, \tag{5.26}$$

i.e.,

$$\cos H_s = -\tan\phi\tan\delta.$$

Hence the difference $(H_t - H_s)$ gives the duration of twilight. In some latitudes at certain times of the year twilight will continue all night. Consider the small circle of altitude JBK. All points on JBK are at a zenith distance (from Z) of 108°. ABD is a parallel of declination $(+\delta)$. Just as in the case of a circumpolar star, if D lies closer to N than J, twilight will not end. If J lies closer to N than D then twilight will end. In mathematical terms, twilight will end if

$$\text{ND} = 90° - \phi - \delta > 18°. \tag{5.27}$$

since $\text{NU} = 90° - \phi$ and $\text{DU} = \delta$. Clearly for latitude 52°, twilight will end only if $\delta < 20°$. Therefore around summer solstice, twilight will continue all night at this latitude.

Only at latitudes less than 48° will twilight end each night. The discussion above has been given in terms of twilight at sunset: clearly similar arguments apply to sunrise.

5.8.2 Rising and setting. If A is any body in Fig. 5.4 then its setting and rising of any body can be considered similarly. When a star reaches the horizon in the west it is said to set and when it appears above the horizon in the east it is said to rise. Times of rising and setting are the instants when the zenith distance is 90°. The instants of rising at any site will be subject to slight variations from this definition since the actual horizon may not correspond to the astronomical horizon. Clearly eqn. (5.26) gives the hour angle of setting (or rising). For setting $0^h < H_s < 12^h$ while for rising $12^h < H_r < 24^h$ where $H_s(H_r)$ is the hour angle of setting (rising). Eqn. (5.26) may be used to calculate the hour angles H_s or H_r for any latitude ϕ.

Of particular interest in connection with rising and setting are the Sun and the Moon. Both these objects have a finite apparent diameter. Sunrise is defined to be the instant when the upper limb of the Sun crosses the horizon. The zenith distance of the Sun's centre is then 90°50′ (allowing for the Sun's angular semi-diameter of 16′ and horizontal refraction of 34′—see Chapter VI). Sunset or Sunrise for sites h feet above sea level are determined by using zenith distances

$$90°50' + 1'{\cdot}17\,h^{1/2}. \tag{5.28}$$

For solar calculations eqn. (5.25) should be used replacing H_t by H_s or H_r and replacing z_t by the zenith distance of the Sun's centre at the time of disappearance or appearance of the upper limb of the Sun.

In the case of the Moon allowance must be made for apparent angular radius and horizontal refraction as in the case of the Sun, and additional allowance must be made for parallax (see Chapter VI). Again moonrise (or set) is determined by the appearance or disappearance of the upper lunar limb. The zenith distance for lunar rising and settings varies between 89°49′ and 89°55′ and again eqn. (5.25) should be used. The correction for height above sea level is the same as for the Sun.

In the case of the Sun it is clear that the systematic motion of the Sun in declination produces a variation in the hour angle of setting (rising). At the equinoxes when $\delta = 0$ the sunset is due west and sunrise due east. As δ increases the Sun sets and rises progressively at more northerly points on the horizon. Sunset and sunrise are at their most northerly at the summer solstice (in the northern hemisphere). Similarly as the Sun's declination decreases the points of sunrise and sunset move south. It should be noted that the fiducial point determining the mean solar day, being on the equator, always sets due west and rises due east. In the case of stars, rising (setting) will always be at the same point on the horizon (apart from the slow changes attributable to precession).

Chapter VI

Correction for Refraction, Parallax and Aberration

In Chapter III coordinate systems were defined on the basis of certain fixed planes, namely the plane of the equator and the plane of the ecliptic. However, the apparent position of an object on the sky will be affected by several phenomena: refraction in the Earth's atmosphere, parallax, aberration, proper motion and precession. Small displacements of the celestial object from its geometrical position will result from each of these phenomena. The effect of precession (and nutation) is the result of a dynamical interaction between the Earth and the remainder of the solar system, such that the planes of the celestial equator and ecliptic cannot be regarded as fixed planes and the zero point of the system of equatorial coordinates must be regarded as continuously altering its position. Proper motion refers to the individual motions in space of the Sun and stars: the Sun and stars are moving with respect to each other and are hence constantly altering their relative positions. Since precession refers to the alteration of the zero of coordinates and proper motion refers to the relative position of stars they will be considered together in the next Chapter (Chapter VII).

Aberration is a kinematic effect of the Earth's motion about the Sun; parallax is a geometrical effect of the Earth's being in different places with respect to the Sun at different times of the year; refraction is purely a property of the nature of the Earth's atmosphere. All of these effects may be treated on the basis that the Sun and the stars retain a fixed relative configuration and that the celestial equator and ecliptic are invariable planes. For this reason they are treated together in this chapter. The physical nature of the effect will be discussed before the displacement from the geometrical position of the celestial object is determined.

The displacement of a celestial object may be determined by consideration of the effect of the phenomenon responsible for the displacement. Displacements will be along great circles. The planes of these great circles, and so their poles, will be determined by the physical nature of the parent phenomenon. The formulae derived in chapters II and III for general displacements in terms of rectangular equatorial coordinates are used. The displacements are treated in this chapter as being of the first order in a Taylor series expansion except in the case of the parallactic displacement of the Moon. Since the displacements are only given correct to the first order, approximations producing a second order error may be substituted into the terms defining the displacement. In consequence if higher order corrections are required some of the results will require reworking from that point where an approximation is made. Corrections to higher orders are given in *Spherical Astronomy* by Woolard and Clemence.

6.1 Refraction in the Earth's Atmosphere

In figure 6.1 aa′ represents the interface between two media. bb′ is the normal to that surface. Light passing through medium 1 is incident on the interface at angle θ and is refracted so as to pass through medium 2 at angle ϕ to the normal. Snells' Law defines the index of refraction μ_{12} between media 1 and 2 to be

$$\mu_{12} = \frac{\sin\theta}{\sin\phi}. \tag{6.1}$$

Figure 6.2 illustrates the situation where light traverses three media.

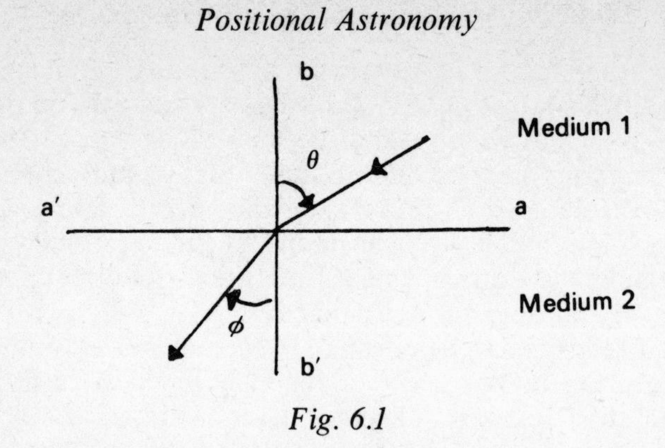

Fig. 6.1

The media are arranged in plane parallel layers. Three refractive indices can be defined,

$$\mu_{12} = \frac{\sin\theta}{\sin\phi},$$
$$\mu_{23} = \frac{\sin\phi}{\sin\psi}, \qquad (6.2)$$
$$\mu_{13} = \frac{\sin\theta}{\sin\psi}.$$

It is inconvenient to refer to refractive indices between media and it is usual to define refractive indices with respect to some

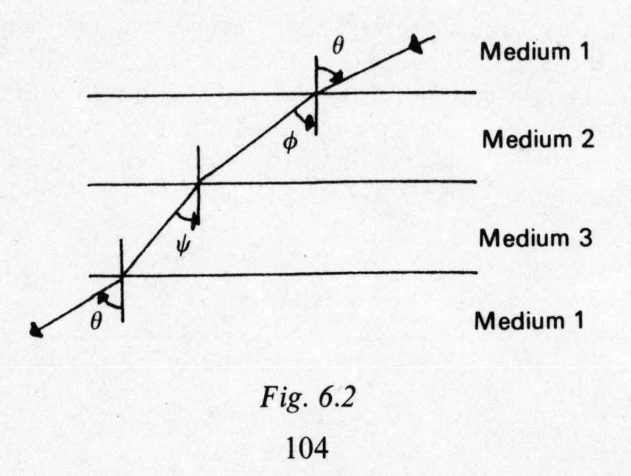

Fig. 6.2

standard medium and to treat them as a material property. A vacuum gives no refraction and in general refractive indices refer to a vacuum as standard. Therefore if medium 1 of Fig. 6.2 is a vacuum the refractive indices of medium 2 and medium 3 may be written,

$$\mu_2 = \frac{\sin\theta}{\sin\phi},$$

$$\mu_3 = \frac{\sin\theta}{\sin\psi} \tag{6.3}$$

$$\mu_{23} = \frac{\mu_3}{\mu_2} = \frac{\sin\phi}{\sin\psi},$$

or

$$\mu_2 \sin\phi = \mu_3 \sin\psi. \tag{6.4}$$

Eqn. (6.4) gives a general result.

Consider now the Earth's atmosphere represented by n plane parallel homogeneous layers as in Fig. 6.3. Any layer i is at height h_i above the bottom of the lowest layer. The

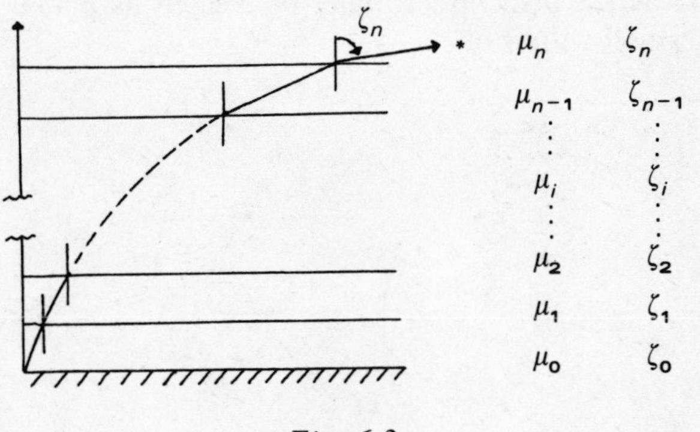

Fig. 6.3

upper layer will be assumed to be a vacuum (i.e. $\mu_n = 1$). The angle of incidence at each interface will be denoted by ζ_i where i refers to the layer above the interface. If light from a star at * is incident at angle ζ_n, it will be refracted so that the light is incident at an angle ζ_{n-1} to the normal at the bottom of the layer $n - 1$ and so on down through the atmosphere. For any two adjacent layers, say i and j in the atmosphere eqn. (6.4) will apply so that

$$\ldots = \mu_i \sin \zeta_i = \mu_j \sin \zeta_j = \ldots, \tag{6.5}$$

and in particular

$$\mu_0 \sin \zeta_0 = \mu_n \sin \zeta_n = \sin \zeta_n. \tag{6.6}$$

In the case of a plane parallel atmosphere ζ_n is just the true zenith distance ζ of the star. Hence

$$\mu_0 \sin \zeta_0 = \sin \zeta = \sin (\zeta_0 + R), \tag{6.7}$$

where

$$\zeta = \zeta_0 + R \tag{6.8}$$

and R is the angle of refraction. Expanding the right hand side of eqn. (6.7) and treating R as a small angle

$$R = (\mu_0 - 1) \tan \zeta_0. \tag{6.9}$$

Eqn. (6.9) gives the variation of an angle of refraction with observed zenith distance. This simple form for R is only true so long as the atmosphere may be treated as plane parallel, i.e. for small values of ζ_0.

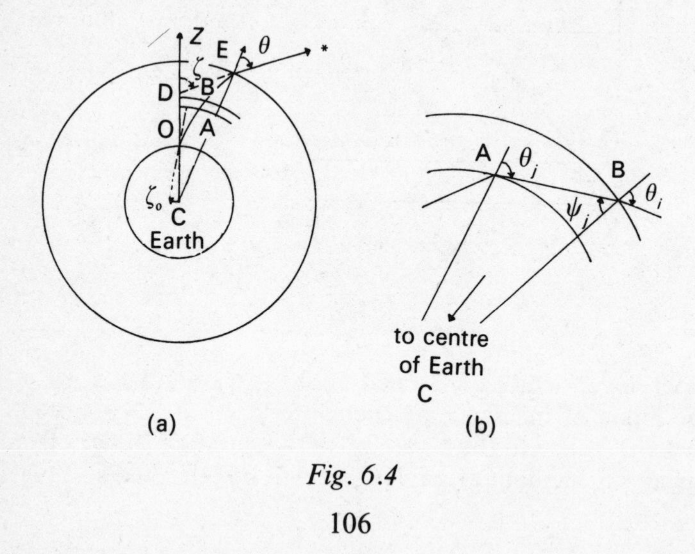

Fig. 6.4

An expression for R may be derived assuming an atmosphere stratified in concentric layers about the earth as in Fig. 6.4. Each layer is assumed to be homogeneous. Suppose a ray of light from a star at * is incident upon the Earth's atmosphere at angle θ to the normal at E. The ray is refracted by the atmosphere and the star is observed from O to have apparent zenith distance ζ_0. Suppose A is at the base of layer j and B is at the base of layer i (see also Fig. 6.4(b)). Then at B eqn. (6.4) may be applied i.e.

$$\mu_i \sin \theta_i = \mu_j \sin \psi_j, \tag{6.10}$$

where θ_i is the angle of incidence and ψ_j is the angle of refraction at the interface. The sine rule for plane triangles can be applied to the triangle CAB to give

$$\frac{r_i}{\sin (180 - \theta_j)} = \frac{r_j}{\sin \psi_j},$$

or

$$r_i \sin \psi_j = r_j \sin \theta_j, \tag{6.11}$$

where $r_i = \text{CB}$, $r_j = \text{CA}$. Combination of eqns. (6.10) and (6.11) give

$$r_i\mu_i \sin \theta_i = r_j\mu_j \sin \theta_j. \tag{6.12}$$

In particular

$$\rho\mu_0 \sin \zeta_0 = r \sin \theta,$$

where r is the radius of the "top" of the atmosphere, $\zeta_0(= \theta_0)$ is the observed zenith distance of the star, μ_0 is the refractive index at ground level and ρ is the radius of the earth. Application of the sine rule for plane triangles to the triangle CDE gives

$$\frac{\rho + h}{\sin \theta} = \frac{r}{\sin (180° - \zeta)},$$

or

$$r \sin \theta = (\rho + h) \sin \zeta, \tag{6.13}$$

where $h = \text{OD}$ and ζ is the true zenith distance ($\text{Z}\hat{\text{D}}\text{E}$).

Combining eqns. (6.12) and (6.13),

$$\rho\mu_0 \sin \zeta_0 = (\rho + h) \sin \zeta,$$

or

$$\left(1 + \frac{h}{\rho}\right) = \frac{\mu_0 \sin \zeta_0}{\sin \zeta} = \frac{\mu_0 \sin \zeta_0}{\sin (\zeta_0 + R)}, \qquad (6.14)$$

where R is the angle of refraction. Although this is a compact expression, it is not a useful expression for practical purposes since it involves both R and h. Both R and h are unknowns and once one is available the other can be determined. It is usual to determine R from observation though in principle h could be worked out were the variation of density with height in the atmosphere known. Details of the necessary calculations can be found in Smart, *Spherical Astronomy* (pp. 62–68) or Woolard and Clemence, *Spherical Astronomy* (pp. 79–84). Refraction on the basis of eqn. (6.14) can be reinterpreted to suggest that for a given zenith distance the observer may be considered to be elevated to a height h. This is the way in which correction for refraction is allowed for in the case of reduction of occultations. Table 6.1 (reproduced by permission of the Controller, HMSO) shows how h varies with ζ.

Table 6.1

Variation of h with ζ

$\zeta^{(0)}$	h(m)	ζ	h	ζ	h
0	0	70	20	88	610
30	0	80	70	90	1540
60	5	84	170	—	—

The value of h/ρ therefore never exceeds 0·000 242 and is frequently much smaller. For a true zenith distance of 90° the angle of refraction—the horizontal refraction—is 34′ (adopted).

Simple expansion of eqn. (6.14) indicates the form of the approximate expressions which may be used.

1st approximation:

$$\left(1 + \frac{h}{\rho}\right)(\sin \zeta_0 + R \cos \zeta_0) = \mu_0 \sin \zeta_0. \qquad (6.15)$$

$$R = \frac{\mu_0 - (1 + h/\rho)}{(1 + h/\rho)} \tan \zeta_0,$$

$$= (\mu' - 1) \tan \zeta_0, \qquad (6.16)$$

where $\mu' = \mu_0/(1 + h/\rho)$. It is clear that eqns. (6.9) and (6.16) are comparable. When ζ is small, R is small and h/ρ is small so that $\mu' \simeq \mu_0$ in this case.

2nd approximation:

$$\left(1 + \frac{h}{\rho}\right)\{\sin \zeta_0(1 - \tfrac{1}{2}R^2) + R \cos \zeta_0\} = \mu_0 \sin \zeta_0,$$

so that

$$R = (\mu' - 1) \tan \zeta_0 + \tfrac{1}{2}(\mu' - 1)^2 \tan^3 \zeta_0. \qquad (6.17)$$

In deriving eqn. (6.17), R^2 has been approximated using eqn. (6.16). Eqn. (6.17) shows that the expansion for R is of the form

$$R = A \tan \zeta_0 + B \tan^3 \zeta_0, \qquad (6.18)$$

where A and B are coefficients depending on ζ. Although A and B depend on ζ, their variation with ζ is small and it is usual to treat A, B as constants. Since μ' cannot be determined independently, the values of A and B cannot be computed but are determined from observations as follows.

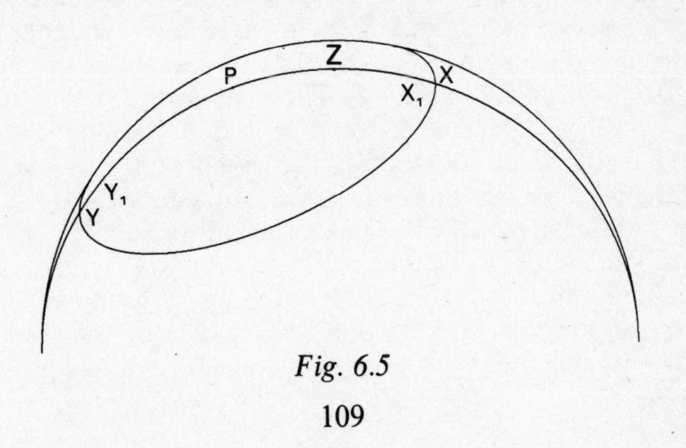

Fig. 6.5

A star of declination δ has upper culmination X and lower culmination Y (see Fig. 6.5). Because of refraction the apparent position of the star is displaced towards the zenith Z, i.e. to X_1 and to Y_1 respectively. Let the observed zenith distance be ζ_1 when the star is apparently at X_1 and ζ_2 when the star is apparently at Y_1. Then

$$\zeta_X = \zeta_1 + R_1,$$

$$\zeta_Y = \zeta_2 + R_2,$$

where $\zeta_{X,Y}$ are the true zenith distances and $R_{1,2}$ are the corresponding angles of refraction. At the time of meridian transit

$$90° = \delta + \zeta_X + 90° - \phi,$$

or

$$\phi - \delta = \zeta_X = \zeta_1 + A \tan \zeta_1 + B \tan^3 \zeta_1, \tag{6.19}$$

where ϕ is the known latitude of the observer. For lower culmination,

$$180° - \phi - \delta = \zeta_Y = \zeta_2 + A \tan \zeta_2 + B \tan^3 \zeta_2. \tag{6.20}$$

Subtraction of eqns. (6.19) from (6.20) gives

$$180° - 2\phi = (\zeta_2 - \zeta_1) + A(\tan \zeta_2 - \tan \zeta_1) + B(\tan^3 \zeta_2 - \tan^3 \zeta_1). \tag{6.21}$$

Observation of a second star will clearly give a similar equation and in principle A and B could be determined. However, it is more usual to consider a large number of stars, determining A and B by a least squares procedure. Woolard and Clemence, *Spherical Astronomy*, give $A = 60''{\cdot}29$ and $B = -0''{\cdot}066\,88$ at S.T.P. These values are not accurate for $\zeta \sim 90°$. For accurate work A and B should be determined from refraction tables. Tables currently in use are:

Greenwich Observations for 1898, Appendix I;
Pub. U.S. Naval Obs., *4* (2nd series), Appendix II;
Refraction tables of Pulkova Observatory (4ed. 1956).

A useful summary of refraction data can be found in *Astrophysical* Quantities (3rd Ed. p. 124, 125). In eqn. (6.9) the value of $(\mu_0 - 1)$ at S.T.P. is usually taken to be $60''{\cdot}4$. For an atmospheric pressure P (in mm Hg) and temperature T(°C) eqn. (6.9) may be written

$$R = 60''\cdot 4 \frac{(P/760)}{1 + T/273^\circ} \tan \zeta_0. \tag{6.22}$$

6.2. Parallax

6.2.1 Stellar Parallax. Because of the motion of the Earth in its orbit around the Sun, a star will be observed in slightly different directions at different times of the year. The situation is illustrated in Fig. 6.6. When the Earth is at E_1 the star X is observed at inclination θ to the Earth–Sun line E_1S. At E_2 the same star is observed at angle θ_2 to the Earth–Sun line SE_2. Assuming that the Sun and other celestial objects occupy fixed places, the effect of the Earth's motion is that any celestial object will appear to occupy different positions on the sky in a periodic way. (The fact that the Sun and stars have motions in space does not alter this definition of parallax since these space motions can be allowed for—see Chapter VII.) Application of the sine rule for plane triangles to the triangle E_1SX gives

$$\frac{d}{\sin \theta_1} = \frac{R_1}{\sin p_1} = \frac{d}{\sin (\theta - p_1)},$$

or

$$\sin p_1 = \frac{R_1}{d} \sin (\theta - p_1), \tag{6.23}$$

where $R_1 = E_1S$ and $\theta = X\hat{S}E_2$. Since R_1/d is a small quantity, p_1 will be a small angle so that eqn. (6.23) may be written in the form

$$p_1 = \frac{R_1}{d} \sin \theta. \tag{6.24}$$

The parallax of the star X is defined to be

$$\pi_1 = \frac{R_1}{d} \tag{6.25}$$

and it is clear that, on the basis of this definition, the parallax

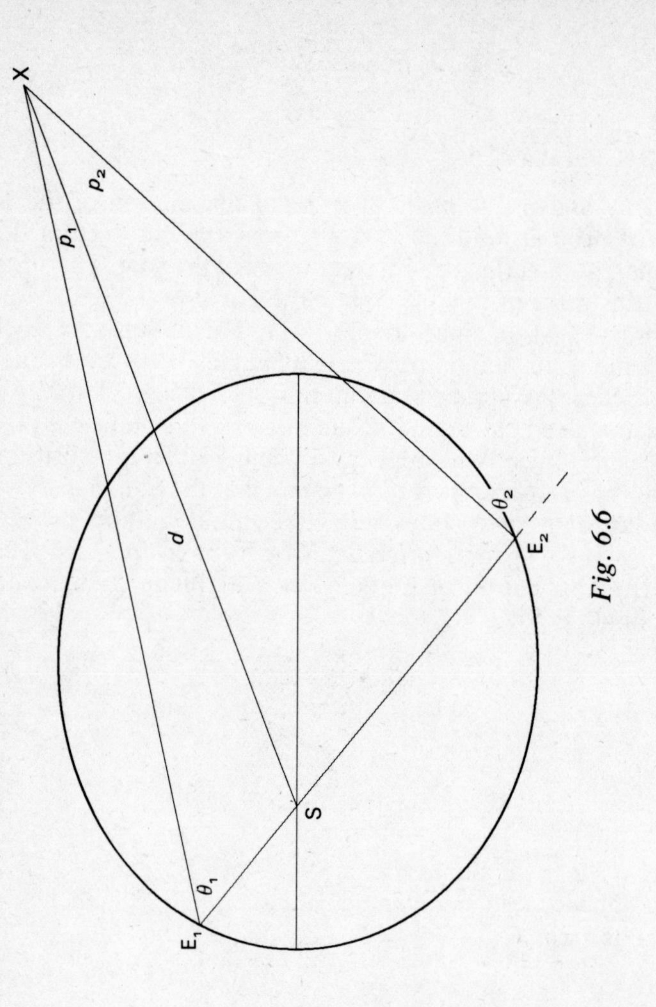

Fig. 6.6

π_1 will vary with the time of year. Therefore, to give a single value for the parallax, stellar parallax is defined to be

$$\pi = \frac{a_E}{d} \tag{6.26}$$

where $2a_E$ is the length of the major axis of the Earth's orbit. π is called the *annual parallax* of the star. Hence

$$p_1 = \left(\frac{R_1}{a_E}\right)\left(\frac{a_E}{d}\right) \sin\theta = \pi\left(\frac{R_1}{a_E}\right) \sin\theta. \tag{6.27}$$

The parallactic displacement is towards the Sun.

Before the distance to any star can be determined the properties of the Earth's orbit must be determined accurately. Measurement of stellar parallax then gives a direct measurement of the distance to a star, though clearly such direct measurement is limited to nearby stars in the neighbourhood of the Sun (say within 200 pc. of the Sun). The largest stellar parallax (for Proxima Centauri) is just less than 1″ (0″·763) and the astronomical unit of length—the *parsec* (pc)—is defined to be the distance of a star such that it would have a parallax of 1″. Therefore, a star's distance in parsecs is the reciprocal of its parallax in seconds of arc. For $a = 1{\cdot}496.10^{13}$ cm, 1 pc = $3{\cdot}086{\cdot}10^{18}$ cm. (Another common measure of stellar distance is the *light year* (l.y.), the distance travelled by a photon in 1 year.

$$1 \text{ l.y.} = 2{\cdot}998.10^{10}\text{ cm s}^{-1} \times 3{\cdot}156.10^{7}\text{ s} = 9{\cdot}462.10^{17}\text{ cm}$$

$$3{\cdot}262 \text{ l.y.} = 1 \text{ pc.})$$

6.2.2 The figure of the Earth. The parallax of a planet may be determined using a baseline drawn between two points on the Earth's surface. For many purposes it is adequate to assume that the Earth is a sphere. In the case of the determination of the distance to, and radius of, the Moon, a better knowledge of the shape of the Earth is required. The shape of the Earth may be approximated to a spheroid of revolution known as the Geoid or Hayford's Spheroid, for the purpose of calculation of parallax corrections. A spheroid is formed by rotation of an ellipse about its minor axis. This representation of the figure of the Earth is of value in geology and geography.

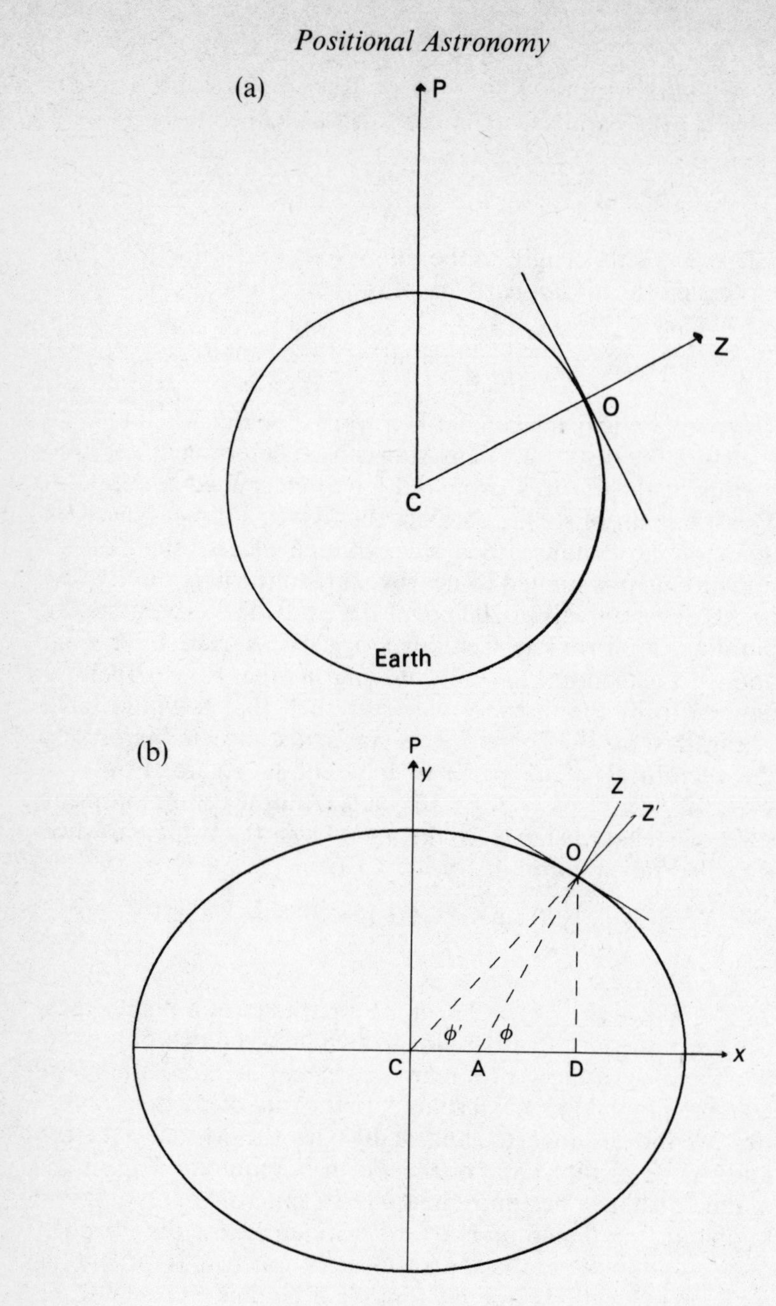

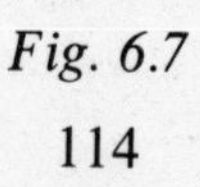

Fig. 6.7

6.2 *Parallax*

Since the tangent plane to the Earth's surface cannot be readily established, the plane of the horizon is defined to be normal to the direction defined by a plumb line. Were the Earth a sphere, either of uniform composition or stratified in shells of uniform composition, the plumb line would define the line joining the centre, C, of the Earth to the zenith, Z as in Fig. 6.7(a). However, the material of the Earth is not distributed in such a simple manner. The direction of the astronomical zenith, determined by plumb line observations is influenced by local irregularities (e.g. mountain masses, subcrustal variation of density) in the distribution of the terrestrial density. The centre of Hayford's Spheroid does not necessarily have its centre at the centre of mass of the Earth and does not necessarily have its poles on the Earth's axis of rotation. However, for the purpose of treating planetary parallax it will be assumed that for astronomical purposes the Earth has its centre of mass at the centre of Hayford's spheroid. In this case the astronomical and geodetic zeniths will be identical, being in the direction of the normal to the tangent plane at the point of observation O. In terms of Fig. 6.7(b) the *astronomical zenith* Z is in the direction AOZ defined by a plumb line. The *geocentric zenith* Z′ is defined to be that point on the celestial sphere intersected by the line CO produced. The angle $Z\hat{O}Z'$ is called the *angle of the vertical* and is denoted by v; the angle ZAD is the *astronomical latitude* ϕ and the angle Z′CD is the *geocentric latitude* ϕ'. Clearly,

$$v = \phi - \phi'. \tag{6.28}$$

Hayford's Spheroid is defined to have a flattening $f = 1/297$ where

$$f = \frac{\rho_0 - \rho_{90}}{\rho_0}, \tag{6.29}$$

where ρ_0 is the length of the Earth's semi-major axis (= Earth's equatorial radius) and ρ_{90} is the length of the semi-minor axis (= Earth's polar radius) of the ellipse of revolution;

$$\begin{aligned} \rho_0 &= 6{,}378{,}388 \text{ m}, \\ \rho_{90} &= 6{,}356{,}912 \text{ m}. \end{aligned} \tag{6.30}$$

Meridians, therefore, have an elliptical shape with equation

$$\frac{x^2}{\rho_0^2} + \frac{y^2}{\rho_{90}^2} = 1, \tag{6.31}$$

where the x-axis is in the direction CD and the y-axis is at right angles to this direction (taking y positive when measured in the direction of the north celestial pole). The eccentricity e of the meridian is defined by

$$e^2 = \frac{\rho_0^2 - \rho_{90}^2}{\rho_0^2} = 2f - f^2. \tag{6.32}$$

For astronomical purposes it is desirable to be able to relate astronomical to geodetic parameters in terms of the flattening f. In Fig. 6.7(b), let the observer O be distant ρ from C. Then

$$x = \rho \cos \phi' = C \cos \phi. \tag{6.33}$$

$$y = \rho \sin \phi' = S \sin \phi. \tag{6.34}$$

where x, y are the rectangular coordinates of O, and C, S are functions defined by eqns. (6.33) and (6.34) respectively.

If the observer is at a height H above the Geoid, then his position in the (x, y) plane will be given by $x' = x + H \cos \phi$, $y' = y + H \sin \phi$ where x, y is the position of an observer with the same astronomical latitude ϕ but situated on the Geoid. Then

$$x' = \rho' \cos \phi'' = (C + H) \cos \phi \tag{6.33a}$$

$$y' = \rho' \sin \phi'' = (S + H) \sin \phi \tag{6.34a}$$

where ρ' is the separation of the observer (at height H) from the centre of the Geoid and ϕ'' is his geocentric longitude.

It will be assumed for the remainder of this Chapter that the observer is situated on the Geoid. The slope of the tangent at O is

$$\frac{\mathrm{d}y}{\mathrm{d}x} = -\frac{x}{y}\frac{\rho_{90}^2}{\rho_0^2} \tag{6.35}$$

The slope of the normal at O is therefore $(y/x)(\rho_0^2/\rho_{90}^2)$ and so

$$\tan \phi = \frac{y}{x}\frac{\rho_0^2}{\rho_{90}^2} = \frac{\rho_0^2}{\rho_{90}^2}\tan \phi',$$

or

$$\tan \phi' = (1 - f)^2 \tan \phi. \tag{6.36}$$

Eqns. (6.33), (6.34) and (6.36) give

$$S = (1 - f)^2 C. \tag{6.37}$$

Again,

$$\begin{aligned}\rho^2 &= \rho^2 \cos^2 \phi' + \rho^2 \sin^2 \phi' = C^2 \cos^2 \phi + S^2 \sin^2 \phi \\ &= \tfrac{1}{2}C^2(1 + \cos 2\phi) + \tfrac{1}{2}S^2(1 - \cos 2\phi), \\ &= \tfrac{1}{2}(C^2 + S^2) + \tfrac{1}{2}(C^2 - S^2)\cos 2\phi, \qquad (6.38) \\ &= C^2[\cos^2 \phi + (1 - f)^4 \sin^2 \phi]. \qquad (6.39)\end{aligned}$$

Substitution of eqns. (6.33) and (6.34) in eqn. (6.31) gives

$$\frac{C^2 \cos^2 \phi}{\rho_0^2} + \frac{S^2 \sin^2 \phi}{\rho_{90}^2} = 1,$$

whence

$$C = \pm \frac{\rho_0}{[\cos^2 \phi + (1 - f)^2 \sin^2 \phi]^{1/2}} \tag{6.40}$$

It is customary to express ρ, C, S in units of ρ_0 and, taking the positive square root, eqn. (6.40) becomes

$$C = [\cos^2 \phi + (1 - f)^2 \sin^2 \phi]^{-1/2}. \tag{6.41}$$

f is a small quantity and ρ, S, C and v can be expressed as expansions in terms of f. The simplest manner of obtaining this series is to use Maclaurin's theorem. If q denotes ρ, S, C or v

$$q = (q)_{f=0} + f\left(\frac{dq}{df}\right)_{f=0} + \frac{1}{2}f^2\left(\frac{d^2q}{df^2}\right)_{f=0} + \dots. \tag{6.42}$$

(*i*) *A series expansion for C, S.*

$$(C)_{f=0} = 1, \quad \left(\frac{dC}{df}\right)_{f=0} = \sin^2\phi = -\tfrac{1}{2}(1 - \cos 2\phi),$$

$$\left(\frac{d^2C}{df^2}\right)_{f=0} = \frac{1}{8}(5 - 8\cos 2\phi + 3\cos 4\phi), \tag{6.43}$$

using eqn. (6.41). Hence

$$C = 1 + \tfrac{1}{2}f + \tfrac{5}{16}f^2 - (\tfrac{1}{2}f + \tfrac{1}{2}f^2)\cos 2\phi + \tfrac{3}{16}f^2\cos 4\phi, \tag{6.44}$$

and similarly,

$$S = 1 - \tfrac{3}{2}f + \tfrac{5}{16}f^2 - (\tfrac{1}{2}f - \tfrac{1}{2}f^2)\cos 2\phi + \tfrac{3}{16}f^2\cos 4\phi. \tag{6.45}$$

These and subsequent expansions are correct to the second order in f.

(*ii*) *A series expansion for* ρ. Using Eqns. (6.39) and (6.41)

$$\rho = \left\{\frac{\cos^2\phi + (1-f)^4\sin^2\phi}{\cos^2\phi + (1-f)^2\sin^2\phi}\right\}^{\frac{1}{2}} \tag{6.46}$$

so that

$$(\rho)_{f=0} = 1, \quad \left(\frac{d\rho}{df}\right)_{f=0} = -\tfrac{1}{2}(1 - \cos 2\phi),$$

$$\left(\frac{d^2f}{df^2}\right)_{f=0} = \tfrac{5}{8}(1 - \cos 4\phi). \tag{6.47}$$

Hence

$$\rho = 1 - \tfrac{1}{2}f + \tfrac{5}{16}f^2 + \tfrac{1}{2}f\cos 2\phi - \tfrac{5}{16}f^2\cos 4\phi. \tag{6.48}$$

where ρ is in units of ρ_0.

(*iii*) *A series expansion for v.*

$$\begin{aligned}\rho\cos v &= \rho\cos(\phi - \phi') \\ &= \rho\cos\phi\cos\phi' + \rho\sin\phi\sin\phi' \\ &= C\cos^2\phi + S\sin^2\phi = C[\cos^2\phi + (1-f)^2\sin^2\phi],\end{aligned} \tag{6.49}$$

since the angle of the vertical $v = \phi - \phi'$. Similarly

$$\rho \sin v = C[1 - (1 - f)^2] \sin \phi \cos \phi. \qquad (6.50)$$

and hence

$$\tan v = \frac{[1 - (1 - f)^2] \sin \phi \cos \phi}{[\cos^2 \phi + (1 - f)^2 \sin^2 \phi]}. \qquad (6.51)$$

Therefore, since

$$(v)_{f=0} = (\tan v)_{f=0} = 0,$$

$$\left(\frac{dv}{df}\right)_{f=0} = \sin 2\phi, \quad \left(\frac{d^2v}{df^2}\right)_{f=0} = \sin 2\phi - \sin 4\phi, \qquad (6.52)$$

$$v = \phi - \phi' = (f + \tfrac{1}{2}f^2) \sin 2\phi - \tfrac{1}{2}f^2 \sin 4\phi.$$

Eqns. (6.44), (6.45) and (6.48) are similar in that they have the general form.

$$q = a_0 + a_1 \cos 2\phi + a_2 \cos 4\phi \qquad (6.53)$$

where q stands for C, S, or ρ; and a_0, a_1, a_2 are constant functions of f. The values of a_0, a_1, a_2 are tabulated in Table 6.2 (reproduced by permission of the Controller of H.M.S.O.).

Table 6.2
Coefficients for the expansions of C, S, ρ

	a_0	a_1	a_2
C	+1·00168705	−0·00168919	+0·00000214
S	+0·99495304	−0·00167783	+0·00000212
ρ	+0·99832005	+0·00168349	−0·00000355

The coefficients in the expansion of v are given in

$$v = 695''.66 \sin 2\phi - 1''.17 \sin 4\phi. \qquad (6.54)$$

Since the astronomical latitude ϕ can be determined, use of these expansions permits the determination of ρ and ϕ'. The introduction of ρ and ϕ' into the problem can be illustrated by the determination of the parallax of a planet and its radius. This treatment is of greater relevance to the Moon.

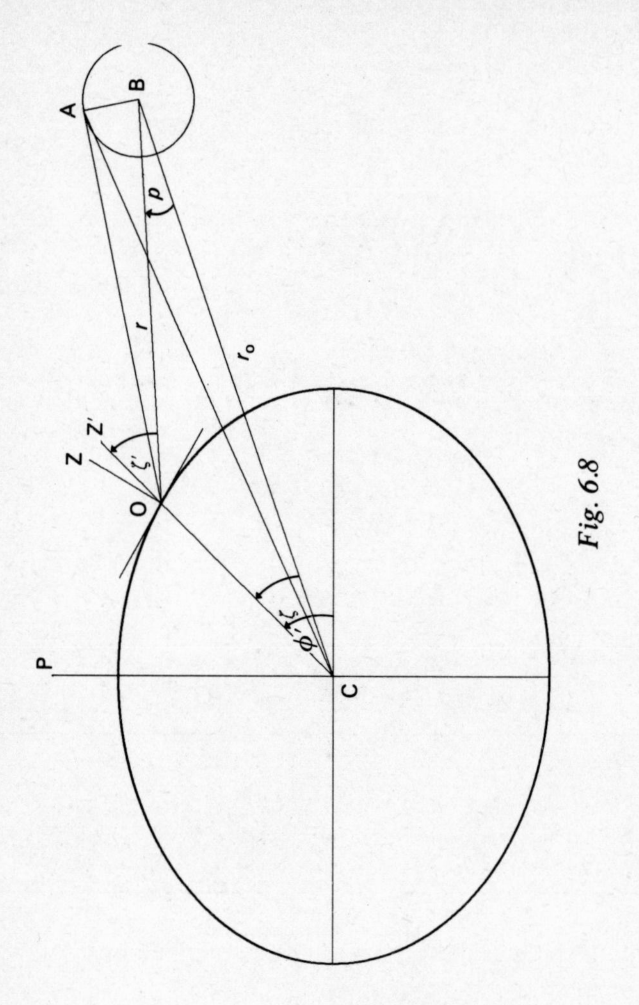

Fig. 6.8

6.2.3 Planetary Parallax

The parallax of a planet is defined in the following way. An observatory O has an astronomical zenith Z and geocentric zenith Z′. The centre of the planet is at B (see Fig. 6.8), distant r_0 from C and r from O. O is distant ρ from C. The parallactic angle C$\hat{\mathrm{B}}$O is denoted by p and from the sine rule for plane triangles

$$\sin p = \frac{\rho}{r_0} \sin \zeta' = \frac{\rho}{r_0} \sin (z - v), \tag{6.55}$$

where z is the observed zenith distance (Z$\hat{\mathrm{O}}$B) of the centre of the planet and v is the angle of the vertical at O. The angle p is a measure of the planetary parallax. However, p varies with time and it is usual to define the *geocentric mean equatorial horizonal parallax* P_0 to be

$$\sin P_0 = \frac{\rho_0}{\bar{r}_0} \tag{6.56}$$

where ρ_0 is the Earth's equatorial radius and $\bar{r}_0$ is the mean value of the separation of Earth and planet. Since it is customary to tabulate the *geocentric equatorial horizontal* parallax P in the Astronomical Ephemeris for each day of the year, the definition of P is given as

$$\sin P = \frac{\rho_0}{r_0}$$

where r_0 is the geocentric distance of the planet for the date in question. P is tabulated at hourly intervals for each day in the Astronomical ephemeris (where P is denoted by π) in the case of the Moon. Hence

$$\sin p = \left(\frac{\rho}{\rho_0}\right)\left(\frac{\rho_0}{r_0}\right) \sin \zeta' = \left(\frac{\rho}{\rho_0}\right) \sin P \sin \zeta', \tag{6.57a}$$

$$= \left(\frac{\rho}{\rho_0}\right)\left(\frac{\rho_0}{\bar{r}_0}\right)\left(\frac{\bar{r}_0}{r_0}\right) \sin \zeta'$$

$$= \left(\frac{\rho}{\rho_0}\right)\left(\frac{\bar{r}_0}{r_0}\right) \sin P_0 \sin \zeta'. \tag{6.57b}$$

Suppose further that Fig. 6.8 represents the Moon whose radius is AB. Let the linear radius of the Moon be R. Let $s(=\mathrm{A\hat{O}B})$ be the angle subtended by the lunar radius at O and $S\ (=\mathrm{A\hat{C}B})$ be the angle subtended by the lunar radius at C. Then

$$\sin s = \frac{R}{r} = \frac{R}{\rho_0}\cdot\frac{\rho_0}{r} = \frac{R}{\rho_0}\cdot\frac{\rho_0}{r_0}\cdot\frac{r_0}{r}$$

$$= \frac{R}{\rho_0}\frac{r_0}{r}\sin P = \frac{r_0}{r}\sin S, \tag{6.58}$$

where

$$\sin S = \frac{R}{r_0} = \frac{R}{\rho_0}\frac{\rho_0}{r_0} = \frac{R}{\rho_0}\sin P. \tag{6.59}$$

The mean angular semi-diameter of the Moon is S_0 where

$$\sin S_0 = \frac{R}{\bar{r}_0} = \frac{R}{\rho_0}\frac{\rho_0}{\bar{r}_0} = \frac{R}{\rho_0}\sin P_0. \tag{6.60}$$

Applying the sine rule for plane triangles to the plane triangle BOC

$$\frac{r}{\sin\zeta} = \frac{r_0}{\sin\zeta'}, \qquad \text{i.e.,}\ \frac{r_0}{r} = \frac{\sin\zeta'}{\sin\zeta}$$

and hence

$$\sin s = \sin S\frac{\sin\zeta'}{\sin\zeta}, \tag{6.61}$$

on using eqn. (6.59). Since ζ' and ζ can be assumed to be known and s can be measured, S (and S_0) can be determined once P (and P_0) are known.

The determination of planetary parallax may be illustrated through the determination of lunar parallax. Ideally two observatories on the same meridian are required for the determination of the lunar parallax. The geometrical conditions are set out in Fig. 6.9. The observatories are at O_1 (geocentric latitude ϕ'_1) and O_2 (geocentric latitude ϕ'_2). The Moon's centre is at B and the geocentric distance of

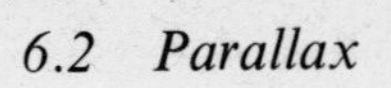

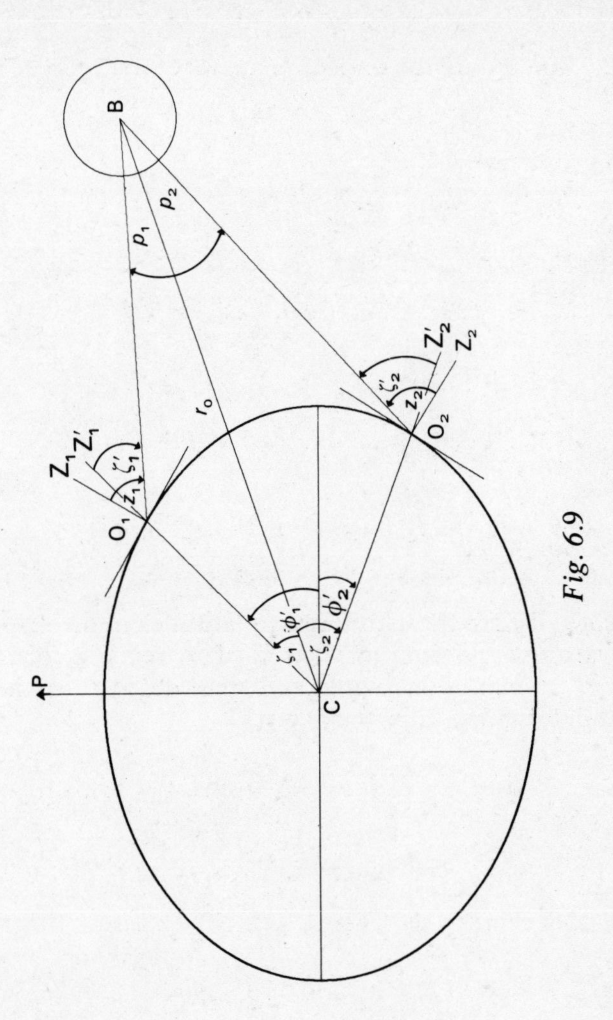
B
p_1
p_2
r_0
Z_1
Z'_1
ζ'_1
z_1
O_1
Z'_2
Z_2
ζ'_2
z_2
O_2
ϕ'_1
ϕ'_2
ζ_1
ζ_2
C
P

Fig. 6.9

B is r_0. The observed zenith distances of B are z_1 and z_2 respectively. Then

$$Z'_1\hat{O}_1B = \zeta'_1 = z_1 - v_1; \qquad Z'_2\hat{O}_2B = \zeta'_2 = z_2 - v_2 \qquad (6.62)$$

where v_1 and v_2 are the respective angles of the vertical. Again

$$O_1\hat{C}B = \zeta_1; \qquad O_2\hat{C}B = \zeta_2. \qquad (6.63)$$

From geometry

$$p_1 = \zeta'_1 - \zeta_1; \qquad p_2 = \zeta'_2 - \zeta_2 \qquad (6.64)$$

$$\begin{aligned} p_1 + p_2 &= (\zeta'_1 + \zeta'_2) - (\zeta_1 + \zeta_2) \\ &= (z_1 - v_1) + (z_2 - v_2) - (\phi'_1 + \phi'_1) = \theta \text{ (say)}, \end{aligned} \qquad (6.65)$$

since

$$(\zeta_1 + \zeta_2) = O_1\hat{C}O_2 = \phi'_1 + \phi'_2,$$

i.e.

$$\begin{aligned} \theta &= (z_1 + z_2) - (v_1 + v_2) - (\phi'_1 + \phi'_2) \\ &= (z_1 + z_2) - (\phi_1 + \phi_2), \end{aligned} \qquad (6.66)$$

where ϕ_1, ϕ_2 are the astronomical latitudes of the respective observatories and eqn. (6.28) is used to relate v, ϕ and ϕ'. Since z_1, z_2 can be measured and since ϕ_1, ϕ_2 are known θ can be determined. Eqn. (6.55) gives

$$\sin p_1 = \frac{\rho_1}{r_0} \sin \zeta'_1 = \frac{\rho_1}{r_0} \sin (z_1 - v_1),$$

$$\sin p_2 = \frac{\rho_2}{r_0} \sin \zeta'_2 = \frac{\rho_2}{r_0} \sin (z_2 - v_2),$$

and

$$p_2 = \theta - p_1,$$

so that

$$\begin{aligned} \sin p_2 &= \sin (\theta - p_1) = \sin \theta \cos p_1 - \cos \theta \sin p_1 \\ &= \frac{\rho_2}{r_0} \sin (z_2 - v_2), \end{aligned}$$

i.e.

$$\sin\theta\cos p_1 = \frac{\rho_1}{r_0}\sin(z_1 - v_1)\cos\theta + \frac{\rho_2}{r_0}\sin(z_2 - v_2)$$

i.e.

$$\tan p_1 = \frac{\rho_1\sin(z_1 - v_1)\sin\theta}{\rho_1\sin(z_1 - v_1)\cos\theta + \rho_2\sin(z_2 - v_2)}. \quad (6.67)$$

Since ρ_1, ρ_2, v_1, v_2 can be determined from the assumed figure of the Earth knowing the astronomical latitudes ϕ_1, ϕ_2 of the observatories, and z_1, z_2 are measured, then θ can be determined from eqn. (6.66). p_1 (and so p_2) can be determined from eqn. (6.67). Eqn. (6.57a) may be rewritten in the form

$$\sin p_1 = \frac{\rho_1}{\rho_0}\sin P\sin(z_1 - v_1), \quad (6.68)$$

from which the geocentric equatorial horizontal parallax P can be determined. Since ζ_1' and p_1 are known ζ_1 can be found from eqn. (6.64) and hence $\sin\zeta_1'/\sin\zeta$ in eqn. (6.61) can be evaluated.

Having found p_1 and p_2, r_0 can be determined. By a series of measurements over a period of time accurate measurements of the Moon's geocentric distance can be made and $\bar{r}_0$ determined.

Direct triangulation of the Moon's distance is now possible. The Apollo landings on the Moon have permitted the setting up of reflectors on the lunar surface. A pulsed laser beam is directed at these reflectors and the reflected beam detected. Accurate timings of the interval between emission of a pulse and reception of the reflected pulse give measurements of lunar distance of an accuracy higher than can be achieved by the more indirect astronomical methods whose principles have been outlined above.

The parallax of planets other than the Moon can be found at a single observatory using a base line determined by the diurnal rotation of the Earth (see next section). The principle is similar to that for the Moon.

6.2.4 The solar parallax. Before any measurement of stellar parallax can be accomplished it is necessary to determine the dimensions of the Earth's orbit. Fig. 6.10 illustrates the orbits of the Earth and an outer planet (e.g. a minor planet or Mars) about the Sun. From the diagram it is clear that when the Earth is at E_1 and the planet is in geocentric *opposition* (or heliocentric *conjunction*) at M_1 the distance between the Earth and the Planet is less than at the (geocentric) opposition when the Earth and planet are at E_2 and M_2 respectively.

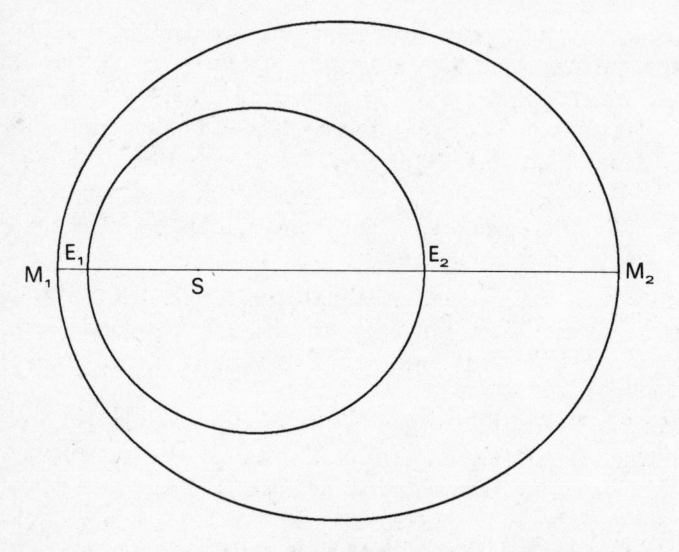

Fig. 6.10

E_1M_1 is the minimum separation of the Earth and planet and consequently the parallactic angles will be at maximum. If the parallax of the planet M is measured at such a favourable opposition by the method outlined for the Moon, a determination of the distance M_1E_1 can be made. In practice measurements from two observatories are unnecessary since two measurements—one before and one after meridian passage at opposition—from a single observatory determined an adequate base line, the base line being determined by the rotation of the observatory as illustrated in Fig. 6.11. The Earth is viewed from a position vertically above the north pole p in the diagram. The rotation of the Earth carries the observatory from O_1 to O_2 in twelve hours. Observations

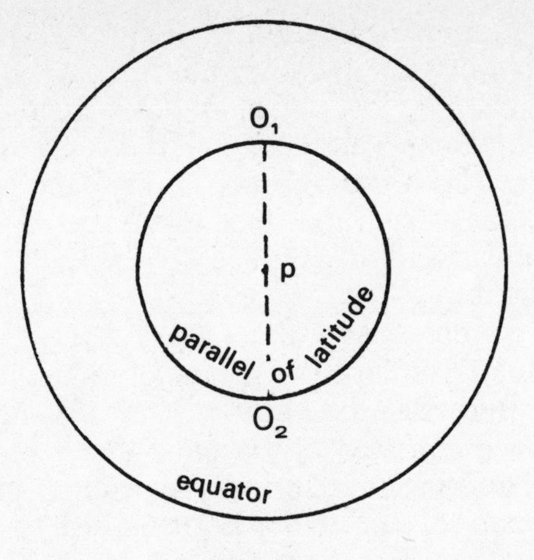

Fig. 6.11

spaced by a 12^h period give the longest attainable baseline O_1O_2. This method is of sufficient accuracy in the case of nearby planets, but of course is inadequate in the case of the Moon. Kepler's 3rd law (see chapter X) can then be used since

$$\left(\frac{a_E}{a_M}\right)^3 = \left(\frac{T_E}{T_M}\right)^2 = \left\{\frac{a}{(a + M_1E_1)}\right\}^3, \tag{6.69}$$

where a_E, a_M are respectively the lengths of the semi-major axes of the orbits of the Earth and planet and similarly, T_E and T_M are the orbital periods. Since orbital periods can be measured very accurately, the error in determination of a depends on the accuracy of measurement of M_1E_1.

Various other methods exist for the determination of solar parallax. The classical method was to utilise a transit of Venus across the solar disk. This method has been superseded by the method outlined above, but was the earliest means of determining the scale of the solar system. It is a complicated method having much in common with the prediction of solar eclipses. The solar parallax may also be derived by measurements of stellar radial velocities. The method, in essence, determines the velocity of the Earth in its orbit around the Sun. If the radial velocity of a star is measured at intervals of 6 months the contribution from the

Earth's velocity will be oppositely directed. Hence the difference of the two measures of radial velocity will give twice the Earth's orbital velocity. The method is complicated by the fact that the Earth's orbit is an ellipse and allowance must be made for the Earth's axial rotation. The solar parallax may also be determined from the constant of aberration once the figure of the Earth and the ellipticity of the Earth's orbit have been determined.

The most convenient and direct method of determining the scale of the solar system is by means of radar. Radio pulses are detected after reflection at the surface of Venus. The time interval between emission of the pulse and reception of the echo gives an accurate measure of the distance of Venus. An equation similar to eqn. (6.64) is then used to deduce the Earth-Sun distance. For astronomical purposes, in view of the differences between the values of the solar parallax obtained by different methods, it has been decided to adopt the value $8''\cdot 80$.

6.3 Aberration

Aberration is a phenomenon which results from the propagation of light at a finite speed. A finite time is required for light to propagate from any celestial body to the Earth. The light emitted by the Sun takes 8 minutes to travel to the Earth and the light from the nearest star takes approximately 4 years to reach the Earth. Since the Earth is moving around the Sun in its orbit, the Earth will move a small, but finite, distance during the time (the light time τ) taken by the light to travel from the Sun to the Earth. In the same way since the stars have space motion with respect to the Sun, the star will not be at the geometric position in space at which it is observed. There is another phenomenon also causing displacement of the star image—namely the displacement resulting from the detection of light by a moving detector—since, clearly, any terrestrial detector shares the motion of the Earth.

The displacement of the actual geometric position of an object from its observed position plus the displacement arising from observation by a moving detector is called *planetary aberration*. The displacement arising from observation by a moving detector is called *stellar aberration.* While it is clear that stars must exhibit planetary aberration, it is impossible to detect it since the absolute motions of the stars are not known. Correction for stellar aberration only is a partial correction for aberration in the case of stars. A full knowledge of the exact geometrical configuration of the stars at any instant cannot be obtained since correction for stellar aberration only allows for the motion of the Earth in its orbit about the Sun and no attempt can be made to allow for the motion of stars relative to the Sun during the light time. However, such a partial correction introduces no inconsistency, since full allowance for planetary aberrations could only be made if the absolute space motion of the Sun and stars could be determined and relativity theory indicates that such an absolute determination is not possible. The relationship between planetary and stellar aberration is illustrated in Fig. 6.12. A celestial object is at P when it emits light. In the

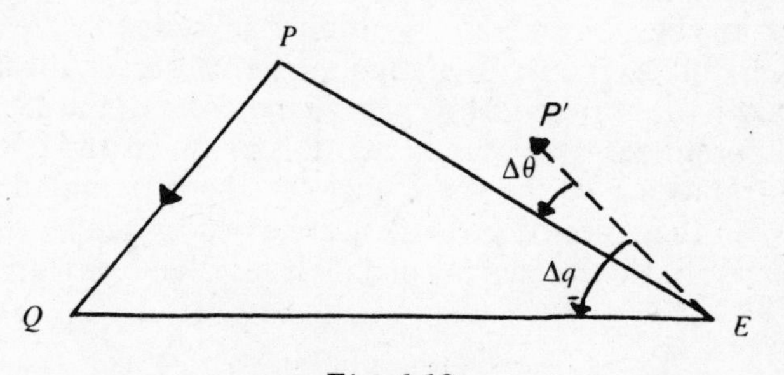

Fig. 6.12

light time the object moves from P to Q. The light is received by the moving observer at E. Because of his motion the object is observed at P′. The direction EP′ is not, in general, in the plane of E, P and Q. The stellar aberration is $P'\hat{E}P = \Delta\theta$ and the planetary aberration is $P'\hat{E}Q = \Delta q$.

While planetary aberration may be ignored in the case of stellar observations this is not the case for observations within the solar system (hence the name planetary aberration). Since the orbits of planets are known, allowance must be made for movement during the light time. This is usually done by assuming that the correction is so small that rectilinear motion can be assumed during the light time. Correction for planetary aberration is made in determining the characteristics of planetary orbits (see Chapter XI).

The discovery of stellar aberration was reported by Bradley in 1728. Bradley was searching for stellar parallax but his endeavours to find a measurable parallax were frustrated by the fact that the star nearest the Sun cannot be observed from Greenwich. Bradley recognised that the stellar displacements he observed could not be the result of parallax since the phenomenon exhibited a different phase relationship with the apparent motion of the Sun. He correctly interpreted the phenomenon as a confirmation of the discovery by Roemer in 1675, that light travels with a finite velocity.

6.3.1 Stellar aberration. The apparent direction in which a celestial object is viewed depends on the speed V with which an observer is moving. A stationary and a moving observer instantaneously at the same point in space would see a celestial object in slightly different positions. In order to treat the problems of reception of light by a moving observer, the theory of relativity is required. However, to the order of the accuracy of measurements which can be made, Newtonian and relativistic kinematics give the same answer. The Newtonian argument will therefore be given first, followed by the relativistic argument. In Fig. 6.13a a ray of light is received by a stationary

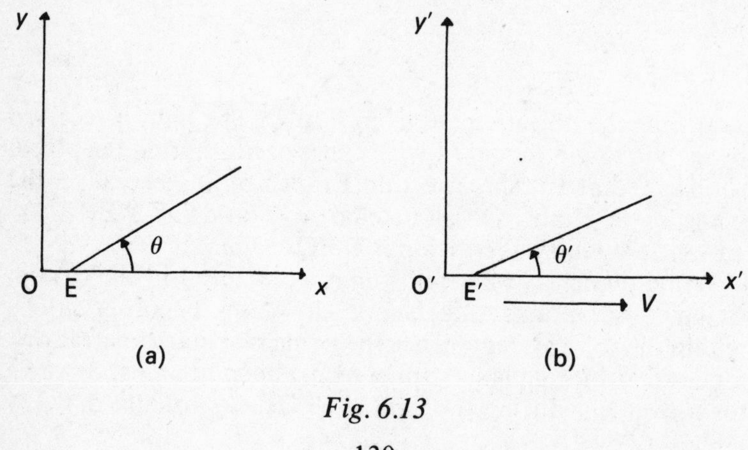

Fig. 6.13

detector at E. In Fig. 6.13b light from the same source is received by a moving detector E′. E′ is fixed in a moving frame of reference O′(x', y'). O′ is moving with uniform speed V parallel to the x-axis of the stationary frame of reference O(x, y). The light ray is observed in direction θ' with respect to the x'-axis by the moving observer E′, while it is observed in the direction θ with respect to the x-axis by the stationary observer E.

For the light ray received at E, E′ let the components of velocity be (u, v), (u', v') with respect to the axes O(x, y), O′(x', y') respectively. Then

$$u = -W\cos\theta, \qquad v = -W\sin\theta$$
$$u' = -W'\cos\theta', \qquad v' = -W'\sin\theta' \tag{6.70}$$

where W, W' is the speed of light with respect to the stationary and moving frames of reference respectively. The addition of velocities in Newtonian kinematics, in the case of a frame of reference moving with velocity V parallel to the x-axis of a stationary frame of reference such that the x coordinate of O′ is increasing with time, is given by

$$u' = u - V,$$
$$v' = v. \tag{6.71}$$

Hence

$$\cot\theta' = \frac{u'}{v'} = \frac{u - V}{v} = \frac{u}{v} - \frac{V}{v} = \cot\theta + \frac{V}{W}\operatorname{cosec}\theta$$
$$= \cot\theta + \frac{V}{c}\operatorname{cosec}\theta, \tag{6.72}$$

where c replaces W as the more customary symbol for the velocity of light. Eqn. (6.72) may be rearranged by multiplying by $\sin\theta\sin\theta'$ to give

$$\sin\theta\cos\theta' - \cos\theta\sin\theta' = \sin(\theta - \theta')$$
$$= \frac{V}{c}\sin\theta', \tag{6.73}$$

which is the more usual expression for astronomical purposes.

Since θ' is the "observed" position of the celestial object the displacement $\theta - \theta'$ from the geometrical position can be calculated from eqn. (6.73).

The addition law for velocities in relativistic kinematics is such that

$$u' = \frac{u - V}{1 - uV/c^2}$$
$$v' = \frac{v}{\beta(1 - uV/c^2)} \tag{6.74}$$

where $\beta = (1 - V^2/c^2)^{\frac{1}{2}}$. Then

$$\cot\theta' = \frac{u'}{v'} = \beta\frac{u - V}{v} = \beta\left(\cot\theta + \frac{V}{c}\operatorname{cosec}\theta\right)$$

using eqn. (6.72)

$$= \cot\theta + \frac{V}{c}\operatorname{cosec}\theta - \frac{V^2}{2c^2}\cot\theta + \ldots. \tag{6.75}$$

Eqn. (6.75) is the relativistic equivalent of the Newtonian expression eqn. (6.72). Since $V/c \sim 10^{-4}$ ($\simeq 20''$), terms in V^2/c^2 will be of the order of 10^{-8} ($\sim 0''\cdot001$) and are not detectable with present techniques. Therefore to the order V/c the Newtonian and relativistic treatments are equivalent. Eqn. (6.73) will therefore be used to estimate the effects of stellar aberration. Eqn. (6.73) may be written

$$\sin(\theta - \theta') = \sin\Delta\theta \simeq \Delta\theta = \kappa\sin\theta \tag{6.76}$$

where κ ($= 20''\cdot47$) is the *constant of aberration* and θ replaces θ' on the right hand side of eqn. (6.76) since $\Delta\theta$ is small.

The term *annual aberration* is often used in the literature. This is simply the aberration to be expected by taking the velocity of the Earth in its orbit relative to the centre of mass (barycentre) of the solar system. In actual practice, because the correction for barycentric motion and other variations in the Earth's orbital elements are small, the annual aberration is based on the heliocentric orbital motion of the Earth, the other effects being neglected.

It is convenient to consider the Earth's orbital motion in terms of two components. One component V_1 is taken to be that velocity perpendicular to the radius vector while the second component V_2 is the velocity perpendicular to the major axis. In magnitude these velocities are

$$V_1 = h/p, \qquad V_2 = eh/p \tag{6.77}$$

where $p = a(1 - e^2)$, $h^2 = n^2a^3p$, e is the eccentricity, a is the semi-major axis and n is the mean angular velocity for the Earth's orbit (see Chapter X, eqns. (10.21, 10.22)). The aberration produced by the Earth's motion can therefore be compounded from two terms.

6.3.2 Diurnal aberration. The rotation of the Earth about its axis introduces a further aberrational effect of the same type as stellar aberration and known as *diurnal aberration.* As the earth rotates from West to East any observatory on the surface of the Earth has a velocity resulting from the rotation. An observatory in latitude ϕ (assuming a spherical Earth) distant ρ from the centre of the Earth will have a velocity

$$V_3 \cos \phi = \frac{2\pi\rho \cos \phi}{s_d}. \tag{6.78}$$

where s_d is the length of the day. The effect of diurnal aberration is small and may be neglected for all except astrometric purposes.

6.3.3 Planetary aberration. Let the velocity of the Earth in its orbit be $\boldsymbol{V}_{\mathrm{E}}$ and the velocity of the planet in its orbit be $\boldsymbol{V}_{\mathrm{P}}$. $\boldsymbol{V}_{\mathrm{P}}$ may be written as the vector difference

$$\boldsymbol{V}_{\mathrm{P}} = \boldsymbol{V}_{\mathrm{E}} - \boldsymbol{V}. \tag{6.79}$$

The motion of the Earth relative to the planet is $\boldsymbol{V}(= \boldsymbol{V}_{\mathrm{E}} - \boldsymbol{V}_{\mathrm{P}})$. If $\boldsymbol{V}$ is directed at an angle q to the line of sight and the planet has a geocentric distance $\rho = c\tau$ where c is the velocity of light and τ is the light time, the linear displacement subtending an angle Δq at the Earth is given by

$$\tan \Delta q = \Delta q = \frac{|V|\tau \sin q}{c\tau}$$
$$= \frac{|V| \sin q}{c} \qquad (6.80)$$

A detailed discussion of planetary aberration will be found in Woolard and Clemence, *Spherical Astronomy*.

6.4 Correction of celestial positions for Refraction, Parallax and Aberration

The phenomena of refraction, parallax and aberration produce displacements of the position of celestial objects. The formulae for a general displacement derived in Chapters II and III will be used to calculate the displacements produced by these phenomena. The displacements are less than 1″ for stellar parallax and less that 20″·5 for aberration. However, displacements for refraction can be as large as 34′ (horizontal refraction) when the object is observed on the horizon. Nevertheless it will be assumed that eqns. (2.27, 28, 29) may be used to express the displacement produced. In the case of the Moon, the parallactic displacement is large and the exact formulae eqns. (2.24, 25, 26) must be used.

6.4.1 The displacement caused by refraction in the Earth's atmosphere. The effect of refraction in the Earth's atmosphere is to cause a displacement of the geocentric position of a celestial object towards the zenith. The displacement of the position of the object is in the plane of the vertical circle containing the geometrical position of the object. In Fig. 6.14 Z is the zenith, P is the north celestial pole, NWS is the astronomical horizon for an observer in astronomical latitude ϕ and UWV is the celestial equator. The geometric position of a celestial object is S and the displacement occasioned by refraction in the Earth's atmosphere causes the object to be observed at S′. The position of the object is therefore displaced through the angle SS′ in the plane of the vertical circle ZS. But SS′ is the angle of refraction R and is given by eqn. (6.9) to be

$$R = (\mu_0 - 1) \tan \zeta_0 \simeq (\mu_0 - 1) \tan \zeta = -d\zeta \qquad (6.9)$$

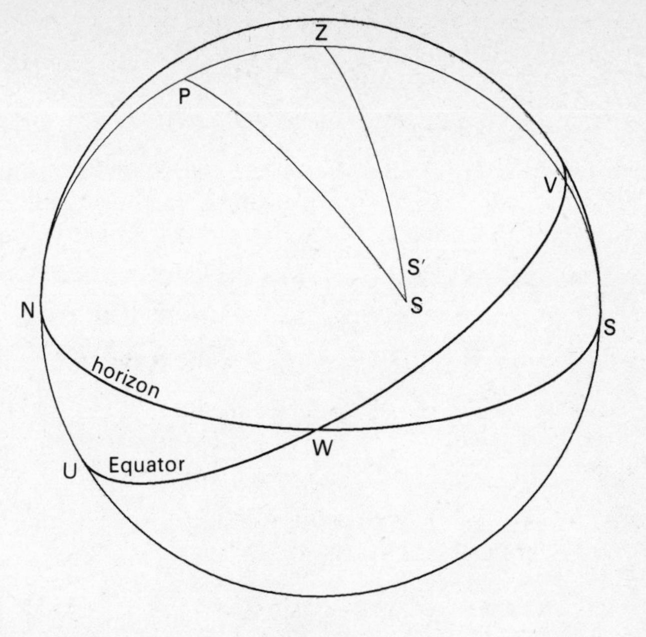

Fig. 6.14

since the effect of refraction is to decrease the zenith distance ζ. ζ replaces ζ_0 since R is assumed small, the difference being of the second order in a first order correction. The more exact formula eqn. (6.18) will not be used since it merely complicates the algebra.

The rectangular equatorial coordinates of S are (x, y, z). The displacements dx, dy, dz produced by refraction are, using eqns. (3.17)

$$\begin{aligned} \mathrm{d}x &= -R(\sin\alpha\sin\eta + \cos\alpha\sin\delta\cos\eta) \\ \mathrm{d}y &= R(\cos\alpha\sin\eta - \sin\alpha\sin\delta\cos\eta) \\ \mathrm{d}z &= R\cos\delta\cos\eta \end{aligned} \qquad (6.81)$$

where (α, δ) are the geocentric equatorial coordinates of the object S and $\eta = \mathrm{P\hat{S}Z}$. The angle of refraction R is taken to be positive since the rotation from S to S′ is anticlockwise if viewed from the pole of ZS on the same side of ZS as P. In the spherical triangle PZS

$$\begin{aligned} &\mathrm{PZ} = 90° - \phi, \qquad \mathrm{PS} = 90° - \delta, \qquad \mathrm{ZS} = \zeta, \\ &\mathrm{S\hat{P}Z} = H = S_t - \alpha, \qquad \mathrm{P\hat{S}Z} = \eta. \end{aligned} \tag{6.82}$$

In eqn. (6.82) S_t denotes the Sidereal Time of observation and H is the hour angle of S. Application of the Cosine, Sine and Transposed Cosine Rules to the spherical triangle PZS gives

$$\begin{aligned} &\cos\zeta = \sin\phi\sin\delta + \cos\phi\cos\delta\cos H, && \text{Cosine Rule} \\ &\sin\zeta\sin\eta = \cos\phi\sin H, && \text{Sine Rule} \\ &\sin\zeta\cos\eta = \sin\phi\cos\delta - \cos\phi\sin\delta\cos H. && \text{Transposed Cosine Rule} \end{aligned} \tag{6.83}$$

Writing $k = (\mu_0 - 1)$, evaluating R using eqn. (6.9) and using eqns. (6.83), eqns. (6.81) become

$$\mathrm{d}x = -k\left[\frac{xz\sin\phi + \cos\phi\{xy\sin S_t - (1 - x^2)\cos S_t\}}{z\sin\phi + (x\cos S_t + y\sin S_t)\cos\phi}\right], \tag{6.84}$$

$$\mathrm{d}y = +k\left[\frac{-yz\sin\phi + \cos\phi\{(1 - y^2)\sin S_t - xy\cos S_t\}}{z\sin\phi + (x\cos S_t + y\sin S_t)\cos\phi}\right], \tag{6.85}$$

$$\mathrm{d}z = +k\left[\frac{(1 - z^2)\sin\phi - z(x\cos S_t + y\sin S_t)\cos\phi}{z\sin\phi + (x\cos S_t + y\sin S_t)\cos\phi}\right]. \tag{6.86}$$

In eqns. (6.84, 85, 86), k is expressed in radians and the equations may only be used for $\zeta < 75°$. For greater zenith distances a more accurate expression must be used for R and the incremental changes. The rectangular equatorial coordinates of the star (x, y, z), the latitude ϕ of the observer and the sidereal time of observation S_t are known. The correction for refraction can be readily worked out from these equations.

Since the correction for refraction involves the sidereal time, the correction can only be made if the sidereal time of observation is known. During the course of an observation the sidereal time will change and so the correction for refraction will alter.

The correction may also be expressed in terms of the equatorial coordinates (α, δ) of the star S. The corrections $d\alpha$, $d\delta$ have already been worked out and are

$$d\alpha = \frac{-y\,dx + x\,dy}{x^2 + y^2}, \qquad d\delta = \frac{dz}{(1 - z^2)^{\frac{1}{2}}}. \tag{3.19}$$

Hence, substituting for dx, dy, dz from eqns. (6.84), (6.85), (6.86), the more usual results:

$$d\alpha = k \tan \zeta \sec \delta \sin \eta, \qquad d\delta = k \tan \zeta \cos \eta, \tag{6.87}$$

are obtained on using eqns. (6.83). Eqns. (6.87) are sometimes modified by using the observed zenith distance ζ_0, the observed declination δ_0 and observed parallactic angle η_0 in place of ζ, δ, η. The changes introduced are of the second order in the determination of the apparent place of the object and are of no consequence when first order displacements are being considered.

6.4.2 The displacement caused by Stellar Parallax. The displacement of the heliocentric position of a celestial object caused by parallax is along the great circle joining the celestial object to the Sun. The magnitude of the displacement is p where p is given by eqn. (6.27),

$$p = \pi\left(\frac{R}{a_E}\right) \sin \theta, \tag{6.27}$$

where π is the parallax of the object (defined by eqn. (6.26)), R is the actual distance of the Earth from the Sun at the time of observation, a_E is the length of the semi-major axis of the Earth's orbit and θ is the angle subtended at the centre of the Earth by the arc of great circle joining the positions of the Sun and object as projected onto the unit celestial sphere.

The displacement p is along the arc of great circle of length θ. R is measured in units of a_E. In Fig. 6.15 the celestial object S has equatorial coordinates (α, δ) and rectangular equatorial coordinates (x, y, z). The Sun is on the ecliptic at a point B whose equatorial coordinates are (A, D). P is the north celestial pole and K is the pole of the ecliptic. In the spherical triangle PSB,

$$\text{PS} = 90° - \delta, \qquad \text{PB} = 90° - \text{D}, \qquad \text{SB} = \theta,$$
$$\text{S}\hat{\text{P}}\text{B} = A - \alpha, \qquad \text{P}\hat{\text{S}}\text{B} = \psi. \tag{6.88}$$

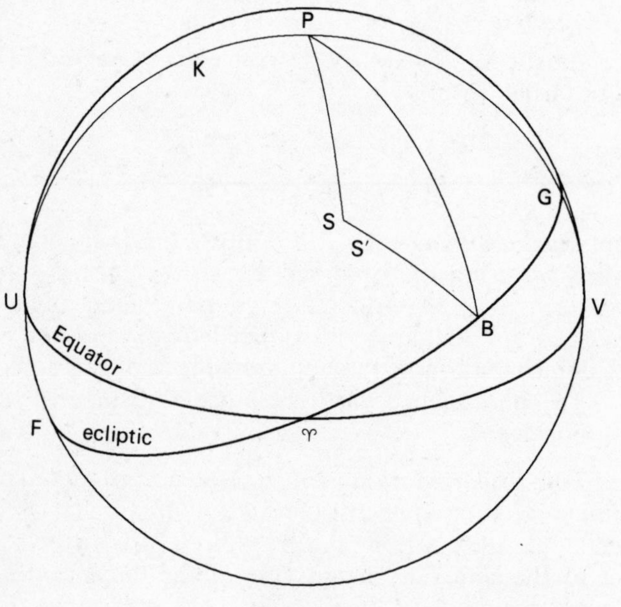

Fig. 6.15

With respect to the pole of the arc of great circle SB in the same hemisphere as the north celestial pole P the displacement p = SS′ is treated as positive if rotation of S to S′ is anti-clockwise. The angle ψ = P$\hat{\text{S}}$B is the angle between the meridian through the geometrical position of the object and the north celestial pole, and the great circle along which the displacement takes place. ψ therefore corresponds with the ψ of eqns. (3.17). Application of the Sine, Cosine and Transposed Cosine Rules to the spherical triangle PSB gives

$$\begin{aligned}
\sin\theta\sin\psi &= \cos D\sin(A-\alpha), && \text{Sine Rule}\\
\cos\theta &= \sin\delta\sin D + \cos\delta\cos D\cos(A-\alpha), && \text{Cosine Rule}\\
\sin\theta\cos\psi &= \cos\delta\sin D - \sin\delta\cos D\cos(A-\alpha). && \text{Transposed Cosine Rule}
\end{aligned} \tag{6.89}$$

The displacements dx, dy, dz in the rectangular equatorial coordinates may be obtained using eqns. (3.17) evaluated using eqns. (6.89),

$$\begin{aligned}
\mathrm{d}x &= -p(y\sin\psi + xz\cos\psi)/(1-z^2)^{\frac{1}{2}}\\
&= -\pi\frac{R}{a_{\mathrm{E}}}(y\sin\theta\sin\psi + xz\sin\theta\cos\psi)/(1-z^2)^{\frac{1}{2}}\\
&= \pi\{+X(1-x^2) - Y(xy) - Z(xz)\}.
\end{aligned} \tag{6.90}$$

Similarly

$$\mathrm{d}y = \pi\{-X(xy) + Y(1-y^2) - Z(yz)\}, \tag{6.91}$$

$$\mathrm{d}z = \pi\{-X(xz) - Y(yz) + Z(1-z^2)\}, \tag{6.92}$$

where in eqns. (6.90, 6.91, 6.92) X, Y, Z are the rectangular equatorial coordinates of the Sun in units of the length of the semi-major axis of the Earth's orbit ($X = R/a_{\mathrm{E}}\cos A\cos D$, $Y = R/a_{\mathrm{E}}\sin A\cos D$, $Z = R/a_{\mathrm{E}}\sin D$). X, Y, Z are tabulated in the Astronomical Ephemeris for each day of the year for (a) the epoch of the year of the ephemeris or the epoch of the next year and (b) the epoch 1950·0. Since x, y, z are readily calculated for any celestial object of known equatorial coordinates the displacements dx, dy, dz may be readily calculated from eqns. (6.90, 91, 92).

In terms of equatorial coordinates the displacements dα, dδ may be readily calculated using eqns. (3.19) and eqns (6.90, 91, 92). The results are

$$\begin{aligned}
\mathrm{d}\alpha &= \pi\sec\delta\{-X\sin\alpha + Y\cos\alpha\},\\
\mathrm{d}\delta &= \pi\{-X\cos\alpha\sin\delta - Y\sin\alpha\sin\delta + Z\cos\delta\},
\end{aligned} \tag{6.93}$$

from which the displacements to the equatorial coordinates may be computed.

6.4.3 The displacement caused by Annual Aberration. The displacement of the heliocentric position of a celestial object caused by annual aberration is along the great circle joining the object to that point on the celestial sphere towards which the Earth's velocity is directed at the instant of observation. Let the magnitude of the Earth's velocity at any instant be denoted by V. The magnitude $\Delta\theta$ of the displacement is

$$\Delta\theta = a = \frac{V}{c}\sin\theta = \kappa\sin\theta \tag{6.76}$$

where $\kappa = V/c$ is the constant of aberration, c is the velocity of light and θ is the length of the arc of great circle joining the celestial object to the point on the celestial sphere intersected by producing the vector instantaneously representing the Earth's velocity. The displacement a is along the arc of great circle of length θ. In Fig. 6.16 U♈V is the celestial

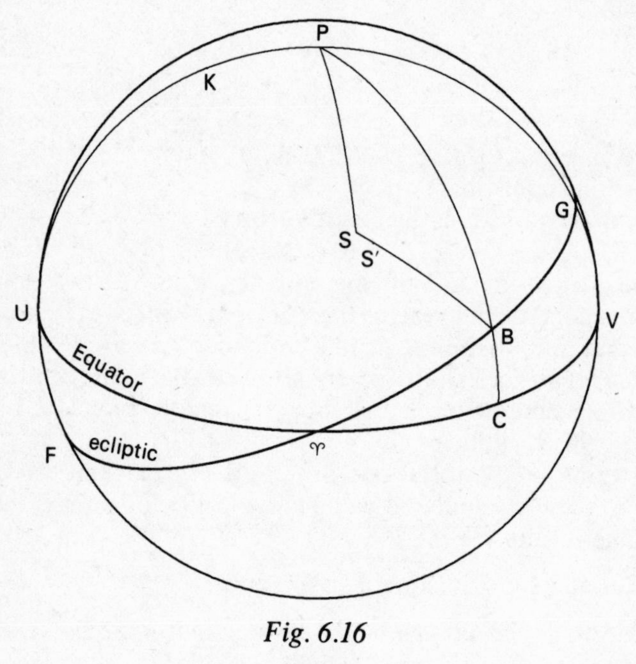

Fig. 6.16

equator, F♈G is the ecliptic, P is the north celestial pole, K is a pole of the ecliptic and ♈ is the Vernal Equinox. The star is at S (heliocentric) and the displacement caused by aberration is SS′ along the arc of great circle SB. B is the point on the sky towards which the Earth's orbital velocity is directed and C is the point, on the celestial equator, on the same meridian as B. Let the equatorial coordinates and the rectangular equatorial coordinates of S be (α, δ), (x, y, z) respectively and let the equatorial coordinates of B be (A, D). In the spherical triangle PSB,

$$\begin{aligned} &\mathrm{PS} = 90^\circ - \delta, \quad \mathrm{PB} = 90^\circ - D, \quad \mathrm{SB} = \theta, \\ &\mathrm{S\hat{P}B} = A - \alpha, \quad \mathrm{P\hat{S}B} = \psi. \end{aligned} \tag{6.94}$$

ψ is the angle $\mathrm{P\hat{S}B}$ between the meridian through the geometrical position of the object and the north celestial pole, and the great circle along which the displacement takes place. ψ therefore corresponds to the ψ of eqns. (3.17). The displacement SS′ is anticlockwise (in the direction of decreasing θ) when viewed from the pole of SB on the same side as the north celestial pole and is therefore taken as positive. It is clear from a comparison of eqns. (6.88) and (6.94) that there is no essential difference between the treatment of annual aberration and stellar parallax. However, the interpretation of the coordinates (A, D) is different. In the case of annual aberration, (A, D) are the equatorial coordinates of the point on the sky towards which the Earth's velocity is instantaneously directed. Let $-X'$, $-Y'$, $-Z'$ be the components of the Earth's velocity with respect to the frame of reference defining rectangular equatorial coordinates (i.e., the Earth has a velocity component $-X'$ parallel to the x-axis, etc.). Thus if V is the velocity of the Earth in its orbit

$$\begin{aligned} -X' &= V \cos A \cos D, \\ -Y' &= V \sin A \cos D, \\ -Z' &= V \sin D. \end{aligned} \tag{6.95}$$

Proceeding in the same manner as that used to determine eqn. (6.90) except that p is replaced by a and use is made of eqn.

(6.95), the displacements dx, dy, dz caused by annual aberration are given by

$$\mathrm{d}x = \frac{1}{c}\{-X'(1 - x^2) + Y'(xy) + Z'(xz)\} \qquad (6.96)$$

$$\mathrm{d}y = \frac{1}{c}\{X'(xy) - Y'(1 - y^2) + Z'(yz)\} \qquad (6.97)$$

$$\mathrm{d}z = \frac{1}{c}\{X'(xz) + Y'(yz) - Z'(1 - z^2)\} \qquad (6.98)$$

Comparison of eqns. (6.90)–(6.91) with eqns. (6.96)–(6.98) indicates the formal similarity between stellar parallax and annual aberration

Use of eqns. (3.19) give the displacements in the equatorial coordinates namely

$$\begin{aligned} \mathrm{d}\alpha &= \frac{\sec\delta}{c}(X'\sin\alpha - Y'\cos\alpha), \\ \mathrm{d}\delta &= \frac{1}{c}(X'\cos\alpha\sin\delta + Y'\sin\alpha\sin\delta - Z'\cos\delta). \end{aligned} \qquad (6.99)$$

The actual motion $(-X', -Y', -Z')$ of the Earth (strictly referred to a coordinate system where origin is the centre of mass of the solar system) has been used in determining the annual aberration since 1960. Prior to that date, for ease of computation, the circular and elliptical components (see section 6.3.1) of the Earth's motion were used. Since the contribution for the elliptical motion only varied slowly with time it was customary to omit any correction for elliptical motion. Consequently the geometrical place quoted before 1960 for a star is in fact the geometrical place plus the small contribution from the elliptical component of the Earth's motion. Before correction for aberration is carried out, it should be ascertained whether or not allowance has been made for this component. If the catalogue places have *not* been corrected the effect of the elliptical component of the Earth's velocity must be removed to find the true heliocentric place for the epoch for which the place is quoted before further corrections are applied. If in eqn. (6.95) V is replaced by V_1 the circular component of the Earth's velocity $(-X', -Y',$

$-Z'$) may be interpreted as the components of V_1. In this case the point B may be identified as that point lying 90° behind the Sun. The spherical triangle ♈BC (see Fig. 6.16) then has the properties

$$\begin{aligned} &\Upsilon\mathrm{B} = \odot - 90°, \qquad \mathrm{BC} = D, \qquad \Upsilon\mathrm{C} = A, \\ &\mathrm{B}\hat{\Upsilon}\mathrm{C} = \varepsilon, \qquad \Upsilon\hat{\mathrm{C}}\mathrm{B} = 90°, \end{aligned} \tag{6.100}$$

where $\odot$ is the longitude of the Sun and ε is the obliquity of the ecliptic. Application of the Cosine, Sine and Transposed Cosine Rules to the Spherical triangle ♈BC gives

$$\begin{aligned} &\sin\odot = \cos A \cos D, \qquad -\cos\odot \sin\varepsilon = \sin D, \\ &-\cos\odot\cos\varepsilon = \sin A \cos D \end{aligned} \tag{6.101}$$

and eqns. (6.96)–(6.98) have the form

$$\begin{aligned} \mathrm{d}x &= \kappa\{(1 - x^2)\sin\odot \\ &\quad + x\cos\odot(y\cos\varepsilon + z\sin\varepsilon)\} \\ \mathrm{d}y &= \kappa\{-(1 - y^2)\cos\odot\cos\varepsilon \\ &\quad - y(x\sin\odot - z\cos\odot\sin\varepsilon)\} \\ \mathrm{d}z &= \kappa\{-(1 - z^2)\cos\odot\sin\varepsilon \\ &\quad - z(x\sin\odot - y\cos\odot\cos\varepsilon)\} \end{aligned} \tag{6.102}$$

where $\kappa = V_1/c$ is the constant of aberration. Again if V in eqn. (6.95) is replaced by V_2 and the point (A, D), i.e. B, considered to be the point on the sky towards which the elliptical component of the Earth's orbital velocity is directed, the longitude of B is then $90° + \omega_E$ where ω_E is the longitude of perihelion (see Chapter XI) for the Earth. In this case the properties of the spherical triangle ♈BC are

$$\begin{aligned} &\Upsilon\mathrm{B} = 90° + \omega_E, \qquad \mathrm{BC} = D, \qquad \Upsilon\mathrm{C} = A, \\ &\mathrm{B}\hat{\Upsilon}\mathrm{C} = \varepsilon, \qquad \Upsilon\hat{\mathrm{C}}\mathrm{B} = 90°. \end{aligned} \tag{6.103}$$

Application of the Cosine, Sine and Transposed Cosine Rules to the Spherical triangle ♈BC gives

$$\begin{aligned} &-\sin\omega_E = \cos A\cos D, \qquad \cos\omega_E\sin\varepsilon = \sin D, \\ &\cos\omega_E\cos\varepsilon = \sin A\cos D \end{aligned} \tag{6.104}$$

and eqns. (6.96, 97, 98) have the form

$$\begin{aligned} dx &= e\kappa\{-(1 - x^2)\sin\omega_E \\ &\quad - x\cos\omega_E(y\cos\varepsilon + z\sin\varepsilon)\}, \\ dy &= e\kappa\{(1 - y^2)\cos\omega_E\cos\varepsilon \\ &\quad - y(-x\sin\omega_E + z\cos\omega_E\sin\varepsilon)\}, \\ dz &= e\kappa\{(1 - z^2)\cos\omega_E\sin\varepsilon \\ &\quad - z(-x\sin\omega_E + y\cos\omega_E\cos\varepsilon)\}. \end{aligned} \tag{6.105}$$

Since ω_E, e remain almost constant for periods of the order of a century and e is small, it is clear why this correction was often neglected.

Chapter VII should be consulted for the representation of aberrational displacements in terms of Day Numbers.

6.4.4 The displacement caused by Diurnal Aberration. Diurnal aberration may be allowed for by the same methods as those outlined in section 6.4.3. The Earth is rotating from west to east and so for the purposes of computing the effect of diurnal aberration the direction of motion of any observer sharing the Earth's rotation will be towards the East point of the horizon. In Fig. 6.17 WUE is the celestial equator, WSE is the horizon, P is the north celestial pole, Z is the zenith for the observer O at the centre of the celestial sphere, W is the West point and E is the East point. PZUS is the observer's meridian. X is a celestial object whose geometric equatorial coordinates are (α, δ) and whose rectangular equatorial coordinates are (x, y, z). The hour angle of X is H. The effect of diurnal aberration is to displace the object X through the arc XX′ in the plane of the great circle XE. The spherical triangle PXE has the properties

$$\begin{aligned} &PX = 90° - \delta, \qquad XE = \theta, \qquad PE = 90, \\ &E\hat{P}X = H - 270°, \qquad P\hat{X}E = \psi. \end{aligned} \tag{6.106}$$

ψ, the angle between the meridian through the star and the great circle in whose plane the displacement takes place, has the same meaning as in eqn. (3.17); θ is the angular distance of X from E. The displacement d caused by diurnal aberration

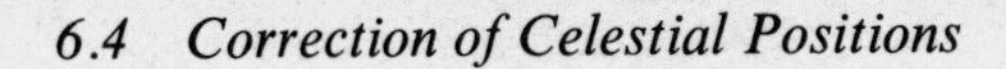

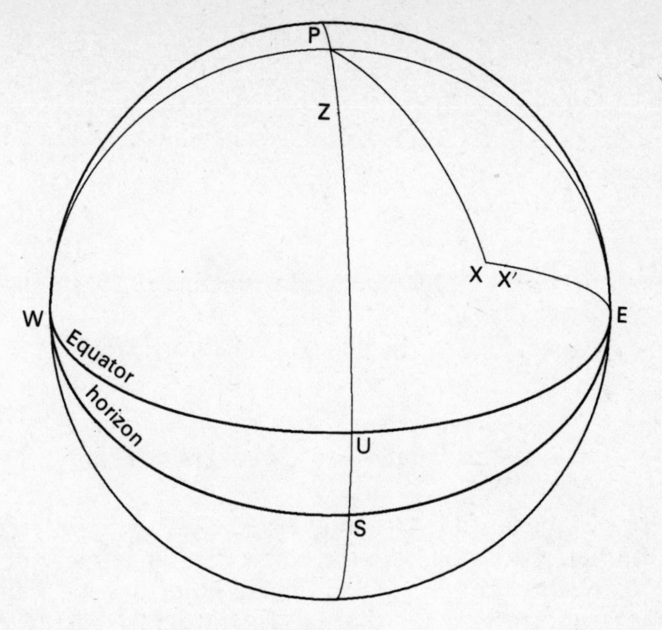

Fig. 6.17

is then

$$d = \frac{V_3}{c}\sin\theta, \tag{6.107}$$

where V_3 is defined by eqn. (6.78). Application of the Sine and Transposed Cosine Rules to the spherical triangle PXE gives

$$\begin{aligned} \sin\theta\sin\psi &= \cos H, && \text{Sine Rule} \\ \sin\theta\cos\psi &= \sin\delta\sin H. && \text{Transposed Cosine Rule} \end{aligned} \tag{6.108}$$

Making use of eqns (3.17) then gives

$$\begin{aligned} \mathrm{d}x &= -d(y\sin\psi + xz\cos\psi)/(1 - z^2)^{\frac{1}{2}} \\ &= -\frac{V_3}{c}(y\cos H + xz\sin\delta\sin H)/(1 - z^2)^{\frac{1}{2}} \\ &= -\frac{V_3}{c}\{xy\cos S_t + (1 - x^2)\sin S_t\} \end{aligned} \tag{6.109}$$

where S_t is the Sidereal Time of observation. Similarly

$$\mathrm{d}y = \frac{V_3}{c}\{(1 - y^2)\cos S_t + xy \sin S_t\}, \qquad (6.110)$$

$$\mathrm{d}z = \frac{V_3}{c} z\{-y \cos S_t + x \sin S_t\}. \qquad (6.111)$$

Eqns. (6.109)–(6.111) may be used with eqns. (3.19) to give

$$\begin{aligned} \mathrm{d}\alpha &= \frac{V_3}{c}\cos H \sec \delta = 0''{\cdot}32 \cos \phi \cos H \sec \delta, \\ \mathrm{d}\delta &= \frac{V_3}{c}\sin H \sin \delta = 0''{\cdot}32 \cos \phi \sin H \sin \delta, \end{aligned} \qquad (6.112)$$

where ϕ is the latitude of the observer. Clearly the correction for diurnal aberration, like the correction for refraction, can only be made once the sidereal time of observation is known. However, correction for diurnal aberration is usually only required for measurement of the highest precision.

6.5 Correction for Parallax in the Case of the Planets

The correction for parallax is large in the case of the Moon, the mean equatorial horizontal parallax P for the Moon being $3422''{\cdot}54$ or nearly $1°$. Eqns. (3.16) must then be employed to find the amount of the displacements in the rectangular equatorial coordinates of the Moon. In the case of the remaining planets, the approximate forms of eqns. (3.17) may be used. However, since a determination of the exact expression must evaluate the term in the approximate expression, the exact form will be derived first.

Since the shape of the Earth must be taken into account when determining planetary parallax, the equatorial and rectangular equatorial coordinates are defined with respect to the geocentric zenith and not the astronomical zenith of the observer. The effect of planetary parallax is such that the zenith distance (with respect to the geocentric vertical) measured by an observer on the surface of the Earth is greater than the zenith distance measured by an observer at the centre of the Earth, as may be seen by inspection of Fig. 6.8 where clearly

$$\zeta' = \zeta + p > \zeta. \tag{6.64}$$

Fig. 6.18 illustrates the effect of planetary parallax when the geocentric zenith is Z′, the north celestial pole is P and M is

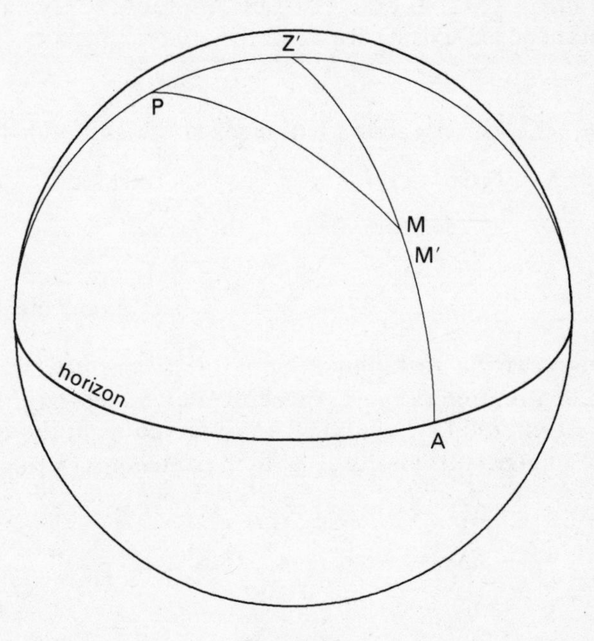

Fig. 6.18

the geocentric position of the Moon's centre. Planetary parallax causes a displacement MM′ along the vertical circle ZMA. The length of the arc PM is $90° - \delta$ and the spherical angle $Z'\hat{P}M = H$, where δ is the declination and H is the hour angle of the Moon's centre. The origin of coordinates is the centre of the Earth. The spherical triangle PZ′M has the properties

$$\begin{aligned} &PZ' = 90° - \phi', \quad Z'M = \zeta, \quad PM = 90° - \delta \\ &Z'\hat{P}M = H, \quad P\hat{M}Z' = \psi, \end{aligned} \tag{6.113}$$

where ϕ' is the geocentric latitude of the observer, ζ (see also

Fig. 6.8) is the zenith distance with respect to the geocentric zenith of the geocentric position of the Moon's centre and ψ has the meaning of ψ in eqns. (3.16). Clearly the displacement caused by parallax must be treated as negative since the displacement from M to M′ is clockwise when viewed from the pole of the vertical circle ZMA on the same side of ZMA as the north celestial pole P. Application of the Cosine, Sine and Transposed Cosine Rules to the spherical triangle PZ′M gives

$$\begin{aligned}
\cos \zeta &= \sin \phi' \sin \delta + \cos \phi' \cos \delta \cos H, && \text{Cosine Rule} \\
\sin \zeta \sin \psi &= \cos \phi' \sin H, && \text{Sine Rule} \\
\sin \zeta \cos \psi &= \sin \phi' \cos \delta - \cos \phi' \sin \delta \cos H. && \text{Transposed Cosine Rule}
\end{aligned} \tag{6.114}$$

If the geocentric rectangular equatorial coordinates of the Moon centre are (x_1, y_1, z_1) when observed from the centre of the Earth and (x_2, y_2, z_2) when observed from the Earth's surface in geocentric latitude ϕ' then from eqns. (3.16)

$$\begin{aligned}
x_2 - x_1 &= -x_1(1 - \cos p) \\
&\quad + \sin p(y_1 \sin \psi + x_1 z_1 \cos \psi)/(1 - z_1^2)^{\frac{1}{2}} \\
y_2 - y_1 &= -y_1(1 - \cos p) \\
&\quad - \sin p(x_1 \sin \psi - y_1 z_1 \cos \psi)/(1 - z_1^2)^{\frac{1}{2}} \\
z_2 - z_1 &= -z_1(1 - \cos p) - \sin p \cos \psi (1 - z_1^2)^{\frac{1}{2}}
\end{aligned} \tag{6.115}$$

where p is the parallax of the Moon at the instant of observation. At any instant the value of ζ may be computed if the rectangular equatorial coordinates of the Moon's centre are known. The first formula of eqn. (6.114) gives

$$\cos \zeta = z_1 \sin \phi' + \cos \phi'(x_1 \cos S_t + y_1 \sin S_t) \tag{6.116}$$

where S_t is the sidereal time of observation. Since $\zeta' = \zeta + p$, eqn. (6.55) gives

$$\sin p = \frac{\rho}{r_0} \sin \zeta' = \frac{\rho}{r_0} \sin (\zeta + p)$$

or

$$\tan p = \frac{(\rho/r_0)\sin\zeta}{1 - (\rho/r_0)\cos\zeta}. \qquad (6.117)$$

Since r_0 may be found from the tabulated values of the horizontal parallax and ρ is known from Geoid calculations (see section 6.2.2) then $\tan p$ and $\sin p$ can be determined. From the final two formulae of eqns. (6.114)

$$\tan\psi = \frac{\cos\phi'(x_1 \sin S_t - y_1 \cos S_t)}{(1 - z_1^2)\sin\phi' - z_1 \cos\phi'(x\cos S_t + y\sin S_t)} \qquad (6.118)$$

Since p and ψ are now known, eqns. (6.115) may be used to compute the displacement caused by parallax in the case of the Moon.

In the case of the remaining planets and the Sun the problem is simpler. In this case the approximate forms may replace eqns. (6.115) namely

$$\begin{aligned} dx &= p(y\sin\psi + xz\cos\psi)/(1 - z^2)^{\frac{1}{2}} \\ dy &= -p(x\sin\psi - yz\cos\psi)/(1 - z^2)^{\frac{1}{2}} \\ dz &= -p\cos\psi(1 - z^2)^{\frac{1}{2}} \end{aligned} \qquad (6.119)$$

where the subscript 1 has been omitted in the case of the rectangular coordinates for the geocentric position. The parallax p may be written in the approximate form

$$p = \frac{\rho}{r_0}\sin\zeta \qquad (6.120)$$

for substitution in eqn. (6.119) without loss of accuracy to the first order. The first of eqns. (6.119) then becomes, on making use of eqns. (6.114),

$$\begin{aligned} dx &= \frac{\rho}{r_0}\{y\cos\phi'(\sin S_t\cos\alpha - \cos S_t\sin\alpha) + xz\sin\phi'\cos\delta \\ &\quad - xz\cos\phi'\sin\delta(\cos S_t\cos\alpha + \sin S_t\sin\alpha)\}/(1 - z^2)^{\frac{1}{2}}, \\ &= \frac{\rho}{r_0}[xz\sin\phi' + \cos\phi'\{-(1 - x^2)\cos S_t + xy\sin S_t\}]. \end{aligned} \qquad (6.121)$$

Similarly

$$\mathrm{d}y = \frac{\rho}{r_0}[yz \sin \phi' + \cos \phi' \{xy \cos S_t - (1 - y^2) \sin S_t\}], \tag{6.122}$$

$$\mathrm{d}z = \frac{\rho}{r_0}[-(1 - z^2) \sin \phi' + \cos \phi' \{xz \cos S_t + yz \sin S_t\}]. \tag{6.123}$$

Again using eqns. (3.19)

$$\begin{aligned} \mathrm{d}\alpha &= -\frac{\rho}{r_0} \cos \phi' \sec \delta \sin H, \\ \mathrm{d}\delta &= -\frac{\rho}{r_0}(\sin \phi' \cos \delta - \cos \phi' \sin \delta \cos H). \end{aligned} \tag{6.124}$$

The time of observation is required before the displacements for planetary parallax can be determined. Eqns. (6.121, 122, 123) are formally similar to eqns. (6.90, 91, 92) for stellar parallax.

To find the displacements $\Delta\alpha$, $\Delta\delta$ in equatorial coordinates for the effect of parallax on the Moon the spherical triangles PZ′M and PZ′M′ (see Fig. 6.18) are considered. By elimination of the spherical angle $P\hat{Z}'M = P\hat{Z}'M'$ and making use of the Cosine, Sine and Transposed Cosine rule applied to each spherical triangle the displacements are found to be given by the following formulae:

$$\begin{aligned} \tan \Delta\alpha &= -\frac{\rho}{r_0} \frac{\sin H \cos \phi'}{\cos \delta - (\rho/r_0) \cos \phi' \cos H}, \\ \frac{\tan \delta + \tan \Delta\delta}{1 - \tan \delta \tan \Delta\delta} &= \frac{\cos H'[\sin \delta - (\rho/r_0) \sin \phi']}{\cos \delta \cos H - (\rho/r_0) \cos \phi'}. \end{aligned} \tag{6.125}$$

$\Delta\alpha = -\Delta H$ is found from the first of these equations and $\Delta\delta$ from the second since $H' = H + \Delta H$ is then known. The eqns. (6.125) are only required in the case of the Moon.

Chapter VII

Precession, Nutation, Proper Motion and the Determination of Star Places of Date

The phenomena of refraction, parallax and aberration cause displacements of the observed place of an object from its geometric position. However, these phenomena do not alter the relative geometrical separations of the actual objects nor do they alter the relative relationship of the frames of reference. For a heliocentric observer, for example, there would be no correction for either parallax or aberration.

The Sun and stars are in relative motion. The relative configuration of celestial objects is therefore continually changing. Fortunately for the development of positional measurements, most celestial objects are too far away for there to be detectable effects of the relative motion. However, such *proper motions* are observed for relatively nearby stars and the determination of this motion is of vital importance in determining the local structure of the Galaxy. The systematic relative motion of celestial objects must be taken into account in determining their observed positions with accuracy.

Again, the Earth is interacting gravitationally with the rest of the solar system. Such interaction is not only responsible for the motion of the Earth in its orbit about the Sun but causes motions of the Earth with respect to fixed axes centred on the Earth. Since the Earth is not a perfectly spherical body composed of spherical shells of uniform density, the Moon and Sun exert couples on the Earth tending to move the plane of its equator. The rest of the solar system perturbs the Earth's orbit. These phenomena have the effect of continuously altering the relative configuration of the (hitherto) fundamental reference planes. This effect can be resolved into two components—a secular component known as *precession* and a periodic component known as *nutation.*

The phenomenon of precession was discovered by Hipparchus in the second century BC. He noted that star positions exhibited a systematic increase in ecliptic longitude but no changes in ecliptic latitude were observed. Proper Motion is a much more recent discovery being first reported in 1718 by Halley, who noted that even having allowed for precession certain bright stars were in positions different from those found by Hipparchus.

Finality cannot be achieved in the determination of the effects of precession and nutation or proper motion. In the case of precession, even if the dynamical interaction of the solar system were fully worked out and improved observational capability were available, small irregular redistributions of mass caused by geophysical phenomena will always occur giving further (small) changes in the relative configurations of the fundamental reference planes. In the case of proper motions refinement of observation will allow the detection of smaller proper motions. However, the likely changes will be small and require considerable refinement of technique in order to be detected. It should be noted that proper motions determined observationally do not give absolute motions in space for celestial objects but motions relative to those objects for which no proper motion can be detected.

7.1 Precession and Nutation

7.1.1 The forces acting on the Earth. The Earth moves under the action of the gravitational forces produced by the remainder of the universe, though the observable effects of such forces are contributed by bodies within the solar system. While a full treatment of the dynamical interaction of the Earth with the rest of the solar system is beyond the scope of this book, a short description of the major effects of such interaction is a necessary preliminary to a discussion of precession and nutation.

The Earth is not spherical nor is its mass distributed symmetrically. In Chapter V the figure of the Earth was approximated to a spheroid. Leaving aside the actual distribution of mass within the Earth, the departure of the figure of the Earth from sphericity has important gravitational consequences. From the point of view of the orbital motion of the Earth about the Sun, the Earth can be treated as a mass point (see also Chapter X); however, from the point of view of the motion of the Earth about its centre of mass, its non-spherical figure is of major importance. The gravitational interactions of the Sun and Moon produce a couple acting on the Earth, such that the Sun is tending to twist the Earth so that its equatorial plane will come into coincidence with the plane of the ecliptic, and the Moon is tending to twist the Earth so that the equatorial plane will come into coincidence with the plane of the lunar orbit. The couples produced by both Moon and Sun are variable. Clearly at those times when the Sun or Moon lies in the plane of the Earth's equator the couple will vanish. Again, since the Earth moves in an elliptical orbit about the Sun and the Moon moves in an elliptical orbit about the Earth and distance separating Earth and Moon or Earth and Sun is variable with time and so the magnitude of the couple also varies periodically for this reason.

The Earth is rotating about its axis. The rotational motion of the Earth, just as in the case of a spinning top, interacts with the gravitational couples in a manner resisting the action of the couples. Continuing the analogy with the spinning top, the rotational axis of the Earth precesses, i.e. the axis of rotation moves around some fixed axis. The couple exerted by the Sun gives a precession of the north celestial pole about the pole of the ecliptic while the couple exerted by the Moon gives a precession of the north celestial pole about the pole of the lunar orbit. The lunar orbit does not define a fixed plane with respect to the ecliptic since the pole of the lunar orbit itself precesses about the pole of the ecliptic. The resolution of the gravitational interaction into these precessional motions is a useful way of visualising the interaction. The precessional motion can be regarded as being composed of two parts—a secular term called the *luni-solar precession* and a periodic term called the *nutation*. The effect of luni-solar precession is to move the Vernal Equinox backwards (i.e., in the sense that ecliptic longitudes increase) along the ecliptic at a rate given by

$$\psi' = 50''{\cdot}370\,8 + 0''{\cdot}005\,0T, \tag{7.1}$$

where T is measured in tropical centuries from 1900·0. The period required for the Vernal Equinox to move through 360° along the ecliptic is therefore of the order of 26 000 years. The nutation includes all the periodic terms.

Clearly there will be periodic terms depending on the solar and lunar longitudes. However, the principal term in the nutation arises from the precession of the pole of the lunar orbit about the pole of the ecliptic and has a period of 18·6 yr. (6798 days) and an amplitude of 9″·210. The nutation itself may be divided into two components—the *nutation in longitude* and the *nutation in the obliquity*. The nutation in longitude describes the periodic terms in the motion of the Vernal Equinox along the ecliptic, while nutation in the obliquity describes the periodic changes in the obliquity of the celestial equator with respect to the ecliptic.

The gravitational interaction of the remainder of the planetary system affects the Earth, Moon, Sun system to the extent that the orbit of the Earth about the Sun suffers perturbation. The effect of the perturbation is such that the Earth does not repeat its path in space with respect to the Sun. Since the plane of the Earth's orbit defines the plane of the ecliptic, the effect of planetary perturbations is to cause variation of the plane of the ecliptic. In practice the effect of the planetary perturbations may be represented by a secular motion of the Vernal Equinox around the celestial equator in such a way that right ascensions are decreased. The effect is known as *planetary precession* and is given by

$$\lambda' = 0''{\cdot}1247 - 0''{\cdot}0188T \tag{7.2}$$

where T is measured in tropical centuries from 1900·0.

The effect of *annual general precession* which combines luni-solar and planetary precession may be expressed in a variety of ways. The *general precession in longitude p* gives the annual secular motion of the Vernal Equinox on the ecliptic, m is defined to be the *general precession in right ascension*, n is defined to be the *general precession in declination* and π is defined to be the annual rate of rotation of the ecliptic arising from planetary precession where,

$$\begin{aligned} p &= 50''{\cdot}2564 + 0''{\cdot}0222T, \\ m &= 3^{s}{\cdot}072\,34 + 0^{s}{\cdot}001\,86T, \\ n &= 20''{\cdot}0468 - 0''{\cdot}0085T, \\ \pi &= 0''{\cdot}4711 - 0''{\cdot}0007T, \\ \varepsilon &= 23°27'08''{\cdot}26 - 46''{\cdot}845T - 0''{\cdot}0059T^2 \\ &\quad + 0''{\cdot}001\,81T^3, \end{aligned} \tag{7.3}$$

where T is measured in tropical centuries from 1900·0 and ε is the obliquity of the ecliptic.

The combined effect of luni-solar and planetary precession is to produce a motion of the Vernal Equinox which may be regarded as resulting from an annual motion ψ' along the ecliptic of a given date and an annual motion λ' along the celestial equator of the same date to give a new position of the Vernal equinox one tropical year later. The new position of the Vernal equinox then forces a redefinition of the plane of the celestial equator and the plane of the ecliptic. It is the purpose of this section to determine the corrections which must be applied to star positions to allow for these changes in the fundamental reference planes

7.1.2 Precession and Nutation. The phenomena contributing to precession and nutation have been described in the previous section: in this section that description will be made more quantitative. The secular effect of luni-solar precession is illustrated in Fig. 7.1. In that diagram the geometric position of a celestial object is S. The ecliptic F$\Upsilon_1\Upsilon_0$G may be regarded as defining a fixed plane. At time t_0 the celestial equator is the great circle $U_0\Upsilon_0V_0$ with north celestial pole P_0. At time t_1 the celestial equator is the great circle $U_1\Upsilon_1V_1$ with north

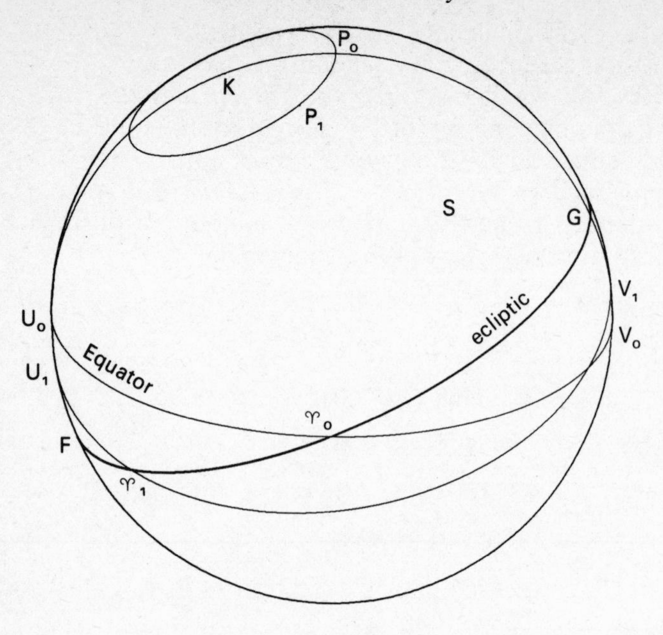

Fig. 7.1

celestial pole P_1. In the interval $t_1 - t_0$ luni-solar precession will move the Vernal Equinox from Υ_0 to Υ_1. If $t_1 - t_0$ is a tropical year then $\Upsilon_0\Upsilon_1$ is the same as ψ', the luni-solar precession in longitude (see eqn. 7.1). Suppose $U_0\Upsilon_0V_0$ is the equator at the beginning of the Besselian Year 19*xy*·0 and $U_1\Upsilon_1V_1$ is the equator for the Besselian Year 19*xz*·0 where $z = y + 1$. Then $U_0\Upsilon_0V_0$ is called the *mean equator for the epoch 19xy·0* and $U_1\Upsilon_1V_1$ is called the mean equator for the epoch 19*xz*·0. The *mean equator of date* is defined by determining the position of the Vernal Equinox of date making allowance only for luni-solar precession.

Fig. 7.1 may also be used to illustrate the combined effect of luni-solar precession and nutation. The nutation in longitude at any instant must be added to the luni-solar precession. It will be presumed that, at some epoch, the position of the mean equator of the epoch and the Vernal equinox at the epoch are known. The change ψ in the position of the Vernal Equinox at any time t subsequent to the fundamental epoch is then the sum of the luni-solar precession in time t and the nutation at time t where

$$\psi = \psi' t + \Delta\psi \tag{7.4}$$

where ψ' is the luni solar precession of eqn. (7.1) and $\Delta\psi$ is the nutation in longitude at time t. The nutation in longitude is given by

$$\Delta\psi = -17''{\cdot}233 \sin \Omega + 0''{\cdot}209 \sin 2\Omega - 1''{\cdot}273 \sin 2L_{\odot} - 0''{\cdot}204 \sin 2L_{☾} + \ldots \tag{7.5}$$

where Ω is the mean longitude of the ascending node of the lunar orbit, $L_{\odot}$ is the mean longitude of the Sun and $L_{☾}$ is the mean longitude of the Moon. The expression for $\Delta\psi$ contains 69 periodic terms with coefficients in excess of $0''{\cdot}0002$. Since the obliquity of the ecliptic ε is altered by the combined effect of precession and nutation, the value of ε must also be evaluated at the instant t for any calculations which refer to that instant. Then

$$\varepsilon = \varepsilon^{\circ} + \Delta\varepsilon \tag{7.6}$$

where ε^0 is the value of ε determined from eqn. (7.3) at the instant t and $\Delta\varepsilon$ is the nutation in the obliquity given by

$$\Delta\varepsilon = 9''{\cdot}210 \cos \Omega - 0''{\cdot}090 \cos 2\Omega + 0''552 \cos 2L_{\odot} + 0''{\cdot}088 \cos 2L_{☾} + \ldots \tag{7.7}$$

where the parameters on the right hand side of eqn. (7.7) are as for eqn. (7.5). The expression for $\Delta\varepsilon$ contains forty periodic terms with coefficients in excess of $0''{\cdot}0002$. If $U_1 ♈_1 V_1$ is the equator determined by allowing for luni-solar precession in the interval t and the nutation at t then $U_1 ♈_1 V_1$ is the *true equator of date*. This definition is not quite precise since allowance for planetary precession must also be made in determining the true equator of date.

The effect of planetary precession is illustrated in Fig. 7.2. At some time t_0 the celestial equator is $U_0 ♈_0 ♈_1 V_0$. Planetary precession does not alter the plane of the celestial equator but perturbs the plane of the Earth's orbit. Let the plane of the ecliptic be defined by the great circle $F_0 ♈_0 G_0$ (with pole K_0) at time t_0 and by the great circle $F_1 ♈_1 G_1$ at some subsequent

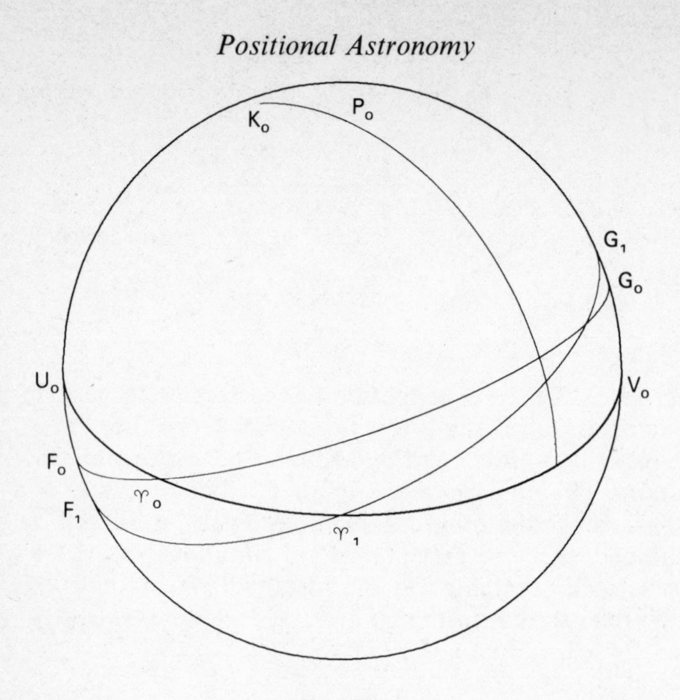

Fig. 7.2

time t_1. The effect of planetary precession is therefore to cause a translation of $♈_0$ to $♈_1$ in the interval $t_1 - t_0$. Planetary precession therefore causes a decrease in the right ascensions of celestial objects. If $t_1 - t_0$ is a tropical year then $♈_0♈_1$ is the planetary precession λ' of eqn. (7.2). Again if $t_1 - t_0$ is a tropical year, $\pi = ♈_0\widehat{B}♈_1$ where π is given in eqn. (7.3) and is the annual rate of rotation of the ecliptic. The longitude Π of B with respect to the Vernal equinox and ecliptic of 1900·0 is given by

$$\Pi = 173°57'{\cdot}06 + 54'{\cdot}77T \qquad (7.8)$$

where T is in tropical centuries after 1900·0.

Having discussed separately the contributions of luni-solar precession, nutation and planetary precession, their combined effect is illustrated in Fig. 7.3. $U_0♈_0V_0$, $F_0♈_0G_0$ are the celestial equator and ecliptic respectively at time t_0, while $U_1♈_1V_1$, $F_1♈_1G_1$ are the celestial equator and ecliptic respectively at

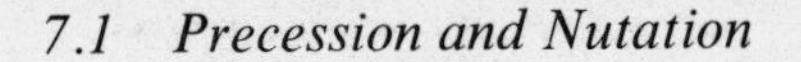

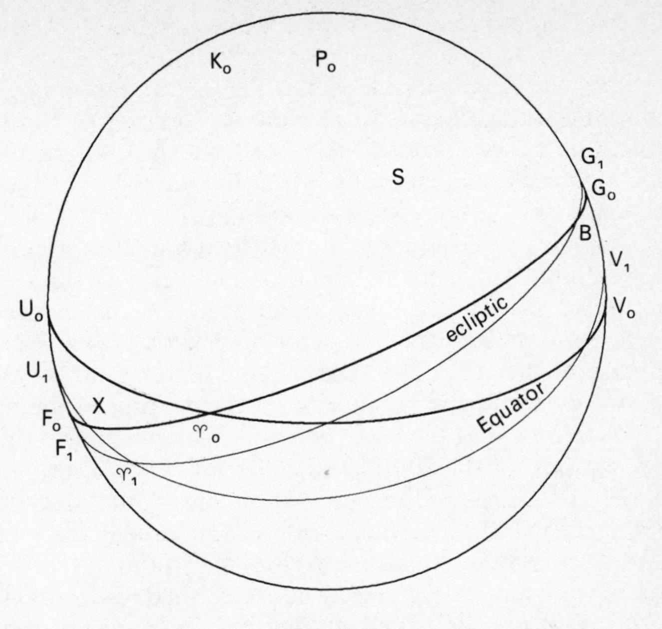

Fig. 7.3

some subsequent time t_1. P_0 is the north celestial pole and K_0 a pole of the ecliptic at time t_0. $♈_0$, $♈_1$ are the Vernal Equinoxes for times t_0, t_1 respectively. The effect of planetary precession in Fig. 7.3 (as in Fig. 7.2) has been exaggerated. The gravitational interaction of the Sun, Moon and Earth causes the plane of the celestial equator to move while the planetary perturbation of the Earth's orbit causes the plane of the ecliptic to move. In consequence of both these movements the Vernal Equinox moves from $♈_0$, the intersection of the celestial equator and ecliptic at time t_0, to $♈_1$, the intersection of the celestial equator and ecliptic at time t_1. The annual effect of the solar system on the fundamental planes combines luni-solar precession, nutation and planetary precession to give the *annual general precession in longitude p*. *p* has components *m* and *n* where *m* is the *annual general precession in right ascension* and *n* is the *annual general precession in declination.*

The concepts discussed above must be capable of realisation in practice. Since the ecliptic is in continual motion and since it requires the elapse of time to establish the apparent position of the Sun on the sky, it is not an easy matter to establish the ecliptic at any instant. The position and motion of the plane of the celestial equator is easier to establish. The motion of the plane of the equator is a continuous one upon which is superimposed a periodic variation. The *mean equatorial* plane can be established as that plane which moves with the secular component of the precessional motion. The *mean equator* therefore moves with the luni-solar precession ψ'. The ascending node of the ecliptic on the mean equator establishes the *mean equinox* and the inclination of the ecliptic to the mean equator is known as the *mean obliquity*. The dynamical theory of the motion of the Earth under the influence of the gravitational attractions of the Sun, Moon and planets allows the determination of the motion of the fundamental reference circles at any time subsequent to some initial instant. The initial values, however, can only be established by observation. The initial instant is called the epoch. The equator, ecliptic and equinox of any subsequent or previous date can then be determined knowing the time elapsed between the epoch and the date in question. The determination of the conditions at epoch is beyond the scope of this book and reference should be made to *Spherical Astronomy* by E. W. Woolard and G. M. Clemence.

Although a definition of annual general precession has been given above an operational definition is required. The definition used by Newcomb for general precession was that it is the motion of the mean equinox of date along the moving ecliptic. Such a definition is not capable of realisation in practical terms. The operational definition is that the *general* precession in longitude (denoted by p) is the longitude of the mean equinox of date referred to the fixed mean equinox and ecliptic of epoch. This definition gives an adequate working measure but is subject to small errors of about $0''\cdot001$ per century.

The constants referring to precession used by the Astronomical Ephemeris have been given in eqns. (7.1), (7.2), (7.3) and (7.8) but are summarised in Table 7.1 for epoch 1900·0.

Table 7.1

(a) Luni-solar precession ψ'	$= 50''\cdot 3708 + 0''\cdot 0050T$	in longitude
(b) planetary precession λ'	$= 0''\cdot 1247 - 0''\cdot 0188T$	in longitude
(c) general precession p	$= 50''\cdot 2564 + 0''\cdot 0222T$	in longitude
(d) precession in right ascension m	$= 3^s\cdot 07234 + 0^s\cdot 00186T$	
(e) precession in declination n	$= 20''\cdot 0468 - 0''\cdot 0085T$	
(f) obliquity of the ecliptic ε	$= 23°27'08''\cdot 26 - 46''\cdot 845T - 0''\cdot 0059T^2 + 0''\cdot 00181T^2$	(7.9)
(g) longitude of the node of the ecliptic on the preceding ecliptic Π	$= 173°57'\cdot 06 + 54'\cdot 77T$	
(h) rate of rotation of ecliptic π	$= 0''\cdot 4711 - 0''\cdot 0007T$	

T is measured in tropical centuries from 1900·0.
Values are annual values except for ε and Π.

These constants are found from continued observation. Dynamical theory gives the variation subsequent to the epoch. Values of p, m, n, Π, π for a current year are given in the Astronomical Ephemeris. The nutation terms $\Delta\psi$ and $\Delta\varepsilon$ are determined from the dynamical theory. The nutation in longitude $\Delta\psi$ (to an accuracy of $0''\cdot 001$) is tabulated in the Astronomical Ephemeris for each day of the year. $\Delta\psi$ represents the correction to be added to convert from the mean equinox to the equinox of date. The equator and equinox of date are found by using the actual motion which has taken place in the interval of time since the epoch. The nutation in obliquity $\Delta\varepsilon$ is treated in a somewhat different way in that it is not tabulated as such but appears as $-B$ in a tabulation of Besselian Day Numbers (see section 7.3.2). The terms making up the nutation have a natural division into long and short period terms. For this purpose a short period term is one with a period of 60 days or less—the short period terms are those depending on the longitudes of the Moon. The sum of the short period terms (accounting for 46 out of 69 terms in $\Delta\psi$ and 24 out of the 40 terms is $\Delta\varepsilon$) are denoted by $d\psi$ and $d\varepsilon$ respectively. These terms are used in specific circumstances; for example, when positions of stars are tabulated at 10-day intervals, the long period terms of the nutation are included. Corrections to the position within the 10-day period can be made using $d\psi$ and $d\varepsilon$. $d\psi$ and $d\varepsilon$ are tabulated in the Astronomical Ephemeris for each day of the year in the same table as the Besselian Day Numbers.

7.2 Reduction of Star Positions for the Equator and Equinox of Date

Star positions are catalogued in terms of right ascension and declination referred to the mean equator and mean equinox of some epoch. However, observations are referred to the equator and equinox appropriate to the instant of observation. (In this book the equator and equinox of date means the equator and equinox appropriate to the instant of observation. This is often called the *true equator* and *true equinox*.) In this section a means of correcting the rectangular equatorial coordinates of a celestial object for the effects of precession and nutation will be developed.

The effect of precession and nutation is to cause a change in the spatial orientation of the plane of the celestial equator and ecliptic. In terms of Fig. 7.3 the equinox has moved from position Υ_0 to some position Υ_1. The problem is to determine the coordinates of the star S with respect to the equator associated with the equinox Υ_1 knowing its coordinates with respect to the equator associated with the equinox Υ_0. The problem may be resolved into two parts—the effect of precession and the effect of nutation, since the methods of treatment differ in each case. The effect of precession is a secular effect increasing with time, and cannot be treated as motion along a great circle whose plane is fixed in space, as was the case for refraction, parallax, proper motion and aberration. It will be treated below in terms of three rotations of the fundamental frame of reference. The effect of nutation will not exceed about 18″ and, to the first order, nutation can be considered in terms of rotation in the plane of a great circle. The effect of nutation will therefore be evaluated in the manner of Chapter VI.

7.2.1 Allowance for precession. In Fig. 7.4 Υ_0, Υ_1 are the Vernal Equinoxes at times t_0, t_1 respectively where Υ_1 is moving secularly from Υ_0 under the influence of precessional effects only. $U_0\Upsilon_0V_0$, $U_1\Upsilon_1V_1$ are the equators at times t_0, t_1

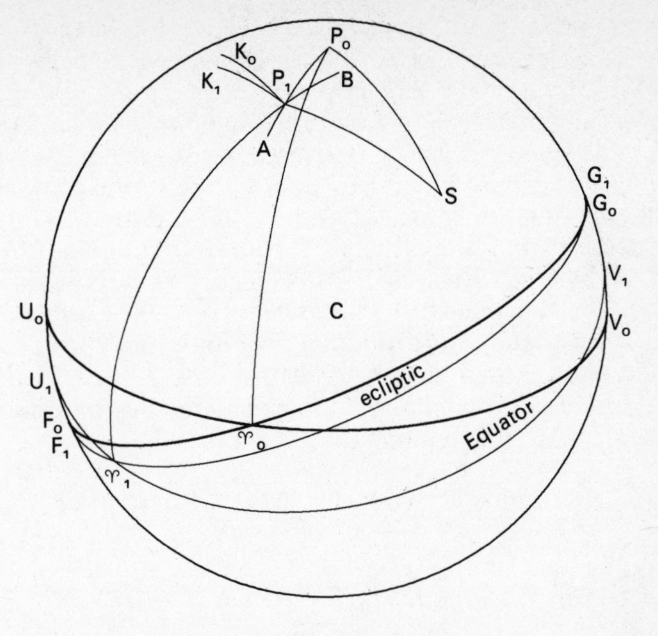

Fig. 7.4

respectively and $F_0♈_0G_0$, $F_1♈_1G_1$ are the corresponding ecliptics. The celestial pole at time t_0 is at P_0 and at time t_1 is at P_1. The pole of the ecliptic is at K_0 at time t_0 and at K_1 at time t_1. P_0P_1A is a great circle (meridian with respect to $U_0♈_0V_0$) through P_0 and P_1. $BP_1♈_1$ is a great circle (meridian with respect to $U_1♈_1V_1$) through P_1 and $♈_1$. Equatorial rectangular axes are defined in the usual way, namely, the z-axis directed towards the instantaneous celestial pole, the x-axis directed towards the instantaneous equinox and the x, y-axes lying in the plane of the celestial equator. The rectangular equatorial frame of reference defined at time t_0 may be brought into coincidence with the rectangular equatorial frame of reference defined at time t_1 by means of three rotations. Let these frames of reference be $C(x_0, y_0, z_0)$ and $C(x_1, y_1, z_1)$ respectively where C is the centre of the celestial sphere. The first rotation is about the z_0-axis, bringing the x_0, z_0 plane into the plane of the great circle P_0P_1A. Rotation is therefore through the angle $\zeta_0 = ♈_0\widehat{P}_0A$. Denote this secondary frame of reference by $C(x', y', z')$ where $z' \equiv z_0$. The frame of reference $C(x', y', z')$ is now rotated about the y'-axis to bring the z'-axis into coincidence with the direction CP_1. The rotation is through the angle $\theta = P_0P_1$. Denote this secondary frame of reference by $C(x^*, y^*, z^*)$ where $y' \equiv y^*$. The great circle PP_1 does not lie in the same plane as the meridian through P_1 and Υ_1. A final rotation of axes about the z^*-axis brings the x^*, z^* plane into the plane of x_1, z_1 defined by the great circle $P_1\Upsilon_1$. Hence $z^* \equiv z_1$. The rotation is through the angle $z = P_0\widehat{P}_1B\ (=\Upsilon_1\widehat{P}_1A)$. The angles ζ_0, θ and z define the transformation. For an initial epoch $1900{\cdot}0 + T_0$ and a final epoch $1900{\cdot}0 + T_0 + T$ the equations, given in the Explanatory Supplement to the Astronomical Ephemeris (p. 30), defining ζ_0, z, θ are

$$
\begin{aligned}
\zeta_0 &= (2304''{\cdot}250 + 1''{\cdot}396T_0)T + 0''302T^2 + 0''{\cdot}018T^3,\\
z &= \zeta_0 + 0''{\cdot}791T^2, \qquad\qquad (7.10)\\
\theta &= (2004''{\cdot}682 - 0''{\cdot}853T_0)T - 0''{\cdot}426T^2 - 0''{\cdot}042T^3,
\end{aligned}
$$

where T_0, T are measured in tropical centuries. Secular changes in the coefficients in terms of order T^2 and higher are ignored. These constants were derived by Newcomb. A table for ζ_0, z, θ is given in the Explanatory Supplement to the Astronomical Ephemeris (pp. 32, 33) for the years 1900·0 to 1980·0.

It was shown in Chapter II (eqn. (2.12)) that if a point has coordinates (a, b), (a', b') with reference to two coordinate systems in the same plane which have the same origin but which are rotated through an angle c with respect to an axis through the origin and perpendicular to the plane that

$$
\begin{aligned}
a' &= a\cos c + b\sin c,\\
b' &= -a\sin c + b\cos c.
\end{aligned}
\qquad (2.12)
$$

In applying eqn. (2.12) it should be remembered that rotations which are anticlockwise when viewed down the axis of rotation are regarded as positive and that the order in which the axes are taken is $x \to y \to z \to x$. Then, applying eqn. (2.12) to each transformation in turn:

1st rotation:

$$
\begin{aligned}
x' &= x_0 \cos(-\zeta_0) + y_0 \sin(-\zeta_0) \\
&= x_0 \cos\zeta_0 - y_0 \sin\zeta_0, \\
y' &= -x_0 \sin(-\zeta_0) + y_0 \cos(-\zeta_0) \\
&= x_0 \sin\zeta_0 + y_0 \cos\zeta_0, \\
z' &= z_0;
\end{aligned}
$$

2nd rotation:

$$
\begin{aligned}
z^* &= z' \cos(+\theta) + x' \sin(+\theta) \\
&= z_0 \cos\theta + x_0 \cos\zeta_0 \sin\theta - y_0 \sin\zeta_0 \sin\theta, \\
x^* &= -z' \sin(+\theta) + x' \cos(+\theta) \\
&= -z_0 \sin\theta + x_0 \cos\zeta_0 \cos\theta - y_0 \sin\zeta_0 \cos\theta, \\
y^* &= y' = x_0 \sin\zeta_0 + y_0 \cos\zeta_0;
\end{aligned}
$$

3rd rotation:

$$
\begin{aligned}
x_1 &= x^* \cos(-z) + y^* \sin(-z) \\
&= x_0 \cos\zeta_0 \cos\theta \cos z - y_0 \sin\zeta_0 \cos\theta \cos z \\
&\quad - z_0 \sin\theta \cos z - x_0 \sin\zeta_0 \sin z - y_0 \cos\zeta_0 \sin z, \\
&= x_0(\cos\zeta_0 \cos\theta \cos z - \sin\zeta_0 \sin z) \\
&\quad + y_0(-\sin\zeta_0 \cos\theta \cos z - \cos\zeta_0 \sin z) \\
&\quad - z_0 \sin\theta \cos z, \qquad (7.11) \\
y_1 &= -x^* \sin(-z) + y^* \cos(-z) = x_0 \cos\zeta_0 \cos\theta \sin z \\
&\quad - y_0 \sin\zeta_0 \cos\theta \sin z - z_0 \sin\theta \sin z \\
&\quad + x_0 \sin\zeta_0 \cos z + y_0 \cos\zeta_0 \cos z, \\
&= x_0(\cos\zeta_0 \cos\theta \sin z + \sin\zeta_0 \cos z) \\
&\quad + y_0(-\sin\zeta_0 \cos\theta \sin z + \cos\zeta_0 \cos z) \\
&\quad - z_0 \sin\theta \sin z, \qquad (7.12) \\
z_1 &= z^* = x_0 \cos\zeta_0 \sin\theta - y_0 \sin\zeta_0 \sin\theta + z_0 \cos\theta. \qquad (7.13)
\end{aligned}
$$

Eqns. (7.11, 12, 13) may be written in the form:

$$\left.\begin{aligned}
x_1 &= x_0X_x + y_0Y_x + z_0Z_x \\
&= x_0 + (X_x - 1)x_0 + y_0Y_x + z_0Z_x, \\
y_1 &= x_0X_y + y_0Y_y + z_0Z_y \\
&= y_0 + x_0X_y + (Y_y - 1)y_0 + z_0Z_y, \\
z_1 &= x_0X_z + y_0Y_z + z_0Z_z \\
&= z_0 + x_0X_z + y_0Y_z + (Z_z - 1)z_0,
\end{aligned}\right\} \quad (7.14)$$

where

$$\left.\begin{aligned}
X_x &= \cos\zeta_0 \cos\theta \cos z - \sin\zeta_0 \sin z, \\
Y_x &= -\sin\zeta_0 \cos\theta \cos z - \cos\zeta_0 \sin z, \\
Z_x &= -\sin\theta \cos z, \\
X_y &= \cos\zeta_0 \cos\theta \sin z + \sin\zeta_0 \cos z, \\
Y_y &= -\sin\zeta_0 \cos\theta \sin z + \cos\zeta_0 \cos z, \\
Z_y &= -\sin\theta \sin z, \\
X_z &= \cos\zeta_0 \sin\theta, \qquad Y_z = -\sin\zeta_0 \sin\theta, \\
Z_z &= \cos\theta.
\end{aligned}\right\} \quad (7.15)$$

The X, Y, Z terms in eqns. (7.14) are direction cosines. Given ζ_0, z, θ eqns. (7.15) may be used to compute the coefficients of x_0, y_0, z_0 in the transformation (7.14) so obtaining x_1, y_1, z_1 the rectangular equatorial coordinates referred to the mean equator and equinox of date. Values of $X_x - 1$, $Y_y - 1$, $Z_z - 1$, Y_x, Z_x, Z_y are tabulated in the Explanatory Supplement to the Astronomical Ephemeris (pp. 36, 37). The remaining values are not tabulated since to the first order in T in eqn. (7.10) $z = \zeta_0$ and so $Y_x = -X_y$, $Z_x = -X_z$ and $Z_y = Y_z$. It is not necessary to make this approximation but the approximation has been used extensively in the past when it was desirable to economise on computing effort. Since,

$$\left.\begin{aligned}
X_x^2 + X_y^2 + X_z^2 &= 1, \\
X_xY_x + X_yY_y + X_zZ_z &= 0, \text{ etc.},
\end{aligned}\right\} \quad (7.16)$$

the inverse relations also hold, namely:

$$
\begin{aligned}
x_0 &= X_x x_1 + X_y y_1 + X_z z_1 = X_x x_1 - Y_x y_1 - Z_x z_1, \\
y_0 &= Y_x x_1 + Y_y y_1 + Y_z z_1 = -X_y x_1 + Y_y y_1 + Z_y z_1, \qquad (7.17) \\
z_0 &= Z_x x_1 + Z_y y_1 + Z_z z_1 = -X_z x_1 + Y_z y_1 + Z_z z_1.
\end{aligned}
$$

The second group of equations in eqns. (7.17) are approximate, but the form has been extensively used. The first group of equations in eqns. (7.17) are exact and would be no problem to use with modern computational resources. For reduction from epoch 1950·0 + T_0 to epoch 1950·0 + T_0 + T where T_0, T are in tropical centuries the following expressions may be used:

$$
\begin{aligned}
X_x - 1 &= -(29\,696 + 26T_0)T^2 - 13T^3, \\
Y_x = -X_y &= -(2\,234\,941 + 1355T_0)T \\
&\quad - 676T^2 + 221T^3, \\
Z_x = -X_z &= -(971\,690 - 414T_0)T \qquad (7.18) \\
&\quad + 207T^2 + 96T^3, \\
Y_y - 1 &= -(24\,975 + 30T_0)T^2 - 15T^3, \\
Y_z = Z_y &= -(10\,858 + 2T_0)T^2, \\
Z_z - 1 &= -(4721 - 4T_0)T^2,
\end{aligned}
$$

where the coefficients are in units of the eighth decimal place; i.e., the first coefficient in the expression for $X_x - 1$ is $2{\cdot}9696 . 10^{-4}$ and so on. By making use of eqns. (7.14) or (7.17) the secular effect of precession can be allowed for exactly. In particular the effect of precession from any standard epoch to the instant of observation could be computed exactly.

7.2.2 *Allowance for nutation.* The effect of nutation can be immediately split into two terms—nutation in longitude and nutation in the obliquity. To the first order, nutation in longitude may be regarded as a rotation through an angle $\Delta\psi$ of the equinox of date in the plane of the ecliptic of date about an axis through its poles. To the same order, nutation in the obliquity may be regarded as a rotation of the plane of the ecliptic about an axis through the equinoxes of date through an angle $\Delta\varepsilon$. The formulae of Chapter II may be applied to this simplified situation. For nutation in the longitude the direction cosines L, M, N of the axis of rotation are

$$\begin{aligned} L &= \cos(90 - \varepsilon)\cos 270^\circ = 0, \\ M &= \cos(90 - \varepsilon)\sin 270^\circ = -\sin\varepsilon, \\ N &= \sin(90 - \varepsilon) = \cos\varepsilon, \end{aligned} \tag{7.19}$$

since the equatorial coordinates of the pole of the ecliptic of date are $\alpha = 270^\circ$, $\delta = 90^\circ - \varepsilon$. Similarly in the case of nutation in the obliquity the direction cosines L, M, N of the axis of rotation (the x-axis since only the motion in obliquity is being considered) are

$$\begin{aligned} L &= \cos 0^\circ \cos 0^\circ = 1 \\ M &= \cos 0^\circ \sin 0^\circ = 0 \\ N &= \sin 0^\circ = 0 \end{aligned} \tag{7.20}$$

The corrections Δx, Δy, Δz for nutation in longitude which must be made to the mean position of date x, y, z are therefore

$$\begin{aligned} \Delta x &= -\Delta\psi(y\cos\varepsilon + z\sin\varepsilon), \\ \Delta y &= \Delta\psi \,.\, x\cos\varepsilon, \\ \Delta z &= \Delta\psi \,.\, x\sin\varepsilon. \end{aligned} \tag{7.21}$$

using eqns. (2.27)–(2.29) of Chapter II. Similarly the corrections for nutation in obliquity have the form

$$\begin{aligned} \Delta x &= 0, \\ \Delta y &= -z\,\Delta\varepsilon, \\ \Delta z &= +y\,\Delta\varepsilon. \end{aligned} \tag{7.22}$$

The convention on the signs of $\Delta\psi$ and $\Delta\varepsilon$ is preserved—while the backwards rotation of the Vernal equinox is clockwise, it reflects as an anticlockwise rotation of the star. The total correction for nutation may be written as

$$\begin{aligned} \Delta x &= -\Delta\psi(y\cos\varepsilon + z\sin\varepsilon), \\ \Delta y &= \Delta\psi\, x\cos\varepsilon - \Delta\varepsilon z, \\ \Delta z &= \Delta\psi\, x\sin\varepsilon + \Delta\varepsilon y. \end{aligned} \tag{7.23}$$

The corrections of eqn. (7.23) must be added to the mean position of date determined by using eqns. (7.14). The true position of date is then obtained. In the Astronomical Ephemeris the nutation in longitude $\Delta\psi$, the obliquity of the ecliptic ε and nutation in obliquity $\Delta\varepsilon$ ($= -B$, Besselian Day Number) are tabulated for each day of the year.

The allowance for precession and nutation may be allowed for in one operation. Eqns. (7.14) may be written in matrix notation as

$$\begin{bmatrix} x_1 \\ y_1 \\ z_1 \end{bmatrix} = \begin{bmatrix} X_x & Y_x & Z_x \\ X_y & Y_y & Z_y \\ X_z & Y_z & Z_z \end{bmatrix} \begin{bmatrix} x_0 \\ y_0 \\ z_0 \end{bmatrix}. \tag{7.14a}$$

Allowance for nutation may be made by introducing another matrix multiplier on the right hand side giving the true coordinates x_1', y_1', z_1', of date as

$$\begin{bmatrix} x_1' \\ y_1' \\ z_1' \end{bmatrix} = \begin{bmatrix} 1 & -\Delta\psi\cos\varepsilon & -\Delta\psi\sin\varepsilon \\ \Delta\psi\cos\varepsilon & 1 & -\Delta\varepsilon \\ \Delta\psi\sin\varepsilon & \Delta\varepsilon & 1 \end{bmatrix}$$
$$\times \begin{bmatrix} X_x & Y_x & Z_x \\ X_y & Y_y & Z_y \\ X_z & Y_z & Z_z \end{bmatrix} \begin{bmatrix} x_0 \\ y_0 \\ z_0 \end{bmatrix}. \tag{7·24}$$

This gives the true position of date (x_1', y_1', z_1') directly from the mean position at epoch (x_0, y_0, z_0) in one stage.

7.3 Reduction of Star Positions from True to Apparent Place of Date

Having established the true place of a star using the techniques outlined in the previous section (7.2) the apparent place of the star must be determined. Star positions are affected by proper motion, parallax, aberration and refraction, and allowance for these phenomena must be made. In Chapter VI expressions for the displacements caused by parallax, aberration and refraction have been derived. In section 7.5.2 of this chapter the method of allowing for proper motion will be discussed. By adding the appropriate correction for these phenomena the true place of an object can be converted to apparent place. Clearly, since the corrections for diurnal aberration and refraction involve the sidereal time of observation, the appropriate correction cannot be made until an approximate time of observation has been established for any individual object.

The procedure may be separated into the following steps.

1. The catalogued position of the object is obtained for a given epoch.
2. The catalogued position is corrected for the proper motion (where appropriate) between the epoch and instant of observation.
3. The corrected position is transformed to the position with respect to the equator and equinox of date. Eqn. (7.24) is used for this purpose.
4. The position of the object for the equator and equinox of date is then corrected for parallax and aberration (where appropriate) using eqns. (6.90, 6.91, 6.92) and (6.96, 6.97, 6.98) respectively. The true position at date gives the initial value of x, y, z and the corrections to these values are then determined.
5. Having corrected for parallax and aberration the equatorial coordinates (α, δ) of the star are found. Knowing the approximate sidereal time of observation the corrections for refraction and diurnal aberration can be determined using eqns. (6.84, 6.85, 6.86) and (6.109, 6.110, 6.111) respectively.

Steps (2) and (3) can be interchanged. The catalogue position at epoch can be transformed to a position with respect to the equator and equinox of date and then allowance made for proper motion in the time interval between the epoch and the instant of observation. In this case the proper motion is referred to the equator and equinox of date.

7.3.1 Approximate formulae for precession. The procedure as described above is straightforward. The formulae, in terms of rectangular equatorial coordinates, are rather cumbersome but do not make great demands on a digital computer. Computation of trigonometric functions only takes place at the beginning and end of the calculation and the entire operation is quick and efficient. However, such a procedure encounters difficulties in practice. The proper motion is often included with the general precession to give the *annual variations* in right ascension and in declination. The time variations of the annual variations are called the *secular variations* and are also quoted. In cataloguing positions, that part of the aberration resulting from the elliptical component of the Earth's motion may not be removed and is included in the star positions since this component is slowly varying. In practice therefore a somewhat different scheme has usually been adopted. Suppose that with respect to the equator and equinox of some epoch t_0 the coordinates of a star are (α_0, δ_0). Suppose that the same star has coordinates (α_1, δ_1) with respect to some other (say later) epoch t_1. Since the change of coordinates is not large a Taylor series expansion for the coordinates could be presumed, namely,

$$\alpha_1 = \alpha_0 + \left(\frac{\mathrm{d}\alpha}{\mathrm{d}t}\right)_0 (t_1 - t_0) + \frac{1}{2}\left(\frac{\mathrm{d}^2\alpha}{\mathrm{d}t^2}\right)_0 (t_1 - t_0)^2 + \dots,$$

$$\delta_1 = \delta_0 + \left(\frac{\mathrm{d}\delta}{\mathrm{d}t}\right)_0 (t_1 - t_0) + \frac{1}{2}\left(\frac{\mathrm{d}^2\delta}{\mathrm{d}t^2}\right)_0 (t_1 - t_0)^2 + \dots. \tag{7.25}$$

Such a scheme has a certain appeal if the higher order terms are not large. In practice it is found that for small values of $(t - t_0)$ ($\not> 10$ yrs) it is sufficient to consider a first order approximation and for most other purposes the inclusion of second order terms is usually sufficient. Expressions for $(\mathrm{d}\alpha/\mathrm{d}t)_0$ and $(\mathrm{d}\delta/\mathrm{d}t)_0$ have first to be determined. Such expressions may be derived by determining first order approximations to eqns. (7.14). If the change occasioned by precession is small then, ζ_0, θ, z are small. Under these circumstances eqn. (7.15) gives

$$\begin{aligned} X_x &= 1, & Y_x &= -(\zeta_0 + z), & Z_x &= -\theta \\ X_y &= (\zeta_0 + z), & Y_y &= 1, & Z_y &= 0 \\ X_z &= \theta, & Y_z &= 0, & Z_z &= 1. \end{aligned} \tag{7.26}$$

and eqns. (7.14) become

$$
\begin{aligned}
x_1 - x_0 &= \mathrm{d}x = -(\zeta_0 + z)y_0 - \theta z_0, \\
y_1 - y_0 &= \mathrm{d}y = (\zeta_0 + z)x_0, \\
z_1 - z_0 &= \mathrm{d}z = \theta x_0.
\end{aligned} \tag{7.27}
$$

Using eqns. (7.27) in eqns. (3.19) it is clear that

$$
\begin{aligned}
\mathrm{d}\delta &= \theta \cos \alpha_0 \\
\mathrm{d}\alpha &= (\zeta_0 + z) + \theta \sin \alpha_0 \tan \delta_0.
\end{aligned} \tag{7.28}
$$

It is clear that to form $(\mathrm{d}\alpha/\mathrm{d}t)_0$, $(\mathrm{d}\delta/\mathrm{d}t)_0$ it is necessary to know the annual change of $(\zeta_0 + z)$ and θ at the epoch t_0. Denote these values by m and n respectively. Then

$$
\begin{aligned}
\left(\frac{\mathrm{d}\alpha}{\mathrm{d}t}\right)_0 &= m + n \sin \alpha_0 \tan \delta_0, \\
\left(\frac{\mathrm{d}\delta}{\mathrm{d}t}\right)_0 &= n \cos \alpha_0.
\end{aligned} \tag{7.29}
$$

m is called the *annual general precession in Right Ascension* and n is called the *annual general precession in Declination*, if the interval over which the variation is considered is one tropical year. m and n are capable of further interpretation. In the following treatment the effect of luni-solar precession will be considered in a first order approximation. The effect of luni-solar precession to this order of approximation may be considered in terms of an ecliptic whose plane is fixed, though the equinox may rotate about its fixed pole. Let us suppose that in the course of a tropical year the equinox rotates through an angle ψ' because of luni-solar precession. Since the rotation is about an axis through the pole of the ecliptic, the rectangular equatorial coordinates of the ecliptic L, M, N (see Chapter II) are therefore

$$
L = 0, \qquad M = -\sin \varepsilon, \qquad N = \cos \varepsilon, \tag{7.19}
$$

and using eqns. (2.27)–(2.29) the increments in the equatorial rectangular coordinates are

$$\begin{aligned} dx &= -\psi'(\sin \varepsilon_0 \,.\, z_0 + \cos \varepsilon_0 \,.\, y_0), \\ dy &= +\psi'(\cos \varepsilon_0 \,.\, x_0), \\ dz &= +\psi'(\sin \varepsilon_0 \,.\, x_0), \end{aligned} \tag{7.30}$$

where x_0, y_0, z_0, ε_0 are evaluated at epoch t_0 for which the rectangular coordinates were established (compare with eqns. (7.21)). It again should be noted that although the motion of the Vernal Equinox is clockwise it reflects as an anticlockwise motion at the star. To the first order, eqns. (7.30) could be used to determine the new coordinates for epoch $t_0 + 1$ yr. The effect of planetary precession may be considered as a reflection of the anticlockwise rotation of the equinox in the plane of the celestial equator about an axis of rotation through the celestial pole. The rotation has a magnitude $-\lambda'$. In this case

$$L = M = 0, \qquad N = 1, \tag{7.31}$$

and so

$$\begin{aligned} dx &= \lambda' y_0, \\ dy &= -\lambda' x_0, \\ dz &= 0. \end{aligned} \tag{7.32}$$

The combined correction is therefore the sums of eqns. (7.30) and (7.32) namely

$$\begin{aligned} dx &= -y_0(\psi' \cos \varepsilon_0 - \lambda') - z_0(\psi' \sin \varepsilon_0), \\ dy &= x_0(\psi' \cos \varepsilon_0 - \lambda'), \\ dz &= x_0(\psi' \sin \varepsilon_0). \end{aligned} \tag{7.33}$$

Eqn. (7.33) may be compared with the approximate eqns. (7.27). It is then clear that

$$\begin{aligned} m &= \text{annual rate of change of } (\zeta_0 + z) = \psi' \cos \varepsilon_0 - \lambda', \\ n &= \text{annual rate of change of } \theta = \psi' \sin \varepsilon_0 . \end{aligned} \tag{7.34}$$

Eqn. (7.34) relates m, n to the luni-solar precession in longitude ψ' and planetary precession λ'. ψ', λ' and ε_0 must be evaluated for the same epoch.

While eqns. (7.14) could be approximated by expressions of order higher than the first, it has proved sufficient in practice to limit expansion to the second order. Indeed it is sufficiently accurate to use the derivative with respect to time of the first order terms in the second order correcting term. Taking the derivatives with respect to time of eqns. (7.29)

$$\left(\frac{\mathrm{d}^2\alpha}{\mathrm{d}t^2}\right)_0 = \frac{\mathrm{d}m}{\mathrm{d}t} + \frac{\mathrm{d}n}{\mathrm{d}t}\sin\alpha_0\tan\delta_0 + n\left\{\cos\alpha_0\tan\delta_0\left(\frac{\mathrm{d}\alpha}{\mathrm{d}t}\right)_0 + \sin\alpha_0\sec^2\delta_0\left(\frac{\mathrm{d}\delta}{\mathrm{d}t}\right)_0\right\},$$

$$\left(\frac{\mathrm{d}^2\delta}{\mathrm{d}t^2}\right)_0 = \frac{\mathrm{d}n}{\mathrm{d}t}\cos\alpha_0 - n\sin\alpha_0\left(\frac{\mathrm{d}\alpha}{\mathrm{d}t}\right)_0. \tag{7.35}$$

The second order terms are known as the secular variations. Secular variations are small and usually quoted as a variation per tropical century. If S_α is the centennial secular variation in Right Ascension and D_δ is the centennial secular variation in Declination then

$$\frac{S_\alpha}{100} = \left(\frac{\mathrm{d}^2\alpha}{\mathrm{d}t^2}\right)_0, \qquad \frac{S_\delta}{100} = \left(\frac{\mathrm{d}^2\delta}{\mathrm{d}t^2}\right)_0. \tag{7.36}$$

Thus to the second order

$$\begin{aligned}\alpha_1 &= \alpha_0 + t\left\{\left(\frac{\mathrm{d}\alpha}{\mathrm{d}t}\right)_0 + \frac{S_\alpha t}{200}\right\}\\ \delta_1 &= \delta_0 + t\left\{\left(\frac{\mathrm{d}\delta}{\mathrm{d}t}\right)_0 + \frac{S_\delta t}{200}\right\}\end{aligned} \tag{7.37}$$

where (α_1, δ_1) are the equatorial coordinates of a celestial object at time t after the epoch when the coordinates were (α_0, δ_0). Different conventions have been adopted in the definition of annual variation and secular variation. The precise assignment of these terms should be carefully checked for each catalogue used. For example, in some catalogues the annual variations include the effect of proper motion while in others the secular variations are already divided by 2.

7.3.2 Day numbers. The reduction of mean to apparent place on the sky may also be carried out using the scheme of section 7.3.1. The following notation will be used:

(α_0, δ_0), equatorial coordinates at a standard epoch t_0,
(α_1, δ_1), equatorial coordinates at epoch t_1,
(α, δ), apparent or true coordinates of date,
τ, fraction of the Besselian year elapsed since epoch t_1.

It is usual to assume that epoch t_1 is the beginning or end of the Besselian year in which the observation is made. For the first half of the year the epoch t_1 is the beginning of the Besselian year t_1 while for the second half of the year the epoch t_1 refers to the beginning of the subsequent Besselian year. The parameters given in the Astronomical Ephemeris assume such a convention.

The catalogue coordinates (α_0, δ_0) for a celestial object at epoch t_0 are either transformed directly to the coordinates (α_1, δ_1) for epoch t_1 using eqns. (7.24) or (7.37) and then correcting for proper motion in the interval $(t_1 - t_0)$ (making due allowance for changes in the proper motion in the interval if significant) using the components of proper motion for epoch t_1. Alternatively the coordinates at epoch t_0 are corrected for proper motion (making due allowance for changes in proper motion in interval if significant) using the components of proper motion for epoch t_0 and then using eqns. (7.24) or (7.37) to transform the corrected coordinates to epoch t_1. The method of making such allowance for proper motion will be discussed in section 7.5.2. Correction must then be made for the effects of proper motion, precession, nutation, aberration, parallax (if any) in the fraction τ of the year that has elapsed since epoch t_1. Suppose the components of proper motion at epoch t_1 are μ_{α_1}, μ_{δ_1} (see section 7.5). Since τ is a short time, a first order correction for precession is allowable, and can be made on the basis of eqns. (7.29). Allowance must also be made for nutation in longitude and in the obliquity. The effects of these phenomena may be determined through eqns. (7.23). From these equations the correction for nutation to the first order is

$$d\alpha = \Delta\psi[\cos\varepsilon_1 + \sin\varepsilon_1 \sin\alpha_1 \tan\delta_1] - \Delta\varepsilon \cos\alpha_1 \tan\delta_1, \qquad (7.38)$$

$$d\delta = \Delta\psi \sin\varepsilon_1 \cos\alpha_1 + \Delta\varepsilon \sin\alpha_1,$$

where ε_1 is the obliquity of the ecliptic at epoch t_1. Correction for aberration must also be made and this has been shown to be

$$d\alpha = \sec\delta_1 \left[\frac{X'}{c_*}\sin\alpha_1 - \frac{Y'}{c_*}\cos\alpha_1\right],$$

$$d\delta = \frac{X'}{c_*}\cos\alpha_1 \sin\delta_1 + \frac{Y'}{c_*}\sin\alpha_1 \sin\delta_1 - \frac{Z'}{c_*}\cos\delta_1, \qquad (6.69)$$

where c_* is the velocity of light. Strictly the correction for aberration should be made with respect to the stellar coordinates of date. However, the error is second order and for most purposes can be neglected. The total correction may be found by summing the individual corrections so that the coordinates at date may be found from

$$\alpha = \alpha_1 + \tau\mu_{\alpha_1} + \tau(m + n\sin\alpha_1\tan\delta_1) + \Delta\psi[\cos\varepsilon_1 + \sin\varepsilon_1\sin\alpha_1\tan\delta_1] - \Delta\varepsilon\cos\alpha_1\tan\delta_1 + \sec\delta_1\left[\frac{X'}{c_*}\sin\alpha_1 - \frac{Y'}{c_*}\cos\alpha_1\right], \qquad (7.39)$$

$$\delta = \delta_1 + \tau\mu_{\delta_1} + \tau(n\cos\alpha_1) + \Delta\psi\sin\varepsilon_1\cos\alpha_1 + \Delta\varepsilon\sin\alpha_1 + \frac{X'}{c_*}\cos\alpha_1\sin\delta_1 + \frac{Y'}{c_*}\sin\alpha_1\sin\delta_1 - \frac{Z'}{c_*}\cos\delta_1, \qquad (7.40)$$

where m, n are the values for epoch t_1. In this form the equations are cumbersome and it is therefore necessary to carry out some rearrangement. The following quantities may be defined:

$$
\begin{aligned}
a &= \frac{m}{n} + \sin\alpha_1 \tan\delta_1, & a' &= \cos\alpha_1, \\
b &= \cos\alpha_1 \tan\delta_1, & b' &= -\sin\alpha_1, \\
c &= \cos\alpha_1 \sec\delta_1, & c' &= \tan\varepsilon_1 \cos\delta_1 - \sin\alpha_1 \sin\delta_1, \\
d &= \sin\alpha_1 \sec\delta_1, & d' &= \cos\alpha_1 \sin\delta_1.
\end{aligned}
\tag{7.41}
$$

In terms of the *star constants* defined in eqn. (7.41), eqns. (7.39) and (7.40) may be written as

$$
\begin{aligned}
\alpha &= \alpha_1 + \tau\mu_{\alpha_1} + a\left(n\tau + n\frac{\Delta\psi}{\psi'}\right) - \Delta\varepsilon\, b \\
&\quad - \frac{Y'}{c_*}c + \frac{X'}{c_*}d + \frac{\lambda'\Delta\psi}{\psi'}, \\
\delta &= \delta_1 + \tau\mu_{\delta_1} + a'\left(n\tau + n\frac{\Delta\psi}{\psi'}\right) - \Delta\varepsilon\,.\,b' \\
&\quad - \frac{Y'}{c_*}c' + \frac{X'}{c_*}d',
\end{aligned}
\tag{7.42}
$$

where

$$\frac{m}{n} = \frac{\psi'\cos\varepsilon_1 - \lambda'}{\psi'\sin\varepsilon_1} = \cot\varepsilon_1 - \frac{\lambda'}{\psi'}\operatorname{cosec}\varepsilon_1,$$

$$\sin\varepsilon_1 = n/\psi'$$

and

$$Z' = Y'\tan\varepsilon_1,$$

on making use of eqns. (7.34). Further coefficients which are independent of stellar position are defined as follows,

$$
\begin{aligned}
A &= n\tau + n\frac{\Delta\psi}{\psi'}, \qquad B = -\Delta\varepsilon, \qquad C = -\frac{Y'}{c_*}, \\
D &= \frac{X'}{c_*}, \qquad E = \frac{\lambda'\Delta\psi}{\psi'}.
\end{aligned}
\tag{7.43}
$$

A, *B*, *C*, *D* and *E* are known as Bessel's Day Numbers.

It is clear that with the above definition of Bessel's Day Numbers the reduction to apparent place is simply,

$$\begin{aligned} \alpha &= \alpha_1 + \tau\mu_{\alpha_1} + Aa + Bb + Cc + Dd + E, \\ \delta &= \delta_1 + \tau\mu_{\delta_1} + Aa' + Bb' + Cc' + Dd'. \end{aligned} \tag{7.44}$$

This form is often used in practice. A, B, C, D, E are tabulated for each day of the year in the Astronomical Ephemeris. The above form is particularly suited for easy computation. It should be noted that the Day Number A is discontinuous in the tabulation at the mid-year point, since in the first half of the year the epoch is the beginning of the Besselian year containing the instant of observation while in the second half of the year the epoch is the beginning of the next Besselian year. It is clear that to the first order, eqn. (7.44) can be rewritten to allow the determination of (α_1, δ_1) from the measured (α, δ).

In some applications, Independent Day Numbers f, g, G, h, H, i may be used in place of Besselian Day Numbers. They are defined by

$$\begin{aligned} f &= \frac{m}{n}A + E, \qquad g \sin G = B, \qquad g \cos G = A, \\ h \sin H &= C, \qquad h \cos H = D, \qquad i = C \tan \varepsilon. \end{aligned} \tag{7.45}$$

The Independent Day numbers are tabulated for each day of the year in the Astronomical Ephemeris. The reduction to apparent place in terms of Independent Day Numbers is therefore

$$\begin{aligned} \alpha &= \alpha_1 + \tau\mu_{\alpha_1} + f + g \sin(G + \alpha_1) \tan \delta_1 \\ &\quad + h \sin(H + \alpha_1) \sec \delta_1, \\ \delta &= \delta_1 + \tau\mu_{\delta_1} + g \cos(G + \alpha_1) \\ &\quad + h \cos(H + \alpha_1) \sin \delta_1 + i \cos \delta_1. \end{aligned} \tag{7.46}$$

The formulation using Independent Day Numbers now has increased value since desk calculators with trigonometric functions are available. In summary the procedure to reduce from mean to apparent place is as follows:

(i) the mean equatorial coordinates (α_0, δ_0) of epoch t_0 are reduced to mean equatorial coordinates of epoch t_1.

(ii) the mean coordinates at epoch t_1 are corrected for proper motion in the interval $t_1 - t_0$ to give (α_1, δ_1).

(iii) the coordinates (α_1, δ_1) are corrected for precession, nutation, proper motion, aberration in the fraction τ of the year since epoch t_1.

(iv) correction for parallax can be made if appropriate.

(v) correction for refraction and diurnal aberration can be determined once the instant of observation has been estimated.

The choice of method used to determine the true place of a celestial object depends on the number of objects for which places are required, the frequency of performance of such calculations and the computing resources available. For infrequent calculations for few stars the method given in this section leading to either eqns. (7.44) or (7.46) is of value. If the calculation is to be carried out repeatedly for large numbers of objects the method outlined at the beginning of this section should be followed. The formulae developed in section (7.2) and Chapter VI should be then used in terms of rectangular equatorial coordinates. In this method formulae may be used to determine ζ_0, θ, z and no reference to tabular material is required.

7.4 Practical Procedure for Allowance of Precession

In practice the procedures outlined in section (7.3) above can be unnecessarily cumbersome if all that is required is the sighting of a telescope. While allowance must be made for precession, it is only necessary that the telescope be sighted so that the required star field is within the field of view though not necessarily at its centre. Allowance for precession can be made by defining parameters $m_{\frac{1}{2}}$, $n_{\frac{1}{2}}$ such that these are the values of m, n at same epoch half way in time between the epoch t_0 of the catalogue and that of date t. Then

$$M = m_{\frac{1}{2}}(t - t_0), \qquad N = n_{\frac{1}{2}}(t - t_0). \tag{7.47}$$

With sufficient accuracy for sighting, the coordinates of the star (α, δ) at date are given by

$$\begin{aligned} \alpha &= \alpha_0 + M + N \sin \tfrac{1}{2}(\alpha + \alpha_0) \tan \tfrac{1}{2}(\delta + \delta_0), \\ \delta &= \delta_0 + N \cos \tfrac{1}{2}(\alpha + \alpha_0), \end{aligned} \tag{7.48}$$

where iteration may be used if essential. For sighting purposes other corrections may be disregarded apart from those relating to errors in the mechanical system for driving the telescope (these may be determined by sighting on a bright star whose coordinates are known and which lies near the field of interest). The stars for observation may then be identified from an identification chart of the desired field. Such approximate methods cannot of course be used for astrometric purposes.

7.5 Proper Motion

7.5.1 The space motion of stars. Stars have up to now been considered to be fixed geometric points on the celestial sphere. However, all stars have their own motions in space. The first evidence of such motion was presented by Halley in 1718 and confirmed by Cassini in 1738. These motions are in part random and in part systematic.

Stars are not distributed throughout the Universe but are aggregated into stellar systems called galaxies. The Sun and all visible stars belong to the Galaxy, which contains about 10^{10}–10^{11} stars. The Galaxy is not spherical in shape, but is flattened toward a plane known as the Galactic plane and is seen on the sky as the Milky Way. The Galaxy is rotating about an axis perpendicular to its plane and the Sun and stars take part in this rotation. The rotation of the Galaxy gives rise to a systematic stellar motion.

Stars may belong to sub-groupings of stars (e.g. globular clusters, open clusters) within the Galaxy and their motion will contain a component arising from the systematic motion of the sub-group. In addition, stars also have a random motion in space which is characteristic of each individual star. No star can therefore be considered to be at rest. However, the effects of such motion are only detectable for those stars which happen to be close neighbours of the Sun in space or which happen to possess high velocities.

Although it has been stated above that stars share in the rotation of the Galaxy the effects of proper motion are treated as though stars moved rectilinearly in straight lines. The period of rotation of the Galaxy is $2{\cdot}4 \, . \, 10^8$ yr and the curvature implied by such rotation is not detectable in periods less than 1000 yr. Again, encounters between stars will cause deflections from a rectilinear path but such encounters are rare (with the same time scale as the period of rotation) and the effects would not be detectable with the timescale available in accurate observational material. The assumption that stars move with a constant velocity in a straight line is therefore useful and does not lead to detectable inaccuracy.

The proper motion of a star is defined to be the angular change per unit time in its position along a great circle of the celestial sphere centred on the Sun. Since the radial component of the star's velocity will not alter the position of the star on the celestial sphere, it is the transverse component, at right angles to the radial velocity, which gives rise to the change in star position. Proper motion in space is considered relative to the Sun. This is a practical definition. The measured proper motion is a relative motion and contains both the intrinsic motions of the star and Sun. However, for an observer interested only in the exact position of a star such a definition is sufficient since the proper motion so defined is capable of accurate measurement. The determination of solar motion cannot be made with precision and so the intrinsic motions of the stars can only be found subject to the same accuracy constraint.

Suppose that, relative to the Sun, a star has velocity V in a direction θ with respect to the line joining Sun and star (see Fig. 7.5). In unit time (usually 1 yr.) suppose that the star moves from A to B with respect to the Sun at S. The star is first observed in the direction of SA and after the elapse of unit time is observed along SB. The angular displacement $A\hat{S}B = \mu$ is called the *proper motion* of the star.

Remembering that μ is the displacement of the star position in unit time, the distance AB' will be $V_T . 1 = V \sin\theta$. Further, if the angle between the star-Sun radial direction changes from θ when the star is at A to $\theta + \Delta\theta$ when the star is at B the properties of the plane triangle ASB give

$$r \sin\mu = V \sin(180° - \theta),$$

or

$$\mu = -\Delta\theta = \frac{V}{r}\sin\theta = \frac{V_T}{r}, \tag{7.49}$$

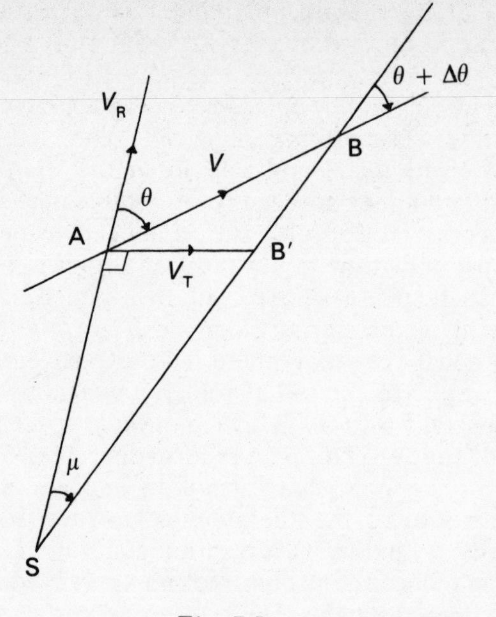

Fig. 7.5

where $r = \text{SB}$ is the distance of the star at B. Since $\Delta\theta$ is the change in θ in unit time μ can be interpreted as the negative of the rate of change of θ with respect to time. If the parallax of the star when at B is π, $r = 1/\pi$ where r is expressed in parsecs and π is in seconds of arc. If μ and π are expressed in seconds of arc and V, V_T in km s^{-1} eqn. (7.49) may be written

$$V_T = V \sin\theta = 4{\cdot}74\frac{\mu}{\pi}. \tag{7.50}$$

Although the rectilinear velocity V of the star may be regarded as a constant with respect to time, its distance from the Sun is not constant. However, the changes produced by such variation of r and θ are not significant except in the case of a very few stars. For such stars the variation of μ may be determined by differentiation of eqn. (7.49) with respect to time;

$$\frac{\mathrm{d}\mu}{\mathrm{d}t} = -\frac{V}{r^2}\sin\theta\frac{\mathrm{d}r}{\mathrm{d}t} + \frac{V}{r}\cos\theta\frac{\mathrm{d}\theta}{\mathrm{d}t},$$

$$= -\frac{\mu}{r}\frac{\mathrm{d}r}{\mathrm{d}t} - \frac{\mu}{r}V\cos\theta = -\frac{2\mu}{r}V_{\mathrm{R}}, \qquad (7.51)$$

$$= -0{\cdot}000\,002\,05\mu\pi V_{\mathrm{R}}, \qquad (7.52)$$

where μ, π are expressed in seconds of arc and V_{R} in km s^{-1}. It is clear from eqn. (7.52) that the *perspective acceleration* $\mathrm{d}\mu/\mathrm{d}t$ will be small unless μ, π, V_{R} are large (e.g., in the case of the nearby Barnard's Star for which $\mu = 10''{\cdot}3$, $\pi = 0''{\cdot}55$, $V_{\mathrm{R}} = -108$ km s^{-1}, the perspective acceleration is only $0''{\cdot}0012$ annually). If perspective acceleration is significant in some interval of time, it is usual to calculate the mean value of μ in the time interval and use this value in all determinations of the observed position of the star.

7.5.2 The displacement of stellar position caused by proper motion. The effect of proper motion is to cause a displacement through an angle μ annually along a great circle whose plane

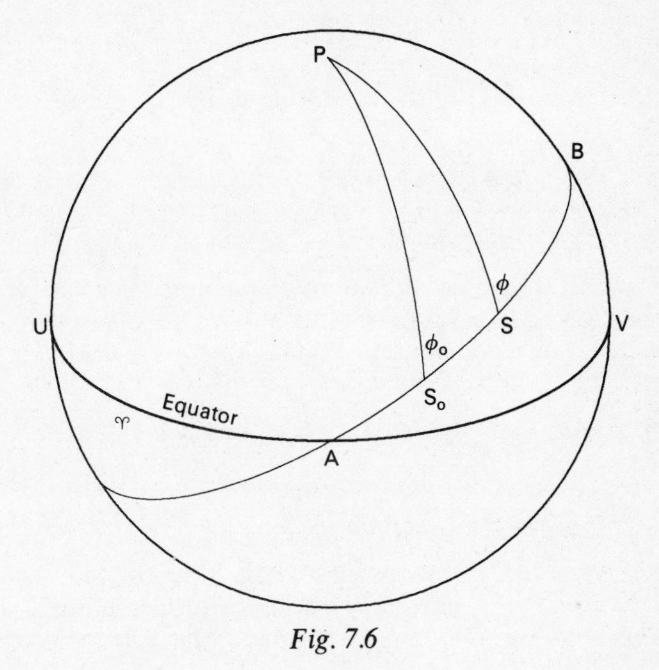

Fig. 7.6

is perpendicular to the vector $\boldsymbol{r} \times \boldsymbol{V}$ where $\boldsymbol{r}$ is the position vector and $\boldsymbol{V}$ is the velocity vector of the star. Because of its proper motion the position of the star moves around this great circle. In this section it will be assumed that perspective acceleration can be neglected and that μ may be assumed constant. In Fig. 7.6 a star has position S_0 at time t_0. At a later time t the star will be observed at S where S_0S is the displacement caused by proper motion in the time interval $t - t_0$. U♈AV is the celestial equator and P is the north celestial pole. Let S_0 have rectangular equatorial coordinates (x_0, y_0, z_0) and equatorial coordinates (α_0, δ_0); let S have rectangular equatorial coordinates (x, y, z) and equatorial coordinates (α, δ). Use of eqns. (3.16) gives,

$$\begin{aligned} x &= x_0 \cos\{\mu(t - t_0)\} \\ &\quad - \sin\{\mu(t - t_0)\}(y_0 \sin\phi_0 + x_0 z_0 \cos\phi_0)/(1 - z_0^2)^{\frac{1}{2}}, \\ y &= y_0 \cos\{\mu(t - t_0)\} \\ &\quad + \sin\{\mu(t - t_0)\}(x_0 \sin\phi_0 - y_0 z_0 \cos\phi_0)/(1 - z_0^2)^{\frac{1}{2}}, \\ z &= z_0 \cos\{\mu(t - t_0)\} + \sin\{\mu(t - t_0)\}(1 - z_0^2)^{\frac{1}{2}} \cos\phi_0, \end{aligned} \tag{7.53}$$

and an application of the Sine Rule to the spherical triangle PS_0S gives

$$\cos\delta \sin\phi = \cos\delta_0 \sin\phi_0,$$

or

$$(1 - z^2)^{\frac{1}{2}} \sin\phi = (1 - z_0^2)^{\frac{1}{2}} \sin\phi_0, \tag{7.54}$$

from which the value of ϕ at any later time t may be determined. For an interval of $(t - t_0) = 1$ year, μ may be treated as a small angle and eqns. (7.53) may then be rewritten in the form

$$\begin{aligned} dx &= \mu_x = -\mu(y_0 \sin\phi_0 + x_0 z_0 \cos\phi_0)/(1 - z_0^2)^{\frac{1}{2}}, \\ dy &= \mu_y = \mu(x_0 \sin\phi_0 - y_0 z_0 \cos\phi_0)/(1 - z_0^2)^{\frac{1}{2}}, \\ dz &= \mu_z = \mu(1 - z_0^2)^{\frac{1}{2}} \cos\phi_0. \end{aligned} \tag{7.55}$$

Indeed, since μ is normally such a small quantity eqns. (7.55) may be used to estimate the effects of proper motion over long intervals of time $(t - t_0)$ in which case μ is replaced by

$\mu(t - t_0)$. Use of eqns. (3.19) then gives

$$\begin{aligned} d\alpha &= \mu_\alpha = \mu \sec \delta_0 \sin \phi_0, \\ d\delta &= \mu_\delta = \mu \cos \phi_0. \end{aligned} \tag{7.56}$$

μ_α, μ_δ are the proper motions in right ascension and declination respectively.

An alternative approximate form of eqns. (7.53) may be obtained by taking the next higher approximation namely

$$\begin{aligned} x &= x_0 + \mu_x(t - t_0) - \tfrac{1}{2}\{\mu(t - t_0)\}^2 x_0, \\ y &= y_0 + \mu_y(t - t_0) - \tfrac{1}{2}\{\mu(t - t_0)\}^2 y_0, \\ z &= z_0 + \mu_z(t - t_0) - \tfrac{1}{2}\{\mu(t - t_0)\}^2 z_0, \end{aligned} \tag{7.57}$$

where μ_x, μ_y, μ_z are given by eqn. (7.55). However, a more useful formulation of eqn. (7.55) is the first order approximation

$$\begin{aligned} \mu_x &= -y_0\mu_\alpha - \frac{x_0 z_0}{(1 - z_0^2)^{\frac{1}{2}}}\mu_\delta, \\ \mu_y &= +x_0\mu_\alpha - \frac{y_0 z_0}{(1 - z_0)^{\frac{1}{2}}}\mu_\delta, \\ \mu_z &= (1 - z_0^2)^{\frac{1}{2}}\mu_\delta, \end{aligned} \tag{7.58}$$

where eqns. (7.56) have been used in eqns. (7.55). This form is useful in practice since μ_α, μ_δ are determined in practice rather than μ, ϕ_0. ϕ_0 may be determined from eqns. (7.56) since μ may be found from the result that

$$\mu_\alpha^2 \cos^2 \delta_0 + \mu_\delta^2 = \mu_x^2 + \mu_y^2 + \mu_z^2 = \mu^2. \tag{7.59}$$

However, eqns. (7.58) are the more direct forms for practical use.

The formulae developed in this section such as eqns. (7.53), (7.55) or (7.58) may be used to determine the displacements caused by proper motion. Such equations would be used in association with the corrections for precession, nutation, parallax, aberration, etc.

7.5.3 Change of μ_α, μ_δ with time. Although it has been assumed that μ is a constant in time, it is clear from the form of eqns. (7.56) that μ_α, μ_δ will change as the position of the star is altered by proper motion. Further, the celestial equator may not be regarded as a fixed plane. Therefore it is necessary to be able to relate μ_α, μ_δ at some time t to their values at an earlier time t_0. It is also necessary to be able to refer μ_α, μ_δ at any instant to the mean equator and equinox of any epoch. The consequent changes in μ_α, μ_δ are small and the values of μ_α, μ_δ for the different circumstances may be related by a Taylor series expansion truncated after the first order term, i.e.,

$$\begin{aligned} \mu_\alpha(t) &= \mu_\alpha(t_0) + (t - t_0)\left(\frac{\mathrm{d}\mu_\alpha}{\mathrm{d}t}\right)_{t_0}, \\ \mu_\delta(t) &= \mu_\alpha(t_0) + (t - t_0)\left(\frac{\mathrm{d}\mu_\delta}{\mathrm{d}t}\right)_{t_0}, \end{aligned} \tag{7.60}$$

where $\mu_{\alpha,\delta}(t)$ refers to either the value of some later time with respect to the same equator and equinox as $\mu_{\alpha,\delta}(t_0)$ or to the value at the same time but referred to the equator and equinox of time t. In the case of no change of equator and equinox the variation of $\mu_{\alpha,\delta}$ may be treated simply by differentiation of eqns. (7.54) and (7.56) whereas if a change of equator and equinox are involved the previous consideration of precession must be included.

(a) μ_α, μ_δ *at different times referred to the same equator and equinox.*

Suppose that the equator and equinox refer to epoch t_0. The results of section 7.5.2 may be used to find the variation in μ_α, μ_δ which results from the displacement caused by proper motion. Differentiation of eqns. (7.56) gives (dispensing with the subscript zero),

$$\begin{aligned} \frac{\mathrm{d}\mu_\alpha}{\mathrm{d}t} &= \frac{\mathrm{d}\mu}{\mathrm{d}t}\sec\delta\sin\phi \\ &\quad + \mu\left\{\tan\delta\sec\delta\sin\phi\frac{\mathrm{d}\delta}{\mathrm{d}t} + \sec\delta\cos\phi\frac{\mathrm{d}\phi}{\mathrm{d}t}\right\}, \\ \frac{\mathrm{d}\mu_\delta}{\mathrm{d}t} &= \frac{\mathrm{d}\mu}{\mathrm{d}t}\cos\phi - \mu\sin\phi\frac{\mathrm{d}\phi}{\mathrm{d}t}. \end{aligned} \tag{7.61}$$

But, $\mathrm{d}\delta/\mathrm{d}t = \mu_\delta$ and $\mathrm{d}\mu/\mathrm{d}t$ is given by eqn. (7.51) (were μ an exact constant $d\mu/dt \equiv 0$). $d\phi/dt$ may be obtained by differentiating eqn. (7.54) when

$$\frac{d\phi}{dt} = \tan\delta\tan\phi\frac{d\delta}{dt} = \mu_\delta\tan\delta\tan\phi. \tag{7.62}$$

Using these results in eqn. (7.61) it is clear that

$$\begin{aligned}\left(\frac{d\mu_\alpha}{dt}\right) &= -\frac{2\mu_\alpha V_R}{r} + 2\mu_\alpha\mu_\delta\tan\delta_0,\\ \left(\frac{d\mu_\delta}{dt}\right)_{t_0} &= -\frac{2\mu_\delta V_R}{r} - \mu_\alpha^2\sin\delta_0\cos\delta_0,\end{aligned} \tag{7.63}$$

where μ_α, μ_δ are evaluated at time $t = t_0$. Hence if μ'_α, μ'_δ are the values of μ_α, μ_δ respectively at time t referred to the equator and equinox of time t_0 then

$$\begin{aligned}\mu'_\alpha &= \mu_\alpha + \left(2\mu_\alpha\mu_\delta\tan\delta_0 - \frac{2\mu_\alpha V_R}{r}\right)(t - t_0),\\ \mu'_\delta &= \mu_\delta - \left(\mu_\alpha^2\sin\delta_0\cos\delta_0 + \frac{2\mu_\delta V_R}{r}\right)(t - t_0).\end{aligned} \tag{7.64}$$

Under usual circumstances the perspective acceleration terms (i.e. those in V_R/r) may be neglected as would have been found were $d\mu/dt \equiv 0$.

(b) μ_α, μ_δ *at the same instant referred to different equators and equinoxes.*

Suppose that the components of proper motion μ_α, μ_δ are known at some instant t_0 referred to the mean equator and equinox of that instant. The components of proper motion may be referred to the mean equator and equinox of another epoch t. Eqns. (7.61) remain valid. However, in this situation there is no perspective acceleration so that $d\mu/dt \equiv 0$, $d\delta/dt$, $d\alpha/dt$ are given by eqns. (7.29). In Fig. 7.7 $U_0\Upsilon_0AV_0$ is the celestial equator for epoch t_0. Proper motion moves the star position in the plane of the great circle ASB. P_0 is the north celestial pole for epoch t_0 and P is the north celestial pole for epoch t. $\phi_0 = P_0\hat{S}B$ and ϕ $(=P\hat{S}B)$ is the equivalent

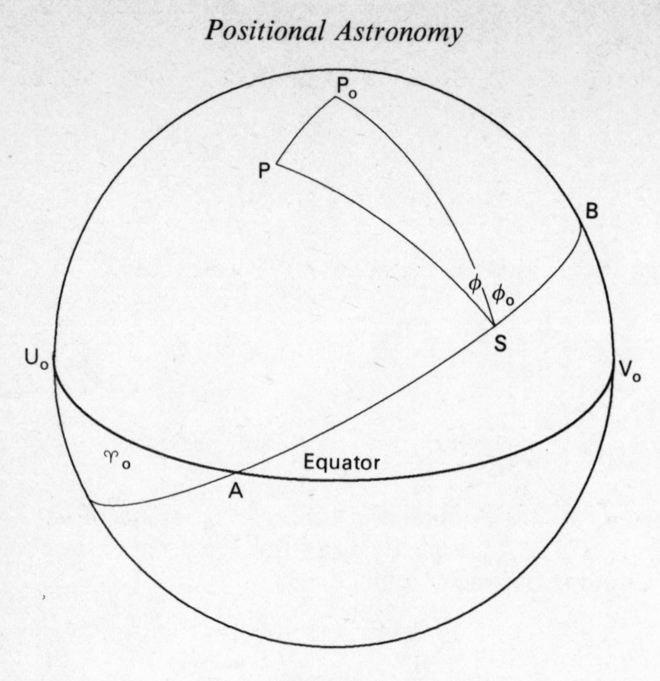

Fig. 7.7

of ϕ_0 at epoch t. The spherical triangle P_0PS has the properties

$P_0S = 90° - \delta_0$, $PS = 90° - \delta$,

$P_0P = \theta$ (see Fig. 7.4 and section 7.2.1)

$P_0\hat{S}P = \phi - \phi_0$,

$$P\hat{P}_0S = \zeta_0 + \alpha_0 \text{ (see Fig. 7.4 and section 7.2.1)} \tag{7.65}$$

The application of the Sine Rule to the spherical triangle P_0SP gives

$$\sin\theta \sin(\zeta_0 + \alpha_0) = \cos\delta \sin(\phi - \phi_0).$$

Since θ is small, and ζ_0 may be neglected by comparison with α_0,

$$\frac{d\phi}{dt} = \frac{d\theta}{dt}\sin\alpha_0 \sec\delta = n \sin\alpha_0 \sec\delta_0, \tag{7.66}$$

to the first order where n is the annual rate of change of θ (see eqn. (7.34)). Using eqns. (7.29), (7.66) and setting $d\mu/dt \equiv 0$, eqn. (7.61) in this case becomes

$$\frac{d\mu_\alpha}{dt} = n(\mu_\alpha \cos \alpha \tan \delta + \mu_\delta \sin \alpha \sec^2 \delta),$$
$$\frac{d\mu_\delta}{dt} = -n\mu_\alpha \sin \alpha. \tag{7.67}$$

If μ''_α, μ''_δ are the components of proper motion referred to epoch t and μ_α, μ_δ are the components of proper motion referred to epoch t_0, then from eqn. (7.60)

$$\mu''_\alpha = \mu_\alpha + n(t - t_0)(\mu_\alpha \cos \alpha_0 \tan \delta_0 + \mu_\delta \sin \alpha_0 \sec^2 \delta_0),$$
$$\mu''_\delta = \mu_\delta - n(t - t_0)(\mu_\alpha \sin \alpha_0). \tag{7.68}$$

Eqns. (7.64) may be used to determine the values of μ_α, μ_δ at any time given their values at one instant for a fixed equator and equinox. Eqns. (7.68) may be used to determine the values of u_α, u_δ at a given instant with respect to the mean equator and equinox of any epoch if the values are known for some epoch.

7.6 A Restatement of the Determination of Star Position

In Chapter VI and the preceding sections of this chapter, corrections which have to be made to the heliocentric position of a star have been discussed. The correction for refraction depends only on the zenith distance at which a star is observed. The corrections for aberration and parallax are periodic in that, if there were no precessional or proper motion displacements, a star would appear to move on the sky over a well determined elliptical track with a period of one year. In a measurement of proper motion the position of a star is determined at two epochs t_0, t_1. The interval $t_1 - t_0$ is made long in order to obtain substantial movement of the star. Let us suppose that the measurements of the star position are corrected for refraction, parallax and aberration appropriate to the times of observation t_0, t_1. Let the coordinates then be (α_0, δ_0), (α_1, δ_1) respectively. The difference in the position simply corresponds to the amount of precession, nutation and proper motion in the interval $t_1 - t_0$. The corrections for precession and nutation are in part determined from observation and as such will contain errors. The proper motion refers to the motion of the Sun, and the intrinsic motion of the star. If allowance is made for precession and nutation in the manner described the remaining difference in star position is made up of proper motion and errors in the determination of the corrections for precession and nutation. Some progress can be made in disentangling the various contributions to the residual difference in position. In the ensuing discussion it will be assumed that the motion of the star is simply a result of its rotation about the centre of the Galaxy. μ_α, μ_δ will be reinterpreted to be simply the annual rate of change of star position referring to the residual motion having made allowance for refraction, aberration, parallax and the known amount of precession and nutation.

7.6.1 Solar motion. Solar motion cannot be determined in any absolute sense. The motion of the Sun can only be established with respect to a specified group of stars. The rectangular equatorial (x, y, z) and equatorial (α, δ) coordinates of a star are related by

$$
\begin{aligned}
x &= r \cos \alpha \cos \delta, \\
y &= r \sin \alpha \cos \delta, \\
z &= r \sin \delta,
\end{aligned}
\tag{3.1a}
$$

where r is the distance of the star. The distance of the star is now of importance and the convention of a unit radius for the celestial sphere has to be dropped. The annual variation of the rectangular equatorial coordinates may be written as

$$
\begin{aligned}
\frac{dx}{dt} &= -r \cos \delta \sin \alpha \, \mu_\alpha - r \sin \delta \cos \alpha \, \mu_\delta \\
&\quad + \cos \delta \cos \alpha \, V_R, \\
\frac{dy}{dt} &= r \cos \delta \cos \alpha \, \mu_\alpha - r \sin \delta \sin \alpha \, \mu_\delta \\
&\quad + \cos \delta \sin \alpha \, V_R, \\
\frac{dz}{dt} &= r \cos \delta \, \mu_\delta + \sin \delta \, V_R,
\end{aligned}
\tag{7.69}
$$

where

$$\mu_\alpha = \frac{d\alpha}{dt}, \qquad \mu_\delta = \frac{d\delta}{dt} \quad \text{and} \quad V_R = \frac{dr}{dt}.$$

If μ_α, μ_δ, V_R and the stellar parallax can be measured then the components dx/dt, dy/dt, dz/dt can be computed for any star. Suppose a group of stars is selected such that the stars may be regarded as having the same (α, δ). If the intrinsic motion of the stars is considered to be random, then the measured proper motion is made up of a random (intrinsic) and a non-random (solar) component. If the motions of the individual stars of the group are added together, the random parts should cancel leaving only the solar component. Then if components of the solar motion are denoted by h_x, h_y, h_z then

$$h_x = -\left(\overline{\frac{dx}{dt}}\right), \qquad h_y = -\left(\overline{\frac{dy}{dt}}\right), \qquad h_z = -\left(\overline{\frac{dz}{dt}}\right) \tag{7.70}$$

where the bar denotes the mean value and the negative sign is introduced since the solar motion will be equal and opposite to the mean motion of the stellar group. The solar velocity is then

$$h = (h_x^2 + h_y^2 + h_z^2)^{\frac{1}{2}} \tag{7.71a}$$

and is directed towards a point whose equatorial coordinates (A, D) are given by

$$\tan A = h_y/h_x, \tag{7.71b}$$

$$\tan D = \frac{h_z}{(h_y^2 + h_x^2)^{\frac{1}{2}}}. \tag{7.71c}$$

The point towards which the solar motion is directed is called the *solar apex*. The point diametrically opposite the solar apex on the celestial sphere is called the *ant-apex*.

However, it is usually the case that the distance r of a group of stars is not known; h_x, h_y, h_z cannot be determined from eqns. (7.69) directly. Eqns. (7.69) may be re-arranged in the form

$$\mu_\alpha \cos\delta = -\frac{1}{r}\left(\frac{dx}{dt}\right)\sin\alpha + \frac{1}{r}\left(\frac{dy}{dt}\right)\cos\alpha,$$

$$\mu_\delta = -\frac{1}{r}\left(\frac{dx}{dt}\right)\cos\alpha\sin\delta$$

$$-\frac{1}{r}\left(\frac{dy}{dt}\right)\sin\alpha\sin\delta + \frac{1}{r}\left(\frac{dz}{dt}\right)\cos\delta, \tag{7.72}$$

$$V_R = \left(\frac{dx}{dt}\right)\cos\alpha\cos\delta + \left(\frac{dy}{dt}\right)\sin\alpha\cos\delta$$

$$+\left(\frac{dz}{dt}\right)\sin\delta.$$

For each star in the group μ_α, μ_δ and V_R can be determined. By summing the contributions of all the stars in the group the quantities $\overline{\mu_\alpha \cos\delta}\,.\,\overline{\mu_\delta}$ and $\overline{V_R}$ can be found. The random contributions to these quantities should cancel if the group of stars is sufficiently large. If

$$X = -\frac{1}{r}\overline{\left(\frac{dx}{dt}\right)}, \qquad Y = -\frac{1}{r}\overline{\left(\frac{dy}{dt}\right)}, \qquad Z = -\frac{1}{r}\overline{\left(\frac{dz}{dt}\right)}, \tag{7.73}$$

then

$$\overline{\mu_\alpha \cos\delta} = +X\sin\alpha - Y\cos\alpha,$$

$$\overline{\mu_\delta} = +X\cos\alpha\sin\delta + Y\sin\alpha\sin\delta - Z\cos\delta, \tag{7.74}$$

$$-\overline{V_R}/r = X\cos\alpha\cos\delta + Y\sin\alpha\cos\delta + Z\sin\delta.$$

Eqns. (7.74) may be solved for X, Y, Z on the assumption that the group of stars have almost the same equatorial coordinates (α, δ) and distance r. In eqn. (7.73) the minus sign is inserted in order that X, Y, Z refer to solar motion. It is usual to determine X, Y, Z by using several groups of stars at the same distance r from the Sun. It is usual to derive X, Y, Z based either on proper motions or on radial velocities through a least squares solution of the pair of equations given by proper motion or the single equation for radial velocities contributed by each group of stars. Clearly

$$\tan A = \frac{rh_y}{rh_x} = \frac{Y}{X}, \qquad \tan D = \frac{rh_z}{r(h_x^2 + h_y^2)^{\frac{1}{2}}} = \frac{Z}{(X^2 + Y^2)^{\frac{1}{2}}}, \tag{7.75}$$

from which the equatorial coordinates (A, D) of the solar apex may be determined. However,

$$\left(\frac{h}{r}\right)^2 = X^2 + Y^2 + Z^2 \tag{7.76}$$

and, as r is unknown, the solar velocity is unknown. However, using radial velocities and the third of eqns. (7.74) values can be obtained for $rX = h_x$, $rY = h_y$ and $rZ = h_z$, from which a determination of $h_\odot$, (A, D) can be obtained using eqns. (7.71).

Hence, in principle, the solar motion can be found. Solar motion is a relative concept and can only be properly defined if the groups of stars used in its determination are also specified. For example, the *standard solar motion* defined with respect to that group of stars for which proper motion and radial velocity data is available gives

$$h = 19{\cdot}7\ \text{km s}^{-1}, \qquad A = 18^{\text{h}}{\cdot}1, \qquad D = +30^\circ\ (1900{\cdot}0), \qquad l = 57^\circ, \qquad b = +22^\circ, \tag{7.77}$$

whereas the *basic solar motion* defined with respect to stars in the solar neighbourhood given by van de Kamp is

$$h = 15{\cdot}4\ \text{km s}^{-1}, \qquad A = 17^{\text{h}}{\cdot}8, \qquad D = +25^\circ, \qquad l = 51^\circ, \qquad b = +23^\circ. \tag{7.78}$$

Comparison of eqns. (7.77) and (7.78) illustrates the effect of different groups of stars on the derivation of h, A, D. Although the concept of solar motion is precise its realisation is not.

If the solar motion is known, a further baseline for measurement of stellar parallax is available. The Sun moves a distance h during the course of a year; the angle θ subtended by such a baseline at a stationary star distant r from the Sun is given by

$$\sin \theta \simeq \theta = \frac{h}{r} \tag{7.79a}$$

Since $1/r$, where the units of r are parsecs, is the annual parallax of the star, θ is also a measure of stellar distance and is called the *secular parallax* of the star. The secular parallax may be written

$$\theta = \frac{\pi h}{4{\cdot}74}, \tag{7.79b}$$

where θ, π are in seconds of arc and h is in km s^{-1}. If h is evaluated using eqn. (7.77)

$$\pi = 0{\cdot}241\, \theta. \tag{7.79c}$$

7.6.2 Intrinsic stellar motions. The determination of solar motion allows an attempt to be made to disentangle the actual motion of the star from the combined measurement of solar

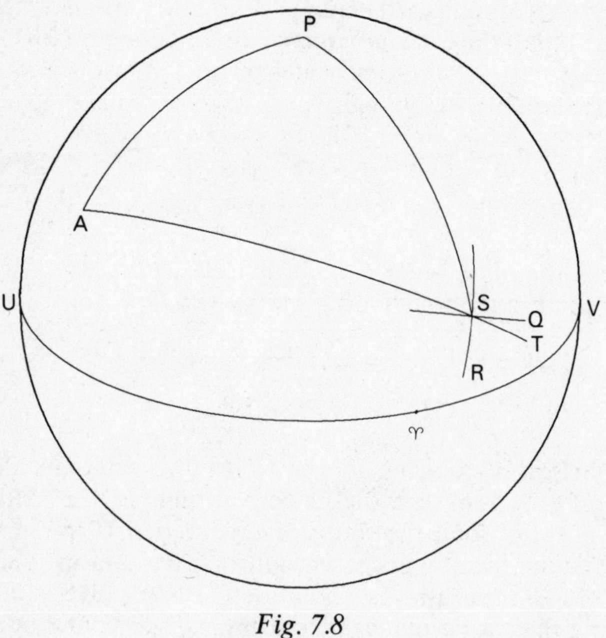

Fig. 7.8

and stellar motion. Let us suppose that the angle between the direction of the solar apex and the direction of the star as viewed from the Sun is λ. In Fig. 7.8 U♈V is the celestial equator and P is the north celestial pole. S is a star. Suppose that the solar motion, reflected in the motion of a star S, is along the great circle AST where A is the solar apex. If the angle between the great circles AST and PS is χ $(=\mathrm{P\hat{S}T})$ then the spherical triangle PAS has the following properties

$$\begin{aligned} &\mathrm{PA} = 90° - D, \qquad \mathrm{PS} = 90° - \delta, \qquad \mathrm{AS} = \lambda, \\ &\mathrm{P\hat{S}A} = 180° - \chi, \qquad \mathrm{A\hat{P}S} = \alpha - A, \end{aligned} \tag{7.80}$$

where (A, D) are the equatorial coordinates of the solar apex and (α, δ) are the equatorial coordinates of the star S. Then, by application of the Cosine and Transposed Cosine Rules to the spherical triangle PAS,

$$\begin{aligned} \cos\lambda &= \sin D \sin\delta + \cos D \cos\delta \cos(\alpha - A), \\ -\sin\lambda\cos\chi &= \sin D \cos\delta - \cos D \sin\delta \cos(\alpha - A). \end{aligned} \tag{7.81}$$

These formulae permit the determination of λ and χ. It should be noticed that the determination of λ and χ are only as good as the determination of (A, D).

If the measured proper motion of the star S is along the great circle SQ where $\mathrm{P\hat{S}Q} = \phi$, components of proper motion υ, τ along the great circle AST and along a great circle SR at right angles to it respectively, may be considered. Then with adequate accuracy

$$\begin{aligned} \upsilon = \mu\cos(\chi - \phi) &= \mu\cos\chi\cos\phi + \mu\sin\chi\sin\phi \\ &= \mu_\alpha\cos\delta\sin\chi + \mu_\delta\cos\chi, \\ \tau = -\mu\sin(\chi - \phi) &= -\mu\sin\chi\cos\phi + \mu\cos\chi\sin\phi \\ &= \mu_\alpha\cos\delta\cos\chi - \mu_\delta\sin\chi. \end{aligned} \tag{7.82}$$

Since the solar motion has no component along SR, τ is entirely a result of intrinsic stellar motion. υ is composed of the parallactic and intrinsic motion of the star. However, knowing the solar motion allows a determination of the parallactic motion. In Fig. 7.9 C_1C_2A is the direction of solar motion. C_1 and C_2 are positions of the Sun at an interval of

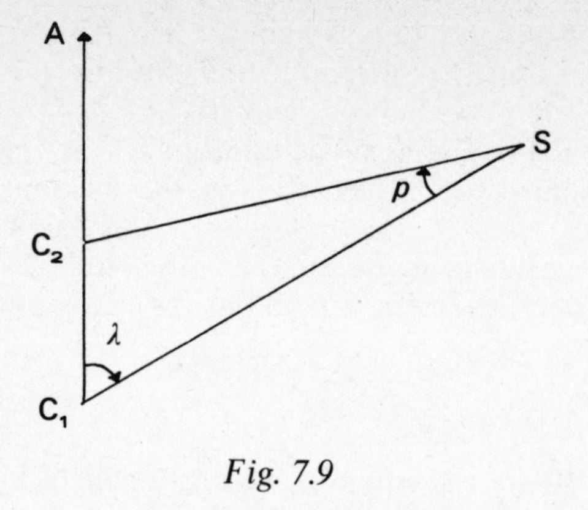

Fig. 7.9

1 year and S is the star. p is the *parallactic motion* of the star resulting from the solar motion i.e. $C_1C_2 = h$. Then if the distance C_2S is r, on using the sine rule for the plane triangle C_1C_2S,

$$\frac{\sin p}{h} = \frac{\sin \lambda}{r}$$

or

$$\sin p = \frac{h}{r} \sin \lambda = \frac{\pi V_\odot}{4{\cdot}74} \sin \lambda. \tag{7.83}$$

Hence, knowing the secular parallax h/r for the star a value of p can be determined and hence $\upsilon - p$ will give the contribution of the intrinsic motion of the star. Clearly such a calculation is hampered by the inaccuracies inherent in the determination of the solar motion.

7.6.3 Motion in the Galaxy. The motion of stars is not random as has been assumed above. Stars share in the rotation of the Galaxy and have some intrinsic velocity dispersion. It will be assumed that the velocity dispersion is small by comparison with the systematic velocity of Galactic rotation and in the following such dispersion will be neglected. The rotation of the Galaxy has to be established by observation and a useful first approximation to Galactic rotation is to assume that the velocity of rotation is only a function of radial distance from the Galactic centre. For the purpose of this discussion the Galaxy will be assumed to be a circular disc of negligible thickness. The velocity of rotation will be directed perpendicularly to the radius vector connecting a star with the centre of the Galaxy. In Fig. 7.10 a star S, the Sun C and the centre of the Galaxy G define the plane of the Galaxy. The solar

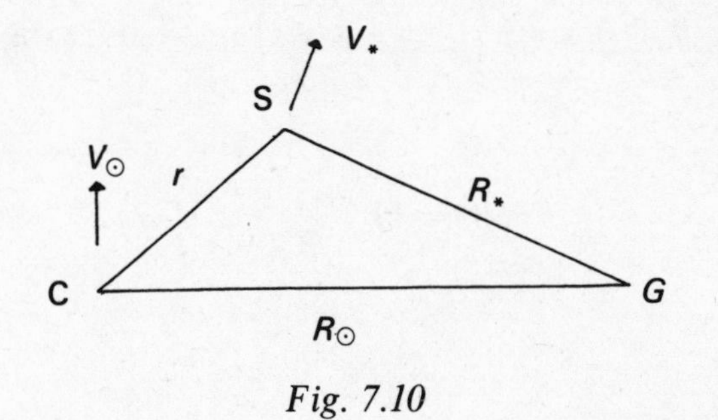

Fig. 7.10

velocity $V_{\odot}$ is directed perpendicularly to the radius GC while the velocity V_* of the star is directed perpendicularly to the radius GS. Since the magnitude of the velocity of rotation is assumed to depend only on distance from the centre, the velocities V_1, V_2 at radii r_1, r_2 may be connected by a Taylor series namely

$$V(r_2) = V(r_1) + \Delta r\left(\frac{\mathrm{d}V}{\mathrm{d}r}\right)_{r_1} + \tfrac{1}{2}\Delta r^2\left(\frac{\mathrm{d}^2V}{\mathrm{d}r^2}\right)_{r_1} + \ldots, \quad (7.84)$$

where $\Delta r = r_2 - r_1$. In this work it is usual to neglect all terms of order greater than the first so that

$$V(r_2) = V(r_1) + \Delta r\left(\frac{\mathrm{d}V}{\mathrm{d}r}\right)_{r_1}. \quad (7.85)$$

The angle $\mathrm{G\hat{C}S}$ is the Galactic longitude l of the star with respect to the Sun. Let the distance CS of the star from the Sun be r and the distances GS $= R_*$ and GC $= R_{\odot}$. Denote the angle $\mathrm{S\hat{G}C}$ by θ. Then

$$\frac{r}{\sin\theta} = \frac{R_{\odot}}{\sin(180° - \theta - l)} = \frac{R_*}{\sin l}. \quad (7.86a)$$

and

$$R_* = R_\odot\left\{1 + \left(\frac{r}{R_\odot}\right)^2 - 2\left(\frac{r}{R_\odot}\right)\cos l\right\}^{\frac{1}{2}}. \qquad (7.86b)$$

If terms in r^2/R^2 and higher powers may be neglected eqns. (7.86) give

$$R_* \sin(\theta + l) = R_\odot\left(1 - \frac{r}{R_\odot}\cos l\right)\sin(\theta + l) = R_\odot \sin l,$$

i.e.,

$$\sin(\theta + l) = \left(1 + \frac{r}{R_\odot}\cos l\right)\sin l,$$

and

$$R_* - R_\odot = -r\cos l. \qquad (7.87)$$

Then the radial velocity V_R of the star S relative to the Sun C is

$$\begin{aligned} V_R &= \left(V_\odot - r\cos l\left(\frac{dV}{dR}\right)_\odot\right)\sin(\theta + l) - V_\odot \sin l \\ &= \left(V_\odot - r\cos l\left(\frac{dV}{dR}\right)_\odot\right)\left(1 + \frac{r}{R_\odot}\cos l\right)\sin l - V_\odot \sin l \\ &= \frac{V_\odot r}{R_\odot}\cos l \sin l - r\cos l \sin l\left(\frac{dV}{dR}\right)_\odot, \end{aligned} \qquad (7.88)$$

using eqns. (7.87) and neglecting the terms in r^2/R. Similarly the transverse velocity V_T of the star with respect to the Sun is given by

$$V_T = -\frac{V_\odot r}{R_\odot}\sin^2 l - r\left(\frac{dV}{dR}\right)_\odot \cos^2 l. \qquad (7.89)$$

If parameters A, B are defined by

$$A = \tfrac{1}{2}\left\{\frac{V_\odot}{R_\odot} - \left(\frac{dV}{dR}\right)_\odot\right\}, \qquad B = \tfrac{1}{2}\left\{-\frac{V_\odot}{R_\odot} - \left(\frac{dV}{dR}\right)_\odot\right\}, \qquad (7.90)$$

the relations

$$V_R = Ar \sin 2l, \qquad V_T = Ar \cos 2l + Br,$$
$$\frac{V_\odot}{R_\odot} = A - B, \qquad \left(\frac{dV}{dR}\right)_\odot = A + B, \tag{7.91}$$

follow immediately.

Were the Galaxy rotating as a solid body with angular velocity of rotation ω then $V/R = dV/dR = \omega$. Hence, from eqn. (7.91), $A = 0$, $B = -V/R$. In the case where V^2 varies inversely as the radius ($V^2 \propto 1/R$—Keplerian rotation) it follows that $A = -3B$. If the parameters A and B are regarded as constants they may be determined for the Galaxy. The *Oort constants* for the Galaxy are

$$A = +0{\cdot}015 \text{ km s}^{-1} pc^{-1} = +0''{\cdot}0032 \text{ yr}^{-1}$$
$$B = -0{\cdot}010 \text{ km s}^{-1} pc^{-1} = -0''{\cdot}0021 \text{ yr}^{-1} \tag{7.92}$$

If the Sun is 10 kpc from the centre of the galaxy $V_\odot$ is 250 km s^{-1}.

The transverse velocity V_T gives rise to the intrinsic motion of the star S. Proceeding by analogy with eqn. (7.50) the proper motion μ' arising from the intrinsic motion of the star may be written:

$$\mu' = \frac{1}{4{\cdot}74}(A \cos 2l + B) = P \cos 2l + Q \tag{7.93}$$

where the units of μ' are seconds of arc and $P = A/4{\cdot}74$, $Q = B/4{\cdot}74$.

The actual distribution of stars in the Galaxy is not that of a disc of negligible thickness and allowance must be made for a non-zero value of Galactic latitude b. It is beyond the scope of this book to derive the formula giving the effect of Galactic rotation in terms of proper motion in right ascension μ'_α and declination μ'_δ. The expressions are of the form

$$\mu'_\alpha = fP + hQ \text{ (with respect to a great circle)}$$
$$\mu'_\delta = gP + jQ \tag{7.94}$$

where f, g, h, j are functions of the galactic coordinates and the relationship between galactic and equatorial coordinates.

For further details the reader should consult Van de Kamp's *Principles of Astrometry*.

Clearly the measured proper motion of a star should be analysed for solar motion, galactic rotation and corrections to the precessional constants. Van de Kamp (in *Principles of Astrometry*) suggests that equations of the type

$$\begin{aligned}
&[X \sin \alpha - Y \cos \alpha] + [fP + hQ] \\
&\quad + [\Delta m \cos \delta + \Delta n \sin \alpha \sin \delta] = \mu_\alpha \cos \delta, \\
&[X \cos \alpha \sin \delta + Y \sin \alpha \sin \delta - Z \cos \delta] \\
&\quad + [gP + jQ] + [\Delta n \cos \alpha] = \mu_\delta,
\end{aligned} \tag{7.95}$$

be solved. The first term in square brackets allows for solar motion (see eqns. (7.74)), the second term allows for galactic rotation (see eqns. (7.94)) while the third term (see eqns. (7.29)) allow for corrections to the precessional terms. These equations are not the most general that could be formulated but they are given here to indicate the basis on which modern astrometry operates. It is clear that the determination of proper motions is an important problem in astrometry since it offers not only a means of determining the motion of stars within the Galaxy but a means of improving the accuracy of precessional corrections. However, the fundamental limitation of the theory underlying these equations should not be forgotten. For example, the model of the Galaxy which has been used is very simple. Modification of that model is likely to give rise to terms which are as important as those from, for example, precession. The detailed analysis is therefore very complicated since at any time the model used for motions in the Galaxy is an approximation to actual conditions. Astrometry is endeavouring to improve its accuracy while at the same time giving definitive information on the structure of the Galaxy.

Chapter VIII

Eclipses and Occulations

The fundamental theory of eclipses and occultations is best illustrated through the treatment of solar eclipses. The theory of occultations and the theory of lunar eclipses can be developed as special cases of solar eclipses. The astronomical importance of solar eclipses and occultations is greater than the importance of lunar eclipses and to some extent this degree of importance is reflected in the degree of accuracy to which the calculations are usually carried out.

A solar eclipse takes place at the time of new Moon when the Moon comes between the Sun and a terrestrial observer. A lunar eclipse takes place at the time of full Moon when the Moon may move into the shadow cone produced by the Earth. Since the plane of the lunar orbit is inclined to the plane of the ecliptic, solar and lunar eclipses can only occur when a new or full Moon is near a node of its orbit. An occultation of a star occurs when, to a terrestrial observer, the star is obscured by the Moon's disc. An occultation of a star by a planet can also occur as can the transit (c.f. an eclipse) of a planet across the solar disc. Since solar eclipses are of fundamental importance an account of the geometrical circumstances of a solar eclipse will now be given.

The conditions for a solar eclipse are illustrated in Fig. 8.1. The centres of the Earth, Sun and Moon are at E, S and M respectively. The shadow produced by the Moon is centred about the line SV_1MV_2. The *umbral* shadow cone has vertex V_2 and generators AV_2, BV_2. The *penumbral* shadow cone has vertex V_1 and generators AV_1, BV_1. The shadow produced by the Moon is a permanent feature but the shadow

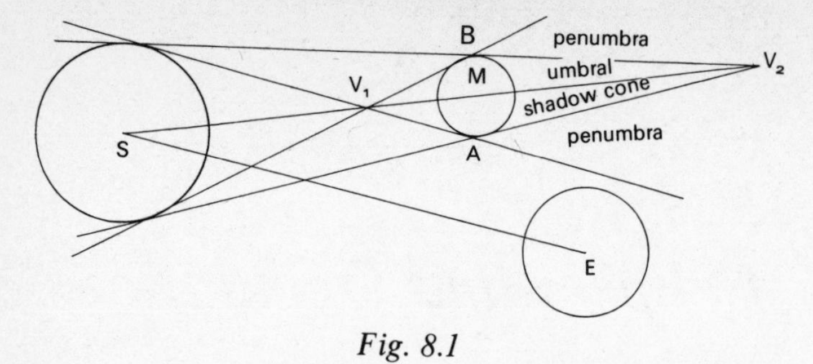

Fig. 8.1

cones can only sweep across the surface of the Earth under very restricted conditions. Since penumbral eclipses of the Sun are of little scientific interest, the following discussion will be confined to umbral eclipses. The lunar orbit is not in the plane of the ecliptic and only when the Moon is new near a node of its orbit on the ecliptic is there the possibility of an eclipse. The angular diameter of the Moon is very close to that of the Sun and the umbral shadow cone of the Moon is only just long enough to intersect the Earth's surface. These two effects mean that solar eclipses are of short duration at totality and that a solar eclipse may only be observed from a small region of the Earth's surface at any given time. Indeed the changes of aspect of the Sun and Moon vary in such a way that it is possible that the vertex V_2 of the umbral shadow does not intersect the Earth's surface. In such circumstances an annular eclipse may be seen. Fig. 8.2 illustrates the relative positions of the Earth's surface and vertex V_2 of the umbral shadow cone for total and annular eclipses. In Fig. 8.2(a) the vertex V_2 would occur inside the Earth. The area enclosed by the curve EFG in which the umbral cone cuts the Earth's surface is that region of the Earth experiencing a total eclipse of the Sun. In Fig. 8.2(b) the vertex V_2 lies outside the Earth. Observers within the region EFG would experience an annular eclipse of the Sun, i.e. the central part of the Sun would be obscured and the obscuring lunar disc would be seen as a dark circular region surrounded by a bright rim of solar photosphere. Annular eclipses of the Sun are most likely to occur when the Earth is at perihelion and the Moon is at apogee. A partial eclipse of the Sun would be observed by an

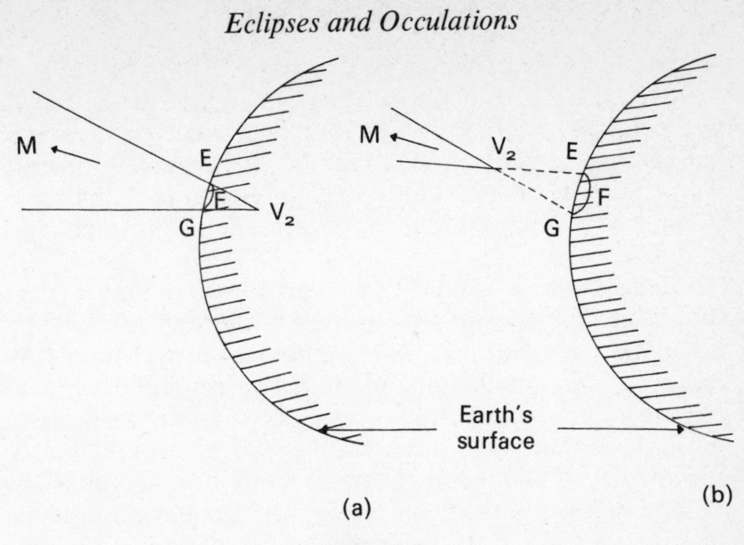

Fig. 8.2

observer within the penumbral shadow cone of the Moon. Again, since the radius of the Moon's penumbral shadow cone is small compared with the dimensions of the Earth, partial eclipses can only be seen from a small region of the Earth's surface. Transits of planets are similar to annular eclipses.

Just as in the case of the Moon, the Earth also produces a shadow cone. When the geometrical circumstances are correct the Moon can pass through the Earth's shadow cones producing an eclipse of the Moon. Eclipses of the Moon can only take place at Full Moon since the Moon is then on the same side of the Earth as its (the Earth's) shadow cone. While there are many similarities between solar and lunar eclipses there are two important differences.

(i) In the case of a solar eclipse the Sun, viewed from the Earth, is being obscured by a third body, the Moon. In the case of a lunar eclipse, the Moon, viewed from the Earth is being shadowed by the body from which the eclipse is being observed.

(ii) The Earth, having a diameter almost 3·7 times that of the Moon, produces an umbral shadow cone at the Moon's distance whose diameter is almost 2·6 the diameter of the Moon and a penumbral shadow cone whose diameter is almost 4·6 the diameter of the Moon.

Because of these differences any eclipse of the Moon may be observed from almost an entire hemisphere of the Earth, lunar eclipses last for a longer time than solar eclipses and, if the relative configurations of the Earth, Sun and Moon are correct, there is the possibility that the Moon may only pass through the penumbral shadow cone of the Earth so giving a penumbral eclipse.

An occultation is similar to a solar eclipse in that it is the Moon that produces the obscuration of the star. Like a solar eclipse the occultation of any particular star may only be observed from a small region of the Earth's surface at any one time. However, because the star may be regarded as being at an infinite distance (and not a finite distance as is the case for the Sun) the fundamental theory is simplified. Occultations by planets have similar properties to occultations by the Moon.

In all calculations regarding eclipses and occultations the time used will be ephemeris time.

8.1 The Calculation of the Conditions of a Solar Eclipse

8.1.1 The determination of the fundamental plane. To determine the conditions of an eclipse the motion of the Moon's shadow in relation to an observer on the Earth must be determined. This motion may be determined by reference to a plane through the Earth's centre which is always perpendicular to the axis of the shadow (i.e. the line SV_1MV_2 of Fig. 8.1 joining the centres of the Sun and Moon). This plane is called the *fundamental plane*. Since the Moon is moving with respect to the line of centres of the Sun and Earth the fundamental plane is continually changing its position. A system of rectangular axes $E(x, y, z)$ are defined with respect to the fundamental plane, E being the centre of the Earth. The z-axis is chosen so that it is parallel to the axis of the shadow. Let the direction of the z-axis be defined by the line ED. The x-axis is taken so that it passes through the point of intersection of the fundamental plane and the Earth's equator (directed positively towards the East). The y-axis (directed positively towards the north) completes the right handed set as shown in Fig. 8.3.

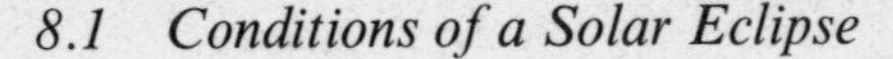

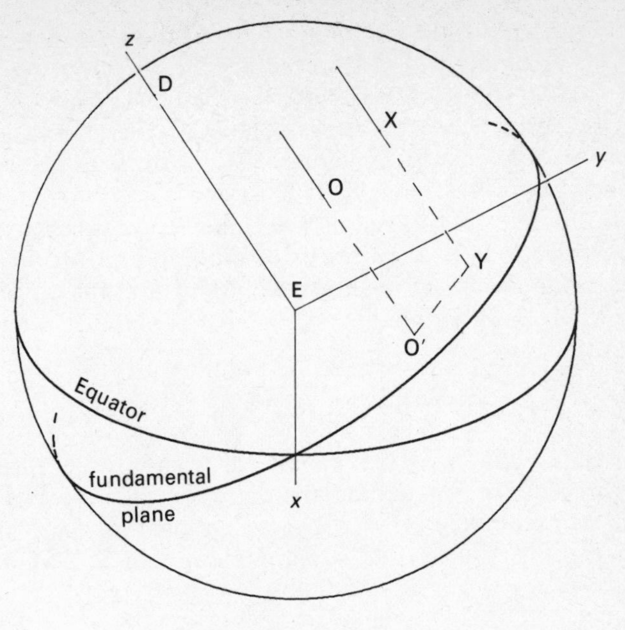

Fig. 8.3

Let the shadow cone cut the Earth's surface in some curve about the point X. The point X projects along the shadow axis (i.e. parallel to the z-axis) to Y on the fundamental plane. The observer is at O on the Earth's surface and O may be projected parallel to the z-axis to become O′ on the fundamental plane. Clearly an eclipse will take place provided O′ lies within the shadowed region about X as projected onto the fundamental plane.

The following notation will be used.

(α, δ): the equatorial coordinates of the Sun.

(α_1, δ_1): the equatorial coordinates of the Moon.

(a, d): the equatorial coordinates of the point in which the z-axis if projected would cut the celestial sphere.

r: the instantaneous geocentric distance of the Sun.

r_1: the instantaneous geocentric distance of the Moon.

$\bar{r}, \bar{r}_1$: the mean geocentric distances of Sun and Moon respectively ($\bar{r} = 1$ A.U.)

k: linear radius of the Moon (in units of the Earth's radius).

$k_\odot$: linear radius of the Sun (in units of the Earth's radius)

x, y, z: rectangular coordinates with respect to the fundamental plane (no subscript denotes solar values; a subscript 1 denotes lunar values).

In Fig. 8.4 rectangular equatorial coordinates are denoted by the frame of reference E(X, Y, Z) where E is the centre of the Earth. The Sun is centred at S and the Moon is centred at M. The projections of the centres of the Sun and Moon on the X, Y plane are S′, M′ respectively. M″ is a point on SS′ such that MM′ = M″S′. Then

$$\begin{aligned} &\Upsilon\hat{\mathrm{E}}\mathrm{S}' = \alpha, \qquad \Upsilon\hat{\mathrm{E}}\mathrm{M}' = \alpha_1, \qquad \Upsilon\hat{\mathrm{M}}'\mathrm{S}' = a, \\ &\mathrm{P}\hat{\mathrm{E}}\mathrm{S} = 90^\circ - \delta, \qquad \mathrm{P}\hat{\mathrm{E}}\mathrm{M} = 90^\circ - \delta_1, \qquad \mathrm{M}''\hat{\mathrm{M}}\mathrm{S} = d, \end{aligned} \tag{8.1}$$

and let G = SM, r = CS, r_1 = CM. It should be noted that the direction of SM defines the direction of the z-axis of the fundamental plane. Hence the assignment of (a, d) in eqn. (8.1). Taking $X-$, $Y-$, $Z-$ components of CS, CM and MS gives

$$G \cos d \cos a = r \cos \delta \cos \alpha - r_1 \cos \delta_1 \cos \alpha_1, \tag{8.2}$$

X-component,

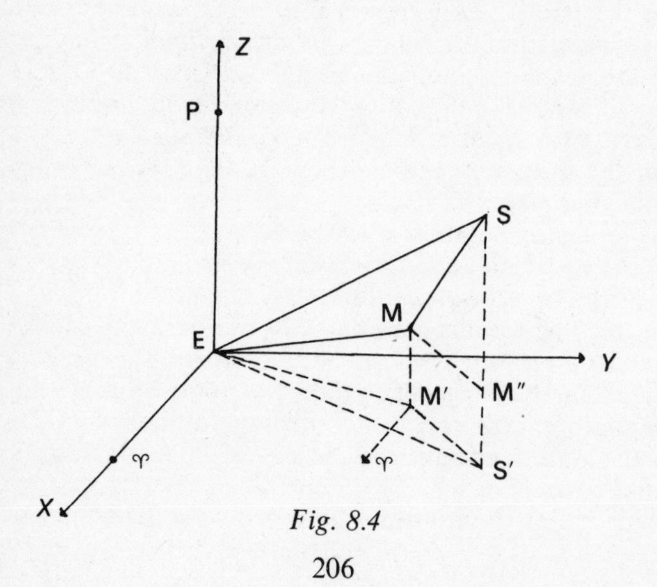

Fig. 8.4

$$G \cos d \sin a = r \cos \delta \sin \alpha - r_1 \cos \delta_1 \sin \alpha_1, \qquad (8.3)$$

Y-component,

$$G \sin d = r \sin \delta - r_1 \sin \delta_1, \text{ Z-component} \qquad (8.4)$$

Defining g, b by

$$g = G/r,\ b = r_1/r; \qquad (8.5)$$

eqns. (8.2, 8.3, 8.4) become,

$$g \cos d \cos a = \cos \delta \cos \alpha - b \cos \delta_1 \cos \alpha_1, \qquad (8.6)$$

$$g \cos d \sin a = \cos \delta \sin \alpha - b \cos \delta_1 \sin \alpha_1, \qquad (8.7)$$

$$g \sin d = \sin \delta - b \sin \delta_1. \qquad (8.8)$$

From Chapter VI, the solar geocentric distance at any time may be expressed in terms of the equatorial horizontal parallax π of the Sun where

$$\sin \pi = \frac{\rho_0}{r}, \qquad (8.9)$$

Hence,

$$b = \frac{r_1}{r} = \frac{\sin \pi}{\sin P} = \frac{\sin \pi_0}{r \sin P} \qquad (8.10)$$

where P is the equatorial horizontal parallax of the Moon, π_0 is the equatorial horizontal parallax of the Sun when the Earth is 1 A.U. distant and r is expressed in A.U. in the final equation of eqns. (8.10). The use of π_0, P and r in A.U. is of value since the Astronomical Ephemeris tabulates P and r (in A.U.). If the orbits of the Earth and Moon are known (and this may be assumed) r (or π), P will be known and consequently b, (α, δ), (α_1, δ_1) can be computed and so g, a and d determined for a range of times spanning the period of the eclipse.

The rectangular coordinates x, y, z of the centre of the Moon can be found with respect to the fundamental plane once a and d have been determined. Consider Fig. 8.5 in which the fundamental plane AB has been projected onto the celestial sphere. The point in which the fundamental plane cuts the celestial equator ♈UV is A where EA is the x-axis of the coordinate system. The z-axis (parallel to the line of centres

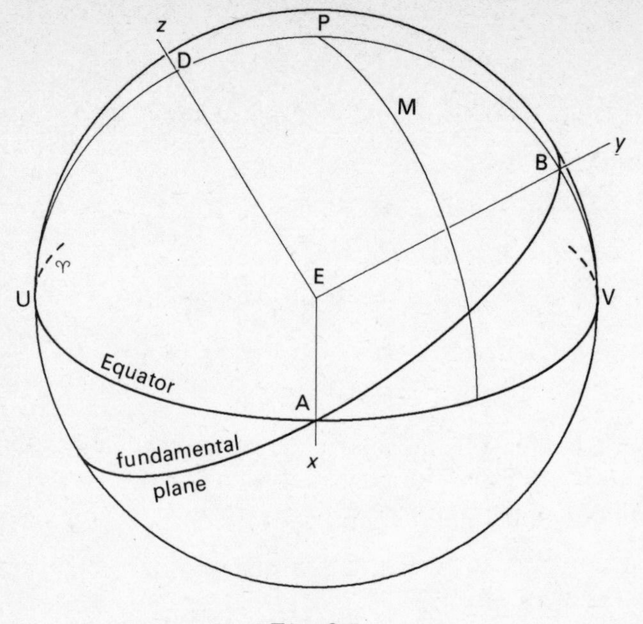

Fig. 8.5

of the Sun and Moon) cuts the celestial sphere in D. The north celestial pole is P and ♈ is the Vernal Equinox, M is the centre of the Moon, ♈U = a, the right ascension of D and UD = d, the declination of D. Since A is 90° from D and P then A must be 90° from all points on the great circle BPD and in particular 90° from U. Hence the right ascension of A is 90° + a. The coordinates of the centre of the Moon with respect to the axes E(x, y, z) in the fundamental plane are

$$x_1 = r_1 \cos A\hat{E}M, \qquad y_1 = r_1 \cos B\hat{E}M, \qquad z_1 = r_1 \cos D\hat{E}M. \tag{8.11}$$

where r_1 is in units of the Earth's equatorial radius. In the spherical triangle PAM

$$PM = 90° - \delta_1, \qquad PA = 90°, \qquad AM = A\hat{E}M, \qquad A\hat{P}M = \alpha_1 - (90° + a). \tag{8.12}$$

Use of the cosine rule gives

$$\cos \mathrm{AM} = \cos \delta_1 \cos \{\alpha_1 - (90° + a)\}$$

or

$$x_1 = r_1 \cos \delta_1 \sin (\alpha_1 - a) \tag{8.13}$$

Similarly in the spherical triangles PBM, PDM

$$\mathrm{PB} = d, \qquad \mathrm{PM} = 90° - \delta_1, \qquad \mathrm{MB} = \mathrm{M\hat{E}B},$$

$$\mathrm{M\hat{P}B} = 180° + a - \alpha_1; \tag{8.14}$$

$$\mathrm{PM} = 90° - \delta_1, \qquad \mathrm{PD} = 90° - d, \qquad \mathrm{DM} = \mathrm{D\hat{E}M},$$

$$\mathrm{D\hat{P}M} = \alpha_1 - a. \tag{8.15}$$

Application of the cosine rules to these spherical triangles gives

$$y_1 = r_1\{\cos d \sin \delta_1 - \sin d \cos \delta_1 \cos (\alpha_1 - a)\}, \tag{8.16}$$

$$z_1 = r_1\{\sin d \sin \delta_1 + \cos d \cos \delta_1 \cos (\alpha_1 - a)\}. \tag{8.17}$$

8.1.2 The Besselian Elements of the eclipse. The Besselian Elements of the eclipse are defined as follows

(i) μ.

The Right Ascension a of the point on the celestial sphere towards which the z-axis of the fundamental plane is directed, can be replaced by μ the ephemeris hour angle of the same point, i.e.

$$\mu = \text{Ephemeris Sidereal Time} - a \tag{8.18}$$

Clearly μ can be calculated once a is known from the known values of the ephemeris sidereal time for a range of times spanning the period of the eclipse. For this set of values the rate of change of μ can be determined by numerical differentiation.

(ii) f_1, f_2.

The angles made by the generators of the shadow cones with the axis of the shadow are denoted by f_1 (penumbral) and f_2 (umbral) i.e. $f_1 = \mathrm{M\hat{V}_1B}$, $f_2 = \mathrm{M\hat{V}_2B}$ with reference to Fig. 8.1. If $\mathrm{V_1M}$ is denoted by i then

$$\sin f_1 = \frac{k_\odot}{G - i} = \frac{k}{i} \tag{8.19}$$

whence,

$$i = \frac{kG}{k_\odot + k}. \tag{8.20}$$

where G = MS (again with reference to Fig. 8.1). Hence

$$\sin f_1 = \frac{k_\odot + k}{G}. \tag{8.21}$$

Again setting $MV_2 = j$,

$$\sin f_2 = \frac{k_\odot}{G + j} = \frac{k}{j}, \tag{8.22}$$

whence,

$$j = \frac{kG}{k_\odot - k}, \tag{8.23}$$

so that,

$$\sin f_2 = \frac{k_\odot - k}{G}. \tag{8.24}$$

By introducing the mean angular diameter S_0 of the Sun, eqns. (8.21) and (8.24) take the form

$$\sin f_1 = \frac{(\sin S_0/\sin \pi_0) + k}{G} = \frac{\sin S_0 + k \sin \pi_0}{gr}$$

$$= 0{\cdot}004\,640\,2/\text{gr}, \tag{8.25}$$

$$\sin f_2 = \frac{\sin S_0 - k \sin \pi_0}{gr} = 0{\cdot}004\,640\,78/\text{gr}, \tag{8.26}$$

since

$$\frac{\sin S}{\sin \pi} = \frac{\sin S_0}{\sin \pi_0} = k_\odot, \qquad \text{(see Chaper VI)}.$$

where S is the angular semi-diameter of the Sun, k (and $k_\odot$) are expressed in units of the Earth's equatorial radius and the units of r are A.U.

Having determined the values of f_1 and f_2 at a range of times spanning the eclipse they can be used to determine the size of region on the Earth's surface intersected by the umbral cone. This is done, approximately, by considering the radius of the circle in which the umbral cone cuts a plane through the observer parallel to the fundamental plane. The geometrical disposition of the Earth, Moon and Sun at the time of eclipse is illustrated in Fig. 8.6.

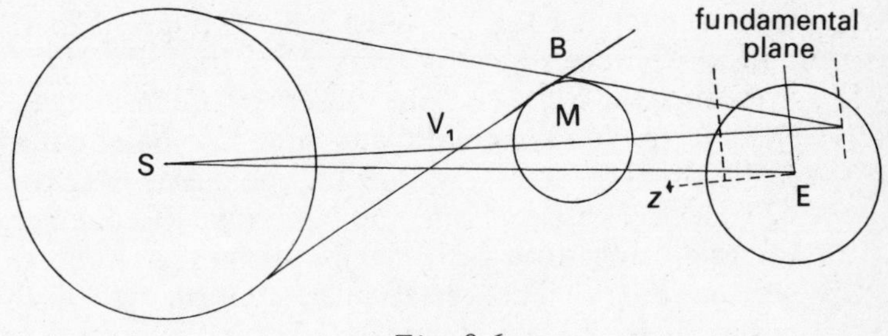

Fig. 8.6

The centre of the Moon M is distant z_1 from the fundamental plane (eqn. 8.17) and so if the distance of V_1 above the fundamental plane is denoted by c_1

$$c_1 = z_1 + k \operatorname{cosec} f_1. \tag{8.27}$$

Similarly if the height of the vertex V_2 above the fundamental plane is c_2 then

$$c_2 = z_1 - k \operatorname{cosec} f_2. \tag{8.28}$$

The units of c_1, c_2, z_1 are those of the Earth's equatorial radius, since k is expressed in these units. It should be noted that for the direction in which the z-axis is chosen c_1 is always positive while c_2 can be negative. If c_2 is positive, the possibility of an annular eclipse must be examined. If c_2 is negative a total eclipse must occur. The radii of the penumbral and umbral cones on the fundamental planes are respectively

$$l_1 = c_1 \tan f_1, \tag{8.29}$$

$$l_2 = c_2 \tan f_2. \tag{8.30}$$

Again l_1 is always positive while l_2 can be positive or negative. A set of values of l_1 and l_2 can be calculated for times spanning the eclipse period.

The elements x_1, y_1, $\sin d$, $\cos d$, μ, l_1, l_2 are given in the Astronomical Ephemeris, calculated at hourly intervals and subtabulated at 10 minute intervals. The hourly variations of x_1, y_1 and l_2 can be obtained from the tabulated values by numerical differentiation. The hourly variations μ', d' of μ, d respectively, are virtually constant. Also $\tan f_1$ and $\tan f_2$ are virtually constant during the period of the eclipse. Therefore one value only of $\tan f_1$, $\tan f_2$, μ', d' is given in the Astronomical Ephemeris.

The coordinates (ξ, η, ζ) of the observer (O in Fig. 8.3) with respect to the fundamental plane must also be obtained. Suppose the observer is at a station in ephemeris longitude ψ_E (longitude reckoned from the Greenwich ephemeris meridian) and geocentric latitude ϕ', distant ρ from the centre of the Earth. The rectangular coordinates of the observer may be found in a manner exactly analogous to that used to find the rectangular coordinates of the Moon with respect to the fundamental plane. In this case the celestial equatorial coordinates (α_1, δ_1) of the Moon are replaced by terrestrial equatorial coordinates (ψ_E, ϕ') of the observer. Fig. 8.5 may be used if M is now interpreted as the geocentric zenith of the observer. In this case since μ is the Greenwich ephemeris hour angle of the point where the z-axis cuts the celestial sphere and ψ_E is the ephemeris longitude of the observer, $M\hat{P}D = \mu - \psi_E$, $MP = 90° - \phi'$, $A\hat{P}M = (\mu - \psi_E) - 90°$ and $B\hat{P}M = 180° - (\mu - \psi_E)$ so that

$$\xi = \rho \cos \phi' \sin (\mu - \psi_E) \tag{8.31}$$

$$\eta = \rho [\sin \phi' \cos d - \cos \phi' \sin d \cos (\mu - \psi_E)], \tag{8.32}$$

$$\zeta = \rho [\sin \phi' \sin d + \cos \phi' \cos d \cos (\mu - \psi_E)]. \tag{8.33}$$

The variation of ξ, η, ζ may be found by differentiation with respect to t,

$$\frac{d\xi}{dt} = \rho \cos \phi' \cos (\mu - \psi_E) \frac{d\mu}{dt}$$

$$= \frac{d\mu}{dt}(-\eta \sin d + \zeta \cos d), \tag{8.34}$$

$$\frac{d\eta}{dt} = \rho\left[+\cos\phi'\sin d\sin(\mu - \psi_E)\frac{d\mu}{dt}\right.$$

$$\left. - \{\sin\phi'\sin d + \cos\phi'\cos d\cos(\mu - \psi_E)\}\frac{dd}{dt}\right],$$

$$= \xi\sin d\frac{d\mu}{dt} - \zeta\frac{dd}{dt}, \tag{8.35}$$

$$\frac{d\zeta}{dt} = \rho\left[-\cos\phi'\cos d\sin(\mu - \psi_E)\frac{d\mu}{dt}\right.$$

$$\left. + \{\sin\phi'\cos d - \cos\phi'\sin d\cos(\mu - \psi_E)\}\frac{dd}{dt}\right],$$

$$= -\xi\cos d\frac{d\mu}{dt} + \eta\frac{dd}{dt}. \tag{8.36}$$

The variation of μ and d with time can be worked out (as pointed out above $d\mu/dt$, dd/dt are nearly constant) by numerical differentiation of tabulated values.

Since the z-axis is parallel to the shadow axis, the projected position of the Moon's centre on the fundamental plane has coordinates (x_1, y_1). The projected position of the observer O′ (see Fig. 8.3) has coordinates ξ, η. To determine the conditions of the eclipse at the observer O we must consider a plane through the observer parallel to the fundamental plane. The coordinates of the observer are ξ, η, ζ and the coordinates of the centre of the shadow on this plane are (x_1, y_1, ζ). Let m denote the separation of observer and shadow centre and let Q be the position angle of the axis of the shadow with respect to the y-axis. It is the convention to measure position angles clockwise ("Eastwards from North") from the observer. In Fig. 8.7 O is the position of the observer and X is the position of the centre of the shadow on the plane through the observer parallel to the fundamental plane. AO is parallel to the y-axis and $A\hat{O}X = Q$ is the position angle of the shadow and $OX = m$ is the separation of the observer from the centre

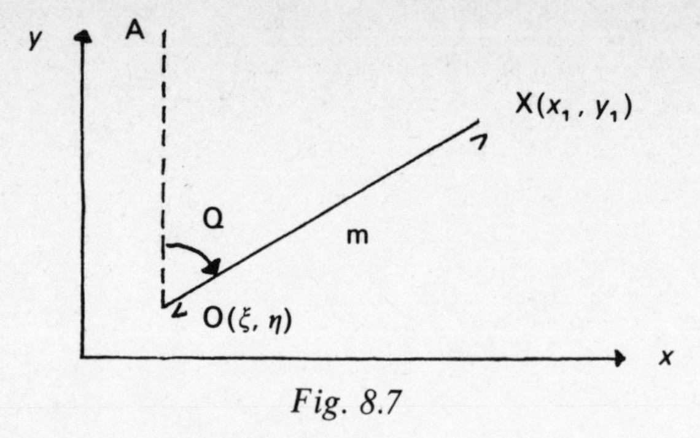

Fig. 8.7

of the shadow. In this case

$$m \sin Q = x_1 - \xi. \qquad (8.37a)$$

$$m \cos Q = y_1 - \eta. \qquad (8.37b)$$

and

$$m^2 = (x_1 - \xi)^2 + (y_1 - \eta)^2. \qquad (8.37c)$$

On a plane through the observer at a height ζ above the fundamental plane the radii of the penumbral and umbral shadows are defined to be respectively,

$$L_1 = l_1 - \zeta \tan f_1 = (c_1 - \zeta) \tan f_1, \qquad (8.38)$$

$$L_2 = l_2 - \zeta \tan f_2 = (c_2 - \zeta) \tan f_2. \qquad (8.39)$$

The sign of L_2 indicates whether or not the eclipse is total or annular. If the eclipse is total V_2 lies inside the Earth and so L_2 is negative (since $\zeta > c_2$); if the eclipse is annular V_2 lies outside the Earth and L_2 will then be positive (since $\zeta < c_2$). L_2 also gives a measure of the maximum distance from the line of central eclipse for a total eclipse still to be seen. An eclipse will only be observable at some station provided that station lies within the shadow. Since the radius of the shadow cone is L_2 for an umbral eclipse the observer would see the eclipse if

$$m < |L_2|. \qquad (8.40)$$

For an observer on the edge of the shadow cone

$$(x_1 - \xi)^2 + (y_1 - \eta)^2 - L_2^2 = 0 \qquad (8.41)$$

In the case of a partial eclipse L_1 would replace L_2.

By making use of the Tabulated Besselian Elements of an eclipse an observer can work out the precise times of contact for his own location. Clearly to get maximum duration of the eclipse one wishes to be as near the centre of the shadow as possible. In practical terms this is often not possible and it is necessary to work out the times of beginning and ending of the eclipse and the time of mid eclipse for a particular location.

An observer on the earth will be at astronomical latitude ϕ, longitude ψ and height H above the Geoid. The geocentric coordinates will then be

$$\rho \sin \phi' = (S + H) \sin \phi, \qquad (6.33a)$$

$$\rho \cos \phi' = (C + H) \cos \phi. \qquad (6.34a)$$

The determination of S and C has been discussed in Chapter VI. The longitude ψ must be converted to the ephemeris longitude ψ_E by increasing it by an amount $1{\cdot}002\,738\Delta T$ where ΔT is the correction from universal to ephemeris time (see Chapter V). In equations (8.31)–(8.33) we must replace $\mu - \psi_E$ by $\mu - \psi - 1{\cdot}002\,738\Delta T$.

8.1.3 Determination of time of eclipse. The time of the beginning or end of umbral eclipse requires accurate calculation. It will be assumed in this section that the conditions for total eclipse are satisfied by the observer. The method of determination is as follows. Define u, v, u', v', m, n by setting

$$u = x_1 - \xi, \qquad u' = x_1' - \xi', \qquad m^2 = u^2 + v^2$$

see also eqn. (8.37c) $\qquad$ (8.42)

$$v = y_1 - \eta, \qquad v' = y_1' - \eta', \qquad n^2 = u'^2 + v'^2$$

An approximate time for the eclipse is assumed, say T_0 (a time near totality determined from information given in the Astronomical Ephemeris). Then let the values of u, u', v, v', m, n at this instant be denoted by

$$u_0, u_0', v_0, v_0', m_0, n_0.$$

At some later time, $T_0 + t$, approximate values of u and v are

$$\begin{aligned} u &= u_0 + tu_0', \\ v &= v_0 + tv_0'. \end{aligned} \qquad (8.43)$$

The time of maximum phase $T_0 + t_{mp}$ is readily determined since the separation of the observer from the centre of the shadow will be a minimum, i.e. m attains its minimum value. By differentiating m^2 with respect to time

$$2m\frac{dm}{dt} = 2u\frac{du}{dt} + 2v\frac{dv}{dt} = 0 \text{ for a minimum;}$$

i.e.,

$$uu' + vv' = 0. \tag{8.44}$$

This may be approximated by

$$(u_0 + t_{mp}u'_0)u'_0 + (v_0 + t_{mp}v'_0)v'_0 = 0,$$

whence,

$$t_{mp} = -\frac{u_0u'_0 + v_0v'_0}{u'^2_0 + v'^2_0} = -\frac{D_0}{n_0^2}, \tag{8.45}$$

where

$$D_0 = u_0u'_0 + v_0v'_0. \tag{8.46}$$

Since D_0 and n_0^2 can be readily evaluated at time T_0 a first estimate to t_{mp} can be obtained. A first estimate for the times of beginning $(T_0 + t_1)$ and end $(T_0 + t_2)$ of totality can also be determined readily through solution of eqn. (8.41) which may be written in the form

$$u^2 + v^2 = L_2^2 \tag{8.47}$$

for two values of t. Eqn. (8.47) may be rewritten

$$(u_0 + tu'_0)^2 + (v_0 + tv'_0)^2 = L_2^2$$

i.e.

$$t^2(u'^2_0 + v'^2_0) + 2t(u_0u'_0 + v_0v'_0) + (u_0^2 + v_0^2) = L_2^2$$

or

$$n_0^2t^2 + 2tD_0 + m_0^2 - L_2^2 = 0. \tag{8.48}$$

Eqn. (8.48) has roots

$$
\begin{aligned}
t_1 &= -\frac{D_0}{n_0^2} - \frac{1}{n_0^2}\{D_0^2 - n_0^2(m_0^2 - L_2^2)\}^{\frac{1}{2}} \\
&= t_{mp} - \frac{1}{n_0^2}\{D_0^2 - n_0^2(m_0^2 - L_2^2)\}^{\frac{1}{2}}, \\
t_2 &= -\frac{D_0}{n_0^2} + \frac{1}{n_0^2}\{D_0^2 - n_0^2(m_0^2 - L_2^2)\}^{\frac{1}{2}} \\
&= t_{mp} + \frac{1}{n_0^2}\{D_0^2 - n_0^2(m_0^2 - L_2^2)\}^{\frac{1}{2}}.
\end{aligned}
\tag{8.49}
$$

The values of t_1 and t_2 can be calculated readily from the known values of D_0, m_0, n_0 and L_2 at time T_0. The duration of the eclipse is clearly

$$
\begin{aligned}
t_d &= (T_0 + t_2) - (T_0 + t_1) = t_2 - t_1 \\
&= \frac{2}{n_0^2}\{D_0^2 - n_0^2(m_0^2 - L_2^2)\}^{\frac{1}{2}}.
\end{aligned}
\tag{8.50}
$$

To obtain greater accuracy values of u, u', v, v' may be computed using the Besselian elements at times $T_0 + t_1$, $T_0 + t_{mp}$ and $T_0 + t_2$ and used in the above manner to obtain new values for the times of the beginning, middle and end of the eclipse.

The foregoing discussion sets out the principles involved in determining the times of the beginning, middle and end of an eclipse. A numerical method based on the algebra as given would be cumbersome. The Besselian elements of the eclipse give x_1, y_1, $\sin d$, $\cos d$, u, l_1, l_2 at a range of times spanning the eclipse while $\tan f_1$, $\tan f_2$, μ', d' are treated as constants. The distance ρ of an observer from the centre of the Geoid can be established using eqns. (6.33a), (6.34a) and ϕ', S, C determined as set out in Chapter VI. ξ, η may be determined from eqns. (8.31, 8.32) using the tabulated Besselian element μ. Then, from the tabulated values of x_1, y_1 values of u, v can be found at the same time intervals as in the tabulation using eqns. (8.42). Finally, using the Besselian elements l_2, $\tan f_2$ and ζ from eqn. (8.33), values of L_2 can be determined. The quantity $(u^2 + v^2 - L_2^2)$ is formed at the times given in the tabulation of the Besselian elements. This quantity will not be zero except at the instant of beginning or end of the eclipse. The zeros may be established using standard numerical procedures for inverse interpolation. Again eqn. (8.44) gives the condition for mid-eclipse. Since the values of u and v are available for a range of times spanning the eclipse, numerical differentiation of the quantity $(u^2 + v^2)$ will give the numerical equivalent of eqn. (8.44). $(\mathrm{d}/\mathrm{d}t)(u^2 + v^2) \equiv 2(uu' + vv')$ will only be zero at the instant of mid-eclipse. Therefore the time corresponding to mid-eclipse can again be found using standard numerical techniques for inverse interpolation. Such a procedure is preferable to use of eqns. (8.48, 8.49), since the arithmetical operations are simpler.

Once the time of beginning and end of the total eclipse have been worked out u and v can be determined at these times. Since the observer is distant L_2 from the centre of the shadow the position angle Q is given by

$$u = L_2 \sin Q, \qquad v = L_2 \cos Q, \qquad \tan Q = u/v, \qquad (8.51)$$

where the values appropriate to beginning or end of the eclipse are used.

It should be noted that all calculated times are in ephemeris time and the universal time is obtained by subtracting the correction ΔT.

In order to calculate the times for a partial eclipse of the Sun, L_2 is replaced by L_1 in the above calculations since an observer within the penumbral shadow cone experiences a partial eclipse.

8.1.4 Magnitude of a solar eclipse. The magnitude of an eclipse is defined to be the fraction of the solar diameter covered by the Moon at the time of greatest phase. Fig. 8.8, drawn for an annular eclipse, illustrates the discs of the Sun and Moon showing the geometrical relationship of the umbral and penumbral cones on a plane parallel to the fundamental plane at the point where the axis of the shadow cone cuts the Earth. S is the centre of the solar disc XX′ and M is the centre of the Moon's disc UU′; V_1, V_2 are the penumbral and umbral vertices respectively. O is the point on the plane PP′ parallel to the fundamental plane, where the shadow axis cuts the Earth's surface. An observer within AA′ will see the annular eclipse. For example the observer at E will see the region YY′ of the solar surface obscured. An observer at B will be in the penumbral shadow cone and will see a partial eclipse of the Sun in that the region ZX′ will not be seen.

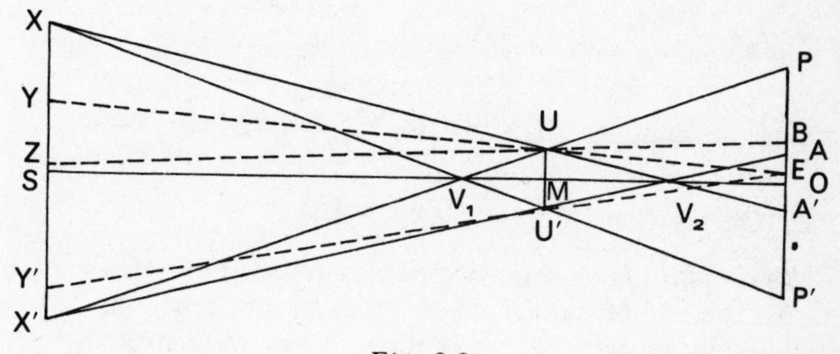

Fig. 8.8

The magnitude M_1 of the penumbral eclipse is defined to be

$$M_1 = \frac{ZX'}{XX'} = \frac{PB}{PA'} \tag{8.52}$$

(using the triangles XX′U, PA′U). The radius of the penumbral cone is $PO = L_1$, while the radius of the umbral cone is $OA = L_2$. The distance (OB) of the observer at B from the shadow axis is m. Then

$$PB = OP - OB = L_1 - m,$$

$$PA' = PO + OA' = L_1 + L_2.$$

Hence

$$M_1 = \frac{L_1 - m}{L_1 + L_2}. \tag{8.53}$$

Again an observer at E is situated in the "umbral" cone and is seeing an annular eclipse. The magnitude of the eclipse M_2

is given by

$$M_2 = \frac{YY'}{XX'} = \frac{YX' - Y'X'}{XX'}$$

$$= \frac{PE}{PA'} - \frac{AE}{PA'} = \frac{PE - AE}{PA'} = \frac{PA}{PA'} \tag{8.54}$$

$$= \frac{OP - OA}{OP + OA'} = \frac{L_1 - L_2}{L_1 + L_2},$$

(using the triangles YUX′, EUP; X′U′Y′, AEU′).

8.2 Occultations of Stars by the Moon

An *occultation* of a star occurs when the Moon's disc passes across the line of sight from observer to star. Since the Moon has a sidereal period of $27\frac{1}{3}$ days, it moves eastwards with respect to the star background at a rate of $13°{\cdot}2$/day or approximately $0{\cdot}55°$/hour. The Moon is therefore constantly obscuring those stars which lie on a band centred on the projection of the lunar orbit on the sky and whose width is the angular diameter of the Moon. Since the Moon has little or no atmosphere, a star disappears suddenly as the Moon's disc just obscures the star and after an interval of time ($\lesssim 1^h$) the star reappears suddenly as the Moon moves on in its orbit. The sudden disappearance and reappearance of a star are called *immersion* and *emersion* respectively.

The observation of occultations was formerly of importance in making corrections to the determination of the Moon's position. This was important so that the observed and the predicted positions of the Moon could be compared with a view to the determination of ΔT, the correction between universal and ephemeris time. Photographic methods have in part superseded the timing of occultations as the means of determining the lunar orbit. However, occultations have again become important in the determination of the positions and structures of radio sources. The immersion and emersion of radio sources can be accurately timed. Therefore knowledge of the Moon's orbit can be used to give positions of these radio sources with great accuracy. Indeed such observations are again making it possible to use occultation measurements to improve the determination of the lunar orbit.

8.2.1 The determination of the fundamental plane. The method of prediction of occultations follows closely the methods used for the determination of solar eclipses with an important reservation. The Sun is at a finite distance from the Earth. The distance of any star from the Earth is very much greater than the solar distance and so the star may be assumed to be at infinity. This means that only the coordinates in the fundamental plane are relevant and the z coordinate ignored. The shadow produced by the Moon in this situation is not a cone, but a cylinder whose generators are parallel to the z-axis which is now defined to be the line of centres of Moon and star. The radius of the cylinder is the radius k of the Moon. The geometrical situation for an occultation is illustrated in Fig. 8.9. The Earth is centred at E and the Moon at M. The line MS is the line of centres of the Moon and stars. The plane perpendicular to the line of centres MS is defined to be the fundamental plane and an axis through the centre of the

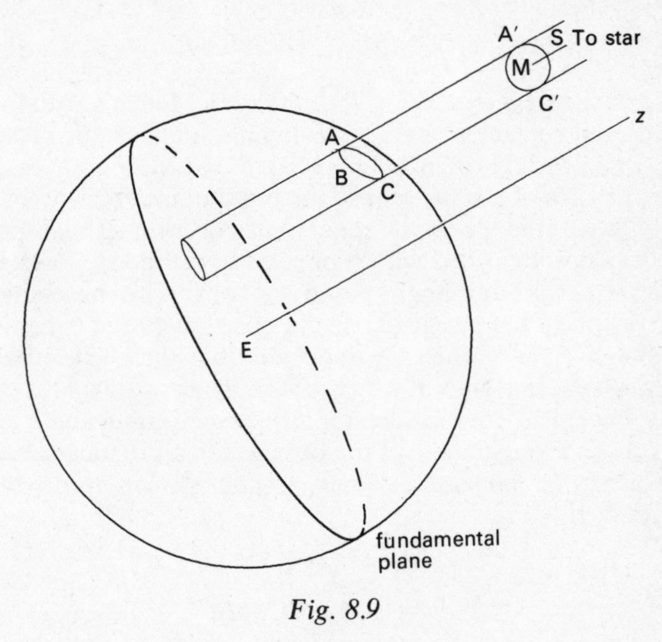

Fig. 8.9

Earth parallel to MS and perpendicular to the fundamental plane is called the z-axis. The generators AA′, CC′ define the generators of the cylinder whose cross section is that of the lunar disc. An observer on the curve ABC will see the star about to be occulted or reappear and the star will be obscured by the Moon for any observer within the curve ABC.

As in the case of a solar eclipse, an x-axis is defined to be that axis through the intersection of the fundamental plane and the celestial equator and is directed positively towards the east. The y-axis (directed positively towards the north) completes the right handed set. Since the star may be assumed to be at infinity there is no point in determining the radius of the shadow cylinder in the manner of solar eclipses, since $f_1 = 0$ and $f_2 = \pi$, and so the elements l_1 and l_2 may be dispensed with. Again, since the star is at infinity the equatorial coordinates of the point on the celestial sphere cut by the z-axis produced are the geometric equatorial coordinates of the star (α_*, δ_*). Hence a, d and μ may be deleted.

For an occultation eqns. (8.13) and (8.16) become

$$x_1 = r_1 \cos \delta_1 \sin (\alpha_1 - \alpha_*), \tag{8.55}$$

$$y_1 = r_1\{\cos \delta_* \sin \delta_1 - \sin \delta_* \cos \delta_1 \cos (\alpha_1 - \alpha_*)\}. \tag{8.56}$$

It is usual to express x_1, y_1 in units of the Earth's equatorial radius and replace r_1 by $1/\sin P$ in eqns. (8.55, 8.56) where P is the equatorial horizontal parallax of the Moon.

In the case of a solar eclipse the point of intersection of the z-axis with the celestial sphere was a geometrical point and was not identified with a physically real body. Therefore an ephemeris hour angle could be used. This necessitated conversion of the longitude of the observer to the ephemeris value. However, in the case of occultations the z-axis cuts the celestial sphere at a point which is identified as a star. Therefore it is convenient to consider the Greenwich hour angle H of the star and longitude ψ of the observer in a practical system. Denoting the ephemeris hour angle and longitude by a subscript E,

$$\begin{aligned} \mathrm{H_E} &= \mathrm{H} + 1{\cdot}002\,738\Delta t, \\ \psi_\mathrm{E} &= \psi + 1{\cdot}002\,738\Delta t. \end{aligned} \tag{8.57}$$

Hence

$$H_E - \psi_E = H - \psi \tag{8.58}$$

and so $H_E - \psi_E$ may be replaced by $H - \psi$. Therefore the coordinates (ξ, η) of the observer in the fundamental plane may be written in the form

$$\xi = \rho \cos \phi' \sin (H - \psi), \tag{8.59}$$

$$\eta = \rho\{\sin \phi' \cos \delta_* - \cos \phi' \sin \delta_* \cos (H - \psi)\}, \tag{8.60}$$

by comparison with eqns. (8.81), (8.32).

In the case of occultations it is usual, though unnecessary, to assume an approximate form for x_1 and y_1 since $\alpha_1 \simeq \alpha_*$ and $\delta_1 \simeq \delta_*$ at the time of an occultation. Eqns. (8.55) and (8.56) then become,

$$x_1 = r_1 \cos \delta_1(\alpha_1 - \alpha_*) \simeq \cos \delta_1 \frac{\alpha_1 - \alpha_*}{P}, \tag{8.61}$$

where P is the equatorial horizontal parallax of the Moon and

$$y_1 \simeq r_1\{\cos \delta_* \sin \delta_1 - \sin \delta_* \cos \delta_1(1 - \tfrac{1}{2}(\alpha_1 - \alpha_*)^2)\}$$

$$\simeq \frac{\delta_1 - \delta_*}{P} + \tfrac{1}{2}x_1(\alpha_1 - \alpha_*) \sin \delta_*. \tag{8.62}$$

The use of P rather than $\sin P$ does not introduce a large error in this order of approximation. The units of x_1, y_1 are in terms of the Earth's equatorial radius and the angular measure is in radians. From these expressions the rates of change of x_1, y_1, namely x'_1, y'_1, may be deduced. However, it is customary to give values of x_1, y_1, ξ, η at time of conjunction in right ascension only. Suppose conjunction occurs at time T_0. At this time $\alpha_1 = \alpha_*$ so that

$$x_1 = 0, \qquad y_1 = Y_1 = r_1 \sin (\delta_1 - \delta_*) \simeq \frac{\delta_1 - \delta_*}{P}. \tag{8.63}$$

Hence,

$$x_1' = \frac{dx_1}{dt} = \frac{dr_1}{dt} \text{ as } \delta_1 \sin(\alpha_1 - \alpha_*) - r_1 \sin\delta_1 \sin(\alpha_1 - \alpha_*)\frac{d\delta_1}{dt}$$

$$+ r_1 \cos\delta_1 \cos(\alpha_1 - \alpha_*)\frac{d\alpha_1}{dt} \tag{8.64}$$

$$= r_1 \cos\delta_1 \alpha_1' \simeq \cos\delta_1 \frac{\alpha_1'}{P} \text{ at time } T_0,$$

where α_1' is the time rate of change of α_1. Again,

$$y_1' = \frac{dy_1}{dt} = \frac{dr_1}{dt}\{\cos\delta_* \sin\delta_1 - \sin\delta_* \cos\delta_1 \cos(\alpha_1 - \alpha_*)\}$$

$$+ r_1\{\cos\delta_* \cos\delta_1 + \sin\delta_* \sin\delta_1 \cos(\alpha_1 - \alpha_*)\}\frac{d\delta_1}{dt}$$

$$+ r_1 \sin\delta_* \cos\delta_1 \sin(\alpha_1 - \alpha_*)\}\frac{d\alpha_1}{dt}$$

$$= \frac{dr_1}{dt}\sin(\delta_1 - \delta_*) + r_1 \cos(\delta_1 - \delta_*)\frac{d\delta_1}{dt}, \text{ at time } T_0,$$

$$\simeq \frac{\delta_1'}{P} - Y_1\frac{P'}{P}, \tag{8.65}$$

where a prime denotes differentiation with respect to t. Hence x_1, y_1, x_1', y_1', can be determined at time T_0. With sufficient accuracy x_1, y_1 vary linearly with time and x', y' are constant during an occultation. Values at any other time near the occultation can then be computed. The time variations of ξ, η may also be derived,

$$\frac{d\xi}{dt} = \xi' = \rho\cos\phi'\cos(H - \psi)\,dH/dt, \tag{8.66}$$

$$\frac{d\eta}{dt} = \eta' = \rho\cos\phi'\sin\delta_*\sin(H - \psi)\frac{dH}{dt} \tag{8.67}$$

Since H changes by 360° in the course of 24^h, dH/dt can be readily determined ($dH/dt = 0{\cdot}2625$ radians/hr) and so ξ' and η' may be determined for any value of H.

In order that a star may be occulted the position of the observer on the fundamental plane must lie within a circle of radius k centred on the position of the Moon's centre projected parallel to the z-axis onto the fundamental plane. The criterion that any observer, whose position is (ξ, η) on the fundamental plane, will see immersion of emersion of a star whose position is (x_1, y_1) on the fundamental plane, is given by

$$(x_1 - \xi)^2 + (y_1 - \eta)^2 = k^2, \qquad (8.68)$$

just as for the times of beginning or end of a solar eclipse (in which case L_1 or L_2 would replace k—see eqn. (8.41)).

8.2.2 The Besselian Elements for an occultation. The Besselian elements for an occultation are defined at the universal time T_0 of conjunction in right ascension (equivalent to T_E in ephemeris time). The elements are T_0, the U.T. of conjunction in RA; H, the Greenwich hour angle of the star at T_0; $Y = Y_1(T_0)$; x', y', the hourly variations in x_1 and y_1 and the equatorial coordinates of the star (α_*, δ_*) referred to the equator and equinox of T_0. The values of x, y, ξ, η at any time $T_E + t$ are then,

$$\begin{aligned} x_1 &= x't & y_1 &= Y + y't, \\ \xi &= \xi_0 + \xi't, & \eta &= \eta_0 + \eta't, \end{aligned} \qquad (8.69)$$

where ξ_0, η_0 are the values of ξ, η at time $T_0(= T_E)$. If

$$\begin{aligned} u &= (x_1 - \xi), & v &= (y_1 - \eta), \\ u' &= (x_1' - \xi'), & v' &= (y_1' - \eta'), \end{aligned} \qquad (8.70)$$

eqn. (8.68), the condition for immersion or emersion, becomes

$$u^2 + v^2 - k^2 = 0. \qquad (8.71)$$

Eqn. (8.71) must then be solved to give the times of immersion and emersion.

8.2.3 Determination of times of immersion and emersion. The times of immersion and emersion for an occultation may be determined by computing the values of u^2 and v^2, forming the quantity $u^2 + v^2 - k^2$ and seeking its zeros by inverse interpolation as outlined in the case of a solar eclipse in section 8.1.3. Eqns. (8.66, 8.67) may be used to compute ξ', η'; ξ, η are derived from eqns. (8.69) using eqns. (8.59, 8.60) to determine ξ_0, η_0. x_1, y_1 are determined using the Besselian elements x', y', Y through eqns. (8.69). u, v are calculated using eqn. (8.71). Since it is customary to express ρ in terms of the Earth's equatorial radius, k is usually expressed in the same units when $k = 0{\cdot}2725$. The quantity $u^2 + v^2 - k^2$ may be readily formed and its zeros sought by the standard methods of inverse interpolation. This method gives preliminary values only for the times of immersion and emersion, as the accuracy obtainable may be insufficient to distinguish between a grazing occultation and a clear miss.

The alternative method of computing the times of solar eclipse can be adapted to the case of occultation prediction. A subscript zero denotes values of u, v, u', v' at time T_0. The correction for time of immersion and emersion may be derived by solving the equation

$$(u_0 + u'_0 t)^2 + (v_0 + v'_0 t)^2 - k^2 = 0 \tag{8.72}$$

for two values of t in the same way as for solar eclipses (see eqn. (8.48)). Two values of t are obtained, namely

$$t_1 = -\frac{D_0}{n_0^2} - \frac{1}{n_0^2}\{D_0^2 - n_0^2(m_0^2 - k^2)\}^{\frac{1}{2}} \tag{8.73a}$$

and

$$t_2 = -\frac{D_0}{n_0^2} + \frac{1}{n_0^2}\{D_0^2 - n_0^2(m_0^2 - k^2)\}^{\frac{1}{2}}, \tag{8.73b}$$

where

$$D_0 = u_0 u'_0 + v_0 v'_0; \qquad m_0^2 = u_0^2 + v_0^2$$

and

$$n_0^2 = u_0'^2 + v_0'^2.$$

The parallel between eqns. (8.73) and eqns. (8.49) suggests that there may be value in adopting the time of closest approach (cf. time of mid-eclipse in the case of a solar eclipse) in order to derive improved times of immersion and emersion. The time of closest approach is $T_E + t'$ where

$$t' = -\frac{D_0}{n_0^2}. \tag{8.74}$$

Values of u, v, u', v' at the time of closest approach may be determined and a further solution for the times of immersion and emersion sought from the equation

$$(u_a + u'_a\Delta t)^2 + (v_a + v'_a\Delta t) - k^2 = 0, \tag{8.75}$$

where a subscript a denotes evaluation at the time of closest approach and the times of immersion and emersion are given by $T_E + t' \mp \Delta t$. The evaluation of u_a, v_a, u'_a, v'_a should follow the method described below.

A more standard method of solution evaluates u, v at times $T_E + t_1$, $T_E + t_2$ here t_1, t_2 have been obtained from eqns. (8.73). The values of u, v at these times will be denoted by $u_1, v_1; u_2, v_2$ respectively. The time of immersion is then defined to be $T_E + t_1 + \Delta t_1$ and the time of emersion is defined to be $T_E + t_2 + \Delta t_2$. The values of u, v at these times are then given by

$$\begin{aligned} u(\Delta t_1) &= u_1 + u'_1\Delta t_1, & v(\Delta t_1) &= v_1 + v'_1\Delta t_1, \\ u(\Delta t_2) &= u_2 + u'_2\Delta t_2, & v(\Delta t_2) &= v_2 + v'_2\Delta t_2. \end{aligned} \tag{8.76}$$

Substitution of these values for u, v into eqn. (8.71) gives

$$\Delta t_1 = \frac{k^2 - u_1^2 - v_1^2}{2(u_1u'_1 + v_1v'_1)}, \qquad \Delta t_2 = \frac{k^2 - u_2^2 - v_2^2}{2(u_2u'_2 + v_2v'_2)}, \tag{8.77}$$

neglecting terms in Δt^2. If necessary new values of u, v can be estimated at times $T_0 + t_1 + \Delta t_1$, $T_0 + t_2 + \Delta t_2$ and the procedure repeated. To indicate how u_1, v_1; u_2, v_2 are evaluated the formulae will be given in terms of u_1, v_1. Similarly formulae can be developed for u_2, v_2. Using eqns. (8.59) and (8.69)

$$u_1 = x_1 - \xi_1 = x't_1 - \rho\cos\phi'\sin h_1,$$

where

$$h_1 = H - \psi + t_s \tag{8.78}$$

and $t_s = 1{\cdot}002\,738t_1$ is the sidereal equivalent of t_1. Using eqns. (8.60, 8.69),

$$v_1 = y_1 - \eta_1 = Y_1 + y't_1 - \rho(\sin\phi'\cos\delta_* - \cos\phi'\sin\delta_*\cos h_1). \qquad (8.79)$$

Again using eqns. (8.66, 8.67) with $(H - \psi)$ replaced by h_1,

$$u'_1 = x'_1 - \xi'_1 = x' - \frac{\mathrm{d}H}{\mathrm{d}t}Q_1 = x'_1 - 0{\cdot}2625Q_1:$$

$$v'_1 = y'_1 - \eta'_1 = y'_1 - \frac{\mathrm{d}H}{\mathrm{d}t}\xi_1\sin\delta_* = y'_1 - 0{\cdot}2625\xi_1\sin\delta_*, \qquad (8.80)$$

where $Q_1 = \rho\cos\phi'\cos h_1$ (and $\mathrm{d}H/\mathrm{d}t(= 2\pi \times 1{\cdot}002\,738/24) = 0{\cdot}2625$. Similarly for u_2, v_2. The values obtained from eqns. (8.79, 8.80) are used in eqn. (8.77).

The position angle $\mathscr{P}$ for occultations is measured eastwards from the north point of the Moon's disc. In Fig. 8.10 the

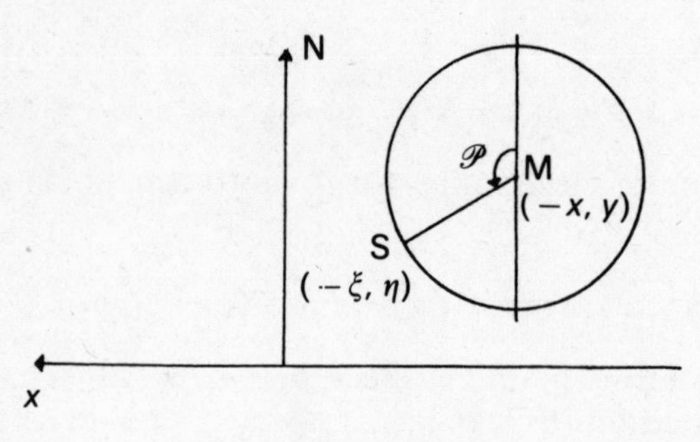

Fig. 8.10

x-axis is taken in an easterly direction and the z-axis is then directed positively downwards perpendicular to the plane of the page. M at $(-x, y)$ is the centre of the moon and S is a star about to be occulted. The coordinates of S at the instant are $(-\xi, \eta)$. The distance MS is the radius of the Moon.

Hence,

$$\cos(180^\circ - \mathscr{P}) = -\cos\mathscr{P} = \frac{y - \eta}{k}$$

$$\text{or} \quad \cos\mathscr{P} = -\frac{v}{k}, \tag{8.81}$$

$$\sin(180^\circ - \mathscr{P}) = \sin\mathscr{P} = -\frac{x - \xi}{k}$$

$$\text{or} \quad \sin\mathscr{P} = -\frac{u}{k}.$$

The position angle of emersion can be calculated similarly. Once values of u and v have been established at the time of immersion or emersion, the position angle can be quickly calculated.

8.2.4 Reduction of occultations. The determination of ephemeris time requires the position of the Moon to be known. Ephemeris time is determined by considering the position of the Moon in its orbit. The lunar ephemeris will give the predicted orbit of the Moon. The accurate timing of occultation will give information on the difference between the actual and predicted places of the Moon. Times of immersion are used since immersion can be timed more accurately than emersion. The displacement of true and observed positions will give the correction ΔT, the difference between ephemeris and universal time. The reduction of occultations requires computation to a higher order of accuracy to that required for prediction. In reduction calculation it is usual to scale all coordinates by the lunar radius. The position of the observer is thus

$$\xi = \frac{\rho}{k}\cos\phi'\sin h, \tag{8.82}$$

$$\eta = \frac{\rho}{k}(\sin\phi'\cos\delta_* - \cos\delta'\sin\delta_*\cos h), \tag{8.83}$$

where h is the hour angle of the star at the time of occultation; i.e.,

$$h = \text{sidereal time at } 0^{\text{h}} \text{ U.T.} + t_s - \psi - \alpha_* \tag{8.84}$$

where ψ is the longitude of the observer, and t_s is the sidereal equivalent of t, the U.T. of immersion. ρ is expressed in units of ρ_0, the equatorial radius of the Earth. Hence ξ, η at the time of immersion can be calculated. Eqns. (8.55, 8.56) for the coordinates (x_1, y_1) of the centre of the Moon may be rewritten in the form

$$x_1 = \frac{\cos \delta_1 \sin (\alpha_1 - \alpha_*)}{k \sin P}, \tag{8.85}$$

$$y_1 = \frac{\cos \delta_* \sin \delta_1 - \sin \delta_* \cos \delta_1 \cos (\alpha_1 - \alpha_*)}{k \sin P}. \tag{8.86}$$

The approximate forms eqns. (8.61, 8.62) with the inclusion of a correction term can be used to replace eqns. (8.85, 8.86) since $(\alpha_1 - \alpha_*)$ and $(\delta_1 - \delta_*)$ are small, namely

$$x_1 = \cos \delta_1 \{(\alpha_1 - \alpha_*) - C_1\}/k \sin P, \tag{8.87}$$

$$y_1 = \{(\delta_1 - \delta_*) - C_2\}/k \sin P + \tfrac{1}{2} x_1 \sin \delta_*(\alpha_1 - \alpha_*), \tag{8.88}$$

where $C_{1,2}$ are the correction terms. A critical table of the correction terms is to be found in the Explanatory Supplement to the Astronomical Ephemeris (p. 299).

ξ, η are derived using the observed times of occultation. x_1, y_1 are calculated using the predicted position of the Moon at the ephemeris time of the occultation. If, therefore, the predicted and observed positions of the Moon's centre are not identical, then

$$\{(x_1 - \xi)^2 + (y_1 - \eta)^2\} - 1 \neq 0, \tag{8.89}$$

if k is taken as unity. The distance between the observer and the predicted position of the Moon's limb on the fundamental plane is the difference between eqn. (8.89) and zero. If s is the geocentric semi-diameter of the Moon, the angular separation of the observer from the predicted position of the Moon's limb is

$$\Delta\sigma = s\{(x_1 - \xi)^2 + (y_1 - \eta)^2\}^{\frac{1}{2}} - s. \tag{8.90}$$

For the purposes of the reduction it is sufficiently accurate to denote the position angle of occultation by

$$\tan \mathscr{P} = \frac{x_1 - \xi}{y_1 - \eta} = \frac{u}{v} \text{(see eqn. (8.81))}. \tag{8.91}$$

If ρ is the position angle of the Moon's direction of motion then

$$\tan \rho = \frac{x'}{y'} \tag{8.92}$$

where x', y' are derived from the predicted lunar orbit. The angle between the direction of motion and the occulted star is therefore $\rho - \mathscr{P}$ and so

$$\begin{aligned} \sin(\rho - \mathscr{P}) &= (x_1 - \xi)\cos\rho - (y_1 - \eta)\sin\rho, \\ \cos(\rho - \mathscr{P}) &= -(x_1 - \xi)\sin\rho - (y_1 - \eta)\cos\rho. \end{aligned} \tag{8.93}$$

If the Moon's position is in error by amounts δL and δB (with respect to the lunar orbit) in longitude and latitude respectively

$$\Delta\sigma = \delta L \cos(\rho - \mathscr{P}) + \delta B \sin(\rho - \mathscr{P}). \tag{8.94}$$

Given a series of occultations of different stars giving $\Delta\sigma$ and $\mathscr{P}$, a least squares analysis will give the corrections δL, δB to the lunar orbit.

8.3 Eclipses of the Moon

Lunar eclipses are simply treated. The geometrical circumstances of a total lunar eclipse are illustrated in Fig. 8.11. The Sun is centred at S, the Earth at E and the Moon at M.

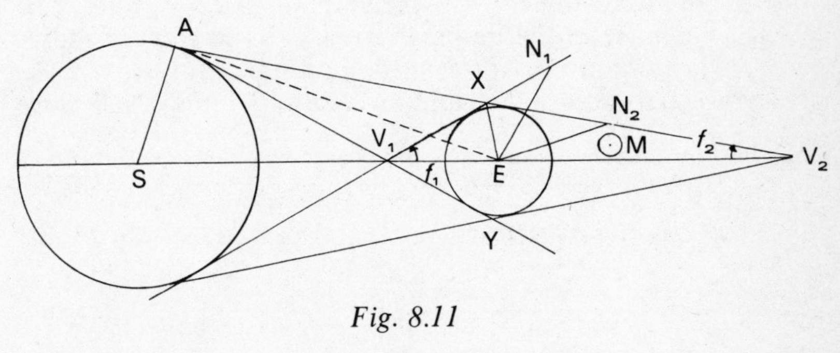

Fig. 8.11

Three types of eclipse may be distinguished.

(i) A penumbral eclipse, or *appulse*, in which the Moon enters the penumbral cone only.
(ii) A partial eclipse in which the Moon enters the umbral cone but is not entirely immersed in the umbral shadow.
(iii) A total eclipse when the Moon is wholly immersed in the umbral shadow.

8.3.1 The determination of the fundamental plane. A lunar eclipse differs from an occultation and a solar eclipse in that the body from which the eclipse is observed—the Earth—is producing the shadow. The fundamental plane assumes a different character. It is still the plane perpendicular to the shadow cone but now passes through the centre of the body producing the shadow. Therefore almost an entire hemisphere the Earth will observe an eclipse of the Moon. The co-ordinates of the observer on the fundamental plane are superfluous. The z-axis perpendicular to the fundamental plane is simply the line joining the Sun to the Earth produced —i.e. the positive direction of z lies on the same side of the Earth as the Moon. The equatorial coordinates of the point in which the z-axis cuts the celestial sphere are therefore

$$a = \alpha + 12^{\text{h}}, \qquad d = -\delta, \tag{8.95}$$

where (α, δ) are the equatorial coordinates of the centre of the Sun.

8.3.2 The "Besselian elements" of a lunar eclipse. The radii of the penumbral and umbral cones at the Moon's distance can be found from simple geometrical considerations. Let f_1 be the semi-angle of the penumbral vertex V_1 and f_2 be the semi-angle of the umbral vertex V_2. The angular radius of the penumbral cone at the Moon's distance is $l_1 = \text{N}_1\hat{\text{E}}\text{V}_2$. The angular radius of the umbral cone at the Moon's distance is $l_2 = \text{N}_2\hat{\text{E}}\text{V}_2$

$$\text{X}\hat{\text{N}}_2\text{E} = l_2 + f_2 = P, \tag{8.96}$$

where P is the equatorial horizontal parallax of the Moon. Also

$$\text{S}\hat{\text{E}}\text{A} = \pi + f_2 = S, \tag{8.97}$$

where π is the equatorial horizontal parallax and S is the angular semi-diameter of the Sun. Hence

$$l_2 = P + \pi - S. \tag{8.98}$$

Again because $Y\hat{A}E = X\hat{A}E$,

$$f_1 = S + \pi,$$

also,

$$l_1 = P + f_1 = P + \pi + S. \tag{8.99}$$

The Earth's atmosphere increases the size of the shadow cones and the effect of the atmosphere is allowed for empirically by increasing the radius of the shadow cones by 2 per cent over the values of eqns. (8.98, 8.99); i.e.,

$$\begin{aligned} l_1 &= 1{\cdot}02(P + \pi + S), \\ l_2 &= 1{\cdot}02(P + \pi - S). \end{aligned} \tag{8.100}$$

The angular distance between the centre of the Moon and the centre of the shadow cones may be denoted by L. At the beginning and end of penumbral eclipse.

$$L_1 = l_1 + s, \tag{8.101a}$$

where s is the angular diameter of the Moon. At the beginning and end of a partial eclipse

$$L_2 = l_2 + s, \tag{8.101b}$$

while at the beginning and end of total eclipse

$$L_3 = l_2 - s. \tag{8.101c}$$

As with solar eclipses and occultations the Besselian elements x_1, y_1 of the centre of the Moon are

$$\begin{aligned} x_1 &= \cos\delta_1 \sin(\alpha_1 - a) \\ y_1 &= \cos d \sin\delta_1 - \sin d \cos\delta_1 \cos(\alpha_1 - a) \end{aligned} \tag{8.102}$$

where (a, d) are given by eqn. (8.95). In this case the distance of the Moon does not matter since the distance is taken up in the determination of the radius of the shadow cones; unit distance has been assumed. The angular distance between the centre of the moon and the centre of the shadow is m, where

$$m^2 = x_1^2 + y_1^2, \tag{8.103}$$

since the centre of the shadow lies on the z-axis.

The Besselian elements are not listed in the Astronomical Ephemeris for lunar eclipses. Clearly since the equatorial coordinates (α, δ) of the Sun and the values of P, π, S, s are tabulated in the Astronomical Ephemeris for each day of the year, it is a trivial calculation to determine x_1, y_1 and L_1, L_2, L_3. Since lunar eclipses are not used to obtain data of high precision (clearly the angular sizes l_1, l_2 of the shadow cones cannot be determined to more than three significant figures because of the empirical allowance for the effect of the Earth's atmosphere) there is no useful purpose in predicting the times of lunar eclipses to high accuracy.

8.3.3 Determination of times of lunar eclipse. Let T_0 be the ephemeris time of some instant near opposition. Denote values of x_1, y_1, by x_0, y_0, when evaluated at time T_0. Let x', y' be the hourly variations of x_1, y_1. For a particular time of contact $T_0 + t$,

$$(x_0 + x't)^2 + (y_0 + y't)^2 = L^2 \tag{8.104}$$

where L is chosen from eqns. (8.101). Eqn. (8.104) may be solved in the usual way by setting

$$n^2 = x'^2 + y'^2, \tag{8.105}$$

$$m_0^2 = x_0^2 + y_0^2, \tag{8.106}$$

and

$$D = x_0 x' + y_0 y', \tag{8.107}$$

then

$$t = -\frac{D}{n^2} \mp \frac{1}{n^2}\{D^2 - n^2(m_0^2 - L^2)\}^{\frac{1}{2}}. \tag{8.108}$$

The upper sign gives first contact and the lower sign gives last contact. The time of greatest obscuration is given by

$$t' = -\frac{D}{n^2}. \tag{8.109}$$

By defining

$$\Delta^2 = \frac{1}{n^2}(n^2 m_0^2 - D^2), \tag{8.110}$$

such that $\Delta > 0$, eqn. (8.108) may be expressed in the form

$$t = -\frac{D}{n^2} \mp \frac{1}{n}(L^2 - \Delta^2)^{\frac{1}{2}}. \qquad (8.111)$$

The time of mid-eclipse can be obtained by solving

$$xx' + yy' = 0 \text{ (see eqn. (8.44))} \qquad (8.112)$$

where x, y are the values of x_1, y_1 at the time of mid eclipse $T_0 + t'$ whence,

$$(x_0 + x't')x' + (y_0 + y't')y' = 0. \qquad (8.113)$$

Again

$$t' = -\frac{1}{n^2}(x_0x' + y_0y') = -\frac{D}{n^2}. \qquad (8.109)$$

The interpretation of Δ is that it is the value of m at time t'. This follows from eqn. (8.103) since

$$\begin{aligned} m^2(t') &= (x_0 + x't')^2 + (y_0 + y't')^2 \\ &= \left(x_0 - x'\frac{D}{n^2}\right)^2 + \left(y_0 - y'\frac{D}{n^2}\right)^2 \\ &= m_0^2 + \frac{D^2}{n^2} - \frac{2D^2}{n^2} \text{ (using eqns. (8.105, 8.107),} \\ &= m_0^2 - \frac{D^2}{n^2} = \Delta^2 \text{ (using eqn. (8.110)).} \qquad (8.114) \end{aligned}$$

Hence Δ is the minimum separation of the centre of the Moon from the centre of the shadow at the time of mid-eclipse. The conditions for eclipse may then be written in terms of Δ;

$$\left.\begin{aligned} &L_1^2 - \Delta^2 < 0 \text{ —no eclipse,} \\ &\left.\begin{aligned} L_1^2 - \Delta^2 > 0 \\ L_2^2 - \Delta^2 < 0 \end{aligned}\right\}\text{—penumbral eclipse,} \\ &\left.\begin{aligned} L_2^2 - \Delta^2 > 0 \\ L_3^2 - \Delta^2 < 0 \end{aligned}\right\}\text{—partial eclipse,} \\ &L_3^2 - \Delta^2 > 0 \text{ —total eclipse.} \end{aligned}\right. \qquad (8.115)$$

The greatest magnitude of the eclipse can be written

$$(L - \Delta)/2s, \tag{8.116}$$

since Δ is the minimum separation of the centres of the shadow and Moon. In eqn. (8.116) L_2 is used for partial eclipses and total eclipses (when the magnitude is unity) and L_1 for penumbral eclipses.

8.4 Determination of the Possibility of Eclipse

It is useful to have a quick guide to the occurrence of eclipses. Conventionally this takes the form of determining the latitude of the centre of the Moon at the time of conjunction (for a solar eclipse) or opposition (for a lunar eclipse). The calculation becomes rather complicated if spherical trigonometry is used and the usual practice is to use plane geometry. This means that the calculation is approximate but is adequate as a guide. The geometrical situation is illustrated in Fig. 8.12 for a solar eclipse but the circumstances for a lunar eclipse can be obtained from a simple reinterpretation of Fig. 8.12. M is the position of the Moon when in conjunction

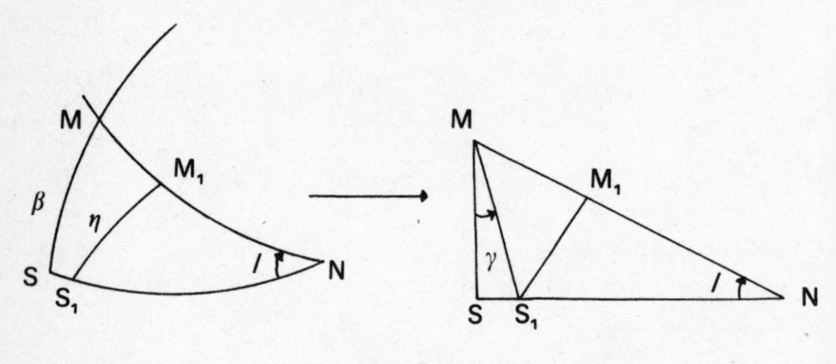

Fig. 8.12

in longitude with the Sun S. Hence SM $= \beta$ is the latitude of the Moon at conjunction. I is the inclination of the lunar orbit to the ecliptic.

At some time after conjunction the Sun has moved to S_1 (at a rate $\dot{\phi}$ radians per unit time) and the Moon to M_1 (at a rate $\dot{\theta}$ radians per unit time). The angular separation of S_1

and M_1 is then η. To establish whether or not an eclipse of some sort is possible the minimum value of η must be determined. To do this the angle $\gamma = S\hat{M}S_1$ is introduced. Treating the figure SMN as a plane triangle,

$$\begin{aligned}
SM &= \beta, \qquad S_1M_1 = \eta, \qquad S\hat{M}S_1 = \gamma\text{-definition}\\
SS_1 &= \beta \tan \gamma, \qquad S_1M = \beta \sec \gamma,\\
S\hat{M}M_1 &= 90° - (I + \gamma),\\
MM_1 &= q\beta \tan \gamma \quad \text{where } q = \dot{\theta}/\dot{\phi}.
\end{aligned}$$

Hence using the cosine rule for plane triangle on S_1MM_1

$$\begin{aligned}
\eta^2 &= \beta^2 \sec^2 \gamma + q^2\beta^2 \tan^2 \gamma - 2q\beta^2 \tan \gamma \sec \gamma\\
&\quad \times \cos \{(90° - (I + \gamma)\}\\
&= \beta^2\{\tan^2 \gamma(q^2 - 2q \cos I + 1)\\
&\quad - 2q \tan \gamma \sin I + 1\}
\end{aligned} \tag{8.117}$$

To find the minimum value of η eqn. (8.117) is differentiated with respect to γ giving

$$\begin{aligned}
2\eta \frac{d\eta}{d\gamma} &= \beta^2\{2 \tan \gamma \sec^2 \gamma(q^2 - 2q \cos I + 1)\\
&\quad - 2q \sec^2 \gamma \sin I\} = 0.
\end{aligned}$$

Hence

$$\sec^2 \gamma = 0,$$

or

$$\tan \gamma = \frac{q \sin I}{q^2 - 2q \cos I + 1}. \tag{8.118}$$

The first solution of eqn. (8.118) is impossible so that,

$$\begin{aligned}
\eta^2_{\min} &= \beta^2\left\{1 - \frac{q^2 \sin^2 I}{q^2 - 2q \cos I + 1}\right\}\\
&= \frac{\beta^2(q \cos I - 1)^2}{(q^2 - 2q \cos I + 1)}.
\end{aligned} \tag{8.119}$$

If a new angle I' is defined by

$$\tan I' = \frac{q \cos I}{q \cos I - 1} \tan I, \tag{8.120}$$

then

$$\eta_{\min} = \beta \cos I'. \tag{8.121}$$

Since the calculation is only approximate and $I \sim 5°$ the expression for $\tan I'$ can be approximated by

$$\tan I' = \frac{q}{q - 1} \tan I. \tag{8.120a}$$

The value of $\eta_{\min}$ can be obtained from eqns. (8.120a), (8.121). The above theory will be applied to solar and lunar eclipses.

8.4.1 Solar Eclipses

Fig. 8.13a is drawn in such a way that the penumbral shadow of the Moon (of centre M) is about to cross the Earth (of centre E). The angular separation of the centre of the Moon from the line of centres SE of the Sun and Earth is $D = \mathrm{S\hat{E}M}$. Hence

$$D = \mathrm{B\hat{E}S} + \mathrm{M\hat{E}B} = \mathrm{B\hat{E}S} + s,$$

$$\mathrm{B\hat{E}S} = \mathrm{V_2\hat{B}E} + \mathrm{B\hat{V}_2E} = P + \mathrm{A\hat{E}S} - \mathrm{E\hat{A}C},$$

$$= P + S - \pi,$$

where s is the angular semi diameter of the Moon and S is the angular semidiameter of the Sun.

Hence,

$$D = P - \pi + s + S. \tag{8.122}$$

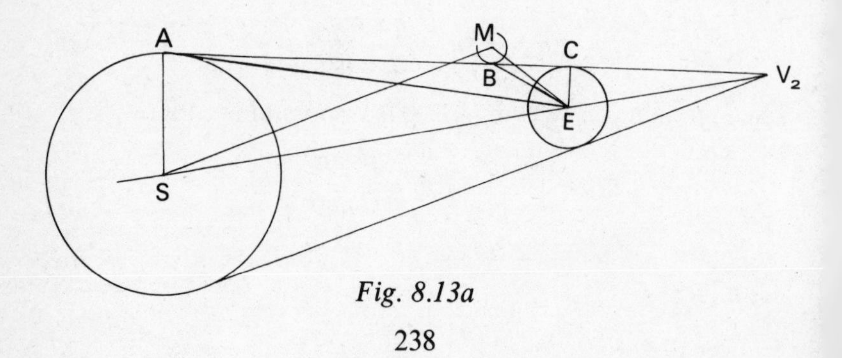

Fig. 8.13a

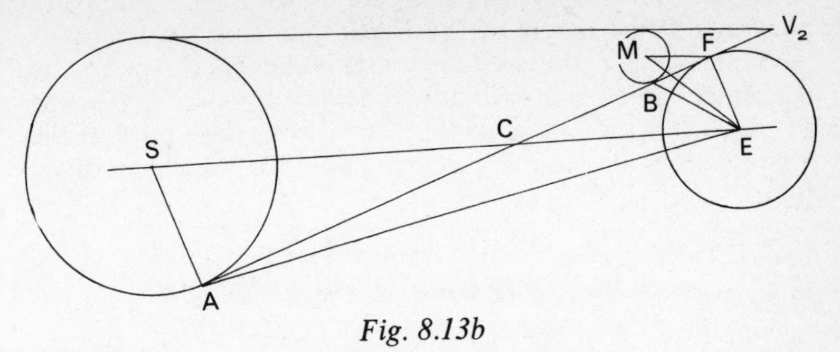

Fig. 8.13b

Hence, from eqn. (8.121), the latitude of the Moon's centre at the time of conjunction in longitude is

$$\beta = (P - \pi + s + S) \sec I'. \tag{8.123}$$

If the actual latitude of the Moon at conjunction is less than that given by eqn. (8.123) penumbral contact will occur. A partial eclipse of the Sun will then occur. For contact with the umbral cone the geometrical conditions are as drawn in Fig. 8.13b. As before,

$$D = \mathrm{M\hat{E}S} = \mathrm{B\hat{E}S} + s,$$

$$P = \mathrm{E\hat{B}F} = \mathrm{B\hat{C}E} + \mathrm{B\hat{E}S} = \pi + \mathrm{S} + \mathrm{B\hat{E}S},$$

whence,

$$D = P - \pi + s - S. \tag{8.124}$$

At conjunction the latitude of the Moon from eqn. (8.121) is

$$\beta = (P - \pi + s - S) \sec I' \tag{8.125}$$

if contact with the umbral cone is possible. In practice it is usual to calculate β from eqn. (8.123). If an eclipse is possible, i.e. if the actual ecliptic latitude of the Moon is less than or equal to the value given by eqn. (8.123), a full investigation of the eclipse is carried out as previously described. Eqn. (8.125) indicates when a total eclipse could be expected but can only be used as a guide and not as a definitive predictor in view of the uncertainties in the derivation of eqn. (8.121).

8.4.2 Lunar eclipses. A lunar eclipse takes place at opposition. Fig. 8.12 is now interpreted such that S and S_1 are positions of the centre of the Earth's shadow cone. Clearly the same analysis follows if the centre of the Sun is replaced by the centre of the Earth's shadow cone. The angular separations L_1, L_2, L_3 of the centre of the Moon from the centre of the Earth's shadow cone are given by eqns. (8.101) which, modified by eqns. (8.100), take the form

$$\begin{aligned} L_1 &= 1{\cdot}02(P + \pi + S) + s, \\ L_2 &= 1{\cdot}02(P + \pi - S) + s, \\ L_3 &= 1{\cdot}02(P + \pi - S) - s. \end{aligned} \tag{8.101}$$

Eqn. (8.121) then gives the ecliptic latitude of the Moon at opposition to be

$$\beta = L_1 \sec I' = \{1{\cdot}02(P + \pi + S) + s\} \sec I', \tag{8.126a}$$

for the penumbral shadow cone to touch the Moon;

$$\beta = L_2 \sec I' = \{1{\cdot}02(P + \pi - S) + s\} \sec I', \tag{8.126b}$$

for the umbral shadow cone to touch the Moon and

$$\beta = L_3 \sec I' = \{1{\cdot}02(P + \pi - S) - s\} \sec I', \tag{8.126c}$$

for a total (umbral) eclipse of the Moon. Of these results only eqns. (8.126a,b) are used when determining the likelihood of an eclipse. Should an eclipse be indicated the precise conditions would be investigated in detail.

8.4.3 Occurrence and repetition of eclipses. In determining the occurrence of eclipses the criteria for establishing whether or not a partial eclipse is possible are used, i.e. eqn. (8.123) in the case of solar eclipses and eqns. (8.126a,b) in the case of lunar eclipses. The conditions for total eclipses are then established from more precise calculation rather than by use of eqn. (8.125) or eqn. (8.126c) for solar and lunar eclipses respectively. Since the Moon moves round the Earth and the Earth round the Sun in elliptical orbits, the Earth–Moon, Earth–Sun distances will vary with time; i.e. P, π, s and S vary with time. The largest values of P, π, s, S occur at perigee/perihelion and the smallest values at apogee/aphelion. However, the Moon can be at perigee while the Earth is at aphelion and vice versa—plus an infinite variety of other combinations. The value of I also changes during the 18·6 year cycle of precession of the node of the lunar orbit about the pole of the ecliptic. The values of P, π, s, S, I and q have been taken from the Explanatory Supplement to the Astronomical Ephemeris (p. 215) and tabulated in Table 8.1. for the purposes of illustration.

Table 8.1

Parameter	Maximum	Minimum	Mean
I	5°18′	4°59′	5°08′
q	16·2	10·9	13·5
sec I′	1·0052	1·0043	1·00472
P	61′27″	53′53″	57′02″·70
π	8″·96	8″·65	8″·80
s	16′45″	14′41″	15′32″·58
S	16′18″	15′46″	15′59″·63

Using these values a range of estimates of D can be obtained above which no eclipse is possible and below which an eclipse is certain. Hence values of β, the ecliptic latitude of the Moon's centre at conjunction, can be computed for the range of values of D. For example in the case of a solar eclipse no eclipse is possible if D is greater than the values given by taking the maximum values of P, s, S and $\sec I'$ at conjunction in longitude. Also of interest are the values of distance ξ of the Sun from the node of the lunar orbit (in the case of solar eclipses) or the distance of the centre of the Earth's shadow from the node of the lunar orbit (in the case of lunar eclipses) at the time of conjunction. Making use of Fig. 8.12 and the four parts Rule on the spherical triangle SMN

$$\sin \xi = \tan \beta \cot I$$

where

$$\xi = \text{SN}, \beta = \text{SM}, I = \text{M}\hat{\text{N}}\text{S and M}\hat{\text{S}}\text{N} = 90°. \quad (8.127)$$

The values of β and ξ are tabulated in Tables 8.2 and 8.3. Table 8.2 refers to solar eclipses while Table 8.3 refers to lunar eclipses. Values of β have been taken from the Explanatory Supplement to the Astronomical Ephemeris (p. 215 in the case of solar eclipses, p. 258 in the case of lunar eclipses).

Table 8.2

Values of β and ξ for the occurrence of solar eclipses

No eclipse	$\beta > 1°34'46''$	$\xi > 17°53'$ $(18°26')$
Eclipse possible	$1°24'36'' < \beta < 1°34'46''$	$15°54' < \xi < 17°53'$
Eclipse certain	$\beta < 1°24'36''$	$\xi < 15°54'$ $(15°23')$

Values of ξ in Table 8.2 are calculated using the mean values of I, and the quantities in parenthesis correspond either to the maximum or minimum value of I.

Values of ξ in Table 8.3 are calculated for the mean value of I, and the values in parenthesis correspond to either the maximum or the minimum value of I.

Table 8.3

Values of β and ξ for the occurrence of lunar eclipse

No penumbral eclipse	$\beta > 1°36'38''$	$\xi > 18°14'$
Penumbral eclipse possible	$1°26'19'' < \beta < 1°36'38''$	$16°14' < \xi < 18°14'$
Penumbral eclipse certain No umbral eclipse	$1°03'46'' < \beta < 1°26'19''$	$11°55' < \xi < 16°14'$
Penumbral eclipse certain Umbral eclipse possible	$0°53'26'' < \beta < 1°03'46''$	$9°58' < \xi < 11°55'$ $(12°17')$
Umbral eclipse certain	$\beta < 0°53'26''$	$\xi < 9°58'(9°39')$

The above values of β are useful in recognising the circumstances in which an eclipse is possible. If the ecliptic latitude of the Moon is examined at time of conjunction, the likelihood of an eclipse can be established. The values of ξ are useful in considering the repeatability of eclipse. Given the range of values of ξ for which a partial eclipse is possible for both solar and lunar eclipses, i.e.

$$\begin{aligned} 15°23' < \xi < 18°26' \quad &\text{for solar eclipses} \\ 9°39' < \xi < 12°17' \quad &\text{for lunar eclipses} \end{aligned} \tag{8.128}$$

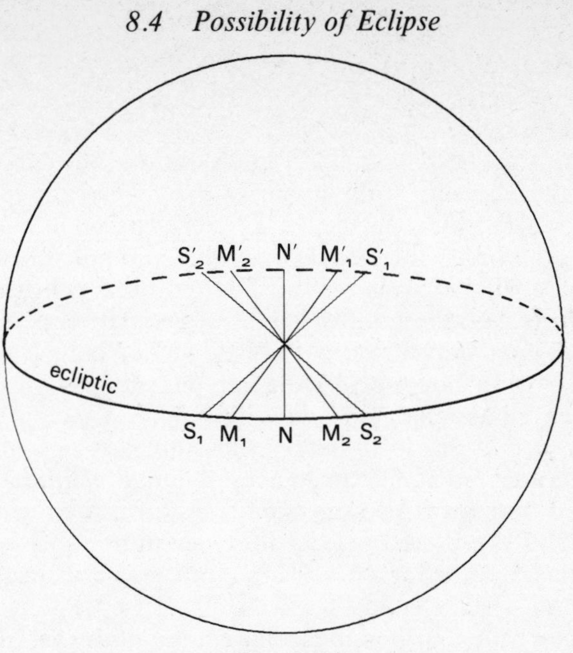

Fig. 8.14

an estimate can be made of the numbers of eclipses to be expected in any given year. In Fig. 8.14 let the plane S_1NN' be the plane of the ecliptic and let

$$\begin{aligned} S_1N = NS_2 = S_1'N' = N'S_2', \\ M_1N = NM_2 = M_1'N' = N'M_2'. \end{aligned} \tag{8.129}$$

S_1N corresponds to a limiting value of ξ defined by (8.128) for solar eclipses while M_1N is the counterpart for lunar eclipses.

The period of the Moon in its orbit about the Earth is 27·3 mean solar days. However, the observed interval between consecutive new moons—the synodic month—is 29·53 mean solar days. The nodes of the Moon's orbit precess once round the ecliptic in 6798·3 days ($\sim$18·6 yr.) so that the interval between successive passages of the Sun through the same node is $\sim$346·62 days. The Sun moves from a node at the rate

$$\left(\frac{360°}{365{\cdot}2422} + \frac{360°}{6798{\cdot}3}\right) \times 29{\cdot}53 = 30°{\cdot}40'/\text{synodic month}.$$

Since the minimum size of S_1S_2 is 30°46′ one new Moon at least must occur when the Sun is within S_1S_2 and similarly when the Sun is within $S'_1S'_2$. Hence at least two solar eclipses are to be expected annually. Suppose that at the time of new Moon the Sun was at the node. At the time of the following full Moon the Sun will be 15°20′ from the node and so be outside the limits for a lunar eclipse. Consequently for that year there will be 2 solar eclipses but no lunar eclipses.

Again 6 synodic months is 177·2 days. The time required for the Sun to move from N to N′ is ~173·3 days. Suppose it was full Moon 2 days *before* the Sun passed through N. Then it will be full Moon again 2 days *after* the Sun passes through N′. The conditions for a lunar eclipse are met on both occasions. Under such circumstances 2 lunar eclipses can be expected that year. Also the conditions for a solar eclipse are satisfied at the preceding and next new moon following the Sun's passage through N and N′. In these circumstances 2 lunar and 4 solar eclipses will occur. Suppose that the occurrence of the first possible solar eclipse of the year (i.e. just inside the upper limit of the range) occurred early in January. After the elapse of 12 synodic months (=354·36 days), a new Moon will occur when the Sun has moved ~368° with respect to the node at the possible eclipse of January. The Sun is now within the limit for which a solar eclipse must occur. At the next full Moon an eclipse of the Moon must occur. However, $12\frac{1}{2}$ lunations =369·13 > 365 days. Therefore in counting possible eclipses in a year one of the eclipses must be dropped. In this case the lunar eclipse would be counted with the following year's eclipses. Accordingly the year in question would have 5 solar and 2 lunar eclipses. Clearly the circumstances could be varied (i.e. the first possible eclipse occurred in the previous year) such that there may be 4 solar and 3 lunar eclipses. But in any period of 365 days there can only be 7 eclipses.

There are certain numerical coincidences of interest. The interval between successive passages of the Moon through a node is 346·62 days. 19 such intervals are 6585·8 days. 223 synodic months are 6585·2 days = 18 yrs, 11 days. This period is called the Saros. If the Sun and Moon were orientated such that on a certain date a solar eclipse took place, then

18 years, 11 days later the same circumstances will prevail and a solar eclipse will again occur. Similarly for lunar eclipses. However, the geographical circumstances will not be exactly reproduced.

Again 235 synodic months = 6939·55 days while 19 tropical years = 6939·60 days. Therefore if on a certain date it is new Moon it will be new Moon on the same date 19 years later. This is known as the Metonic cycle.

8.5 Eclipse Information given in the Astronomical Ephemeris

8.5.1 Solar eclipses. As stated previously the Astronomical Ephemeris tabulates the Besselian elements of the eclipse at 10-minute intervals of ephemeris time during the eclipse. However, additional information is given such that an appraisal of whether or not an eclipse can be observed at a given geographical location can be quickly made. For each eclipse the Elements of the eclipse are listed, namely: the ephemeris time of conjunction in right ascension and at that time; the equatorial coordinates; the hourly variations of the coordinates; the equatorial horizontal parallax and true semi-diameter of the Sun and Moon. The Circumstances of the Eclipse—namely, the beginning and end of the eclipse, the beginning and end of central eclipse and central eclipse at local apparent noon—are given with respect to ephemeris time and geographical location.

However, a more pictorial representation of the circumstances of a solar eclipse can be obtained from the eclipse map which is given in the Astronomical Ephemeris. A specimen map is reproduced in Fig. 8.15 (by permission of the Controller, HMSO) for the total solar eclipse of 1973 June 30. The map summarises the conditions of the eclipse. For the eclipse illustrated in Fig. 8.15 the path of totality is clearly evident and is fairly wide. A partial eclipse is visible outside the path of totality and the magnitude of the eclipse declines to zero on the loci marking the northern and southern limits of the eclipse. It is clear from the diagram that in the U.K. only the southern fringe of England had a chance of seeing a very minor partial eclipse on this occasion. The other natural terminations of the eclipse are the beginning of eclipse at

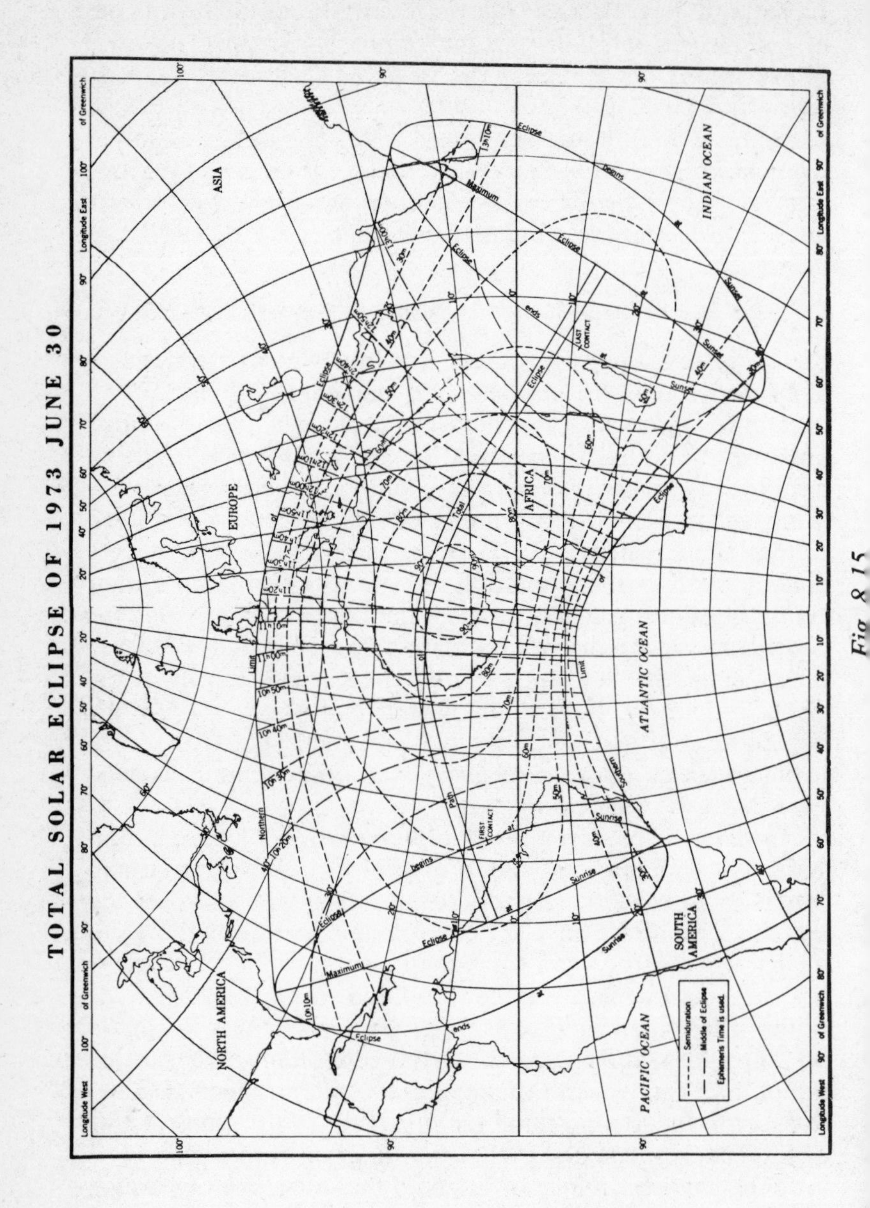
TOTAL SOLAR ECLIPSE OF 1973 JUNE 30
ASIA
EUROPE
AFRICA
NORTH AMERICA
SOUTH AMERICA
INDIAN OCEAN
ATLANTIC OCEAN
PACIFIC OCEAN
Eclipse begins at Sunset
Eclipse ends at Sunrise
Maximum Eclipse
Northern Limit of Total Path
FIRST CONTACT
LAST CONTACT
Semiduration
Middle of Eclipse
Ephemeris Time is used.
Longitude West of Greenwich
Longitude East of Greenwich

Fig. 8.15

sunset and its ending at sunrise. Within these loci some sort of eclipse can be seen. However, for scientific purposes, a solar eclipse is only useful if the first contact occurs at or after dawn. Indeed the site most useful scientifically is that having the longest duration of totality. On the map are two other loci, namely the loci joining places that have the same semi-duration of the partial phase and the loci of places having mid-eclipse at the same ephemeris time. For example: Catania in Sicily would have mid-eclipse at about 11^h40^m (ephemeris time) with the beginning of the eclipse about 60^m prior. From the relative position of Catania with respect to the path of totality and the northern limit of eclipse one might expect an eclipse of magnitude between one quarter and one third.

However, if serious eclipse observations are planned in the path of totality the general information available from a map is insufficient. In the Astronomical Ephemeris there is a table of the longitude (ephemeris) and latitude of the central line of the eclipse as a function of ephemeris time for each eclipse. While maximum duration of the eclipse may be expected about the time of conjunction in right ascension, the table gives the variation of the duration of the total eclipse on the centre line of the eclipse path. The northern and southern limits for totality (or annularity) are also given. This information permits the selection of a site on or near the centre of the path of the total eclipse giving as long a period as possible for observation. Having obtained the relevant information for site selection from the table, a detailed calculation using the Besselian Elements can be made for a given site if necessary.

8.5.2 Lunar eclipses. The information available for lunar eclipses is less detailed than for solar eclipses. The Elements of the Eclipse are as for the solar case (except that the ephemeris time at which they are listed refers to *opposition* and not conjunction). The circumstances of the eclipse give the ephemeris times of entering and leaving the penumbra and umbra and mid-eclipse. The position angles of first and last contacts are given and the location of sites at which the Moon is in the zenith at these instants. The magnitude of the eclipse is given. Since lunar eclipses are not used for any scientific measurement and can be observed from a whole hemisphere of the Earth at a time, it is sufficient to give an outline description of the geographical areas from which the eclipse can be observed. If more detail is required the necessary information can be obtained elsewhere in the Ephemeris.

8.6 Occultations by Planets and Transits

8.6.1 Occultations of stars by planets. Occultations of stars by planetary bodies other than the Moon cannot be determined with the same accuracy as lunar occultations because the positions of the planets in their orbits are not known with the same precision as in the lunar case and, also, because of their small angular size, their diameters are known less accurately. Again because of small planetary angular size such occultations are rarer. However, the procedure is as for lunar occultations, the coordinates and radius of the planet being used in place of the corresponding lunar values. Eqn. (8.124) may be used to calculate the angular separation D for an occultation where the star replaces the Sun (i.e. $\pi = S = 0$) and the planet replaces the moon. Hence

$$D = P_p + S_p \tag{8.130}$$

where P_p is the equatorial horizontal parallax of the planet and S_p is its angular semi-diameter. In practice it is usual to consider the situation at the time of conjunction in right ascension of planet and star. The angular separation is then in declination only. If the difference in declination is $\Delta\delta$ an investigation of the possibility of occultation is worthwhile if

$$\Delta\delta \sin\theta < (P_p + S_p) \tag{8.130a}$$

where θ is the position angle of the direction of motion of the planet with respect to a meridian. Condition (8.130a) follows since $\Delta\delta \sin\theta$ is the minimum separation of planet centre and star.

8.6.2 Transits of Mercury. Transits of the planets Mercury and Venus across the disc of the Sun are similar to annular eclipses of the Sun. The shadow cones produced by these planets have their vertices in the immediate neighbourhood of their respective planets. Hence, as viewed from the Earth, the effect of a transit is similar to that of an annular eclipse (see Fig. 8.2b where M now stands for the centre of Mercury or Venus). However, because these planets are very much further from the Earth than the Moon the radii of their shadow cones at the Earth's distance are large. In the case of Mercury the radius of its umbral shadow cone at the Earth's distance is about 10^6 km while for Venus the equivalent parameter is about $2{\cdot}6 . 10^5$ km. Clearly these radii are very much larger than the Earth's radius ($\sim 6{\cdot}4 . 10^3$ km). It therefore follows that, provided the conditions for a transit are met, a large fraction of the Earth's sunlight hemisphere will be able to see some part of the transit.

The method of calculation of the conditions of transit can be carried out in the same manner as the computation of annular solar eclipses (see section 8.1). However, because of their large shadow cones at the Earth's distance, more approximate methods can be used. Transits of Venus will not be discussed in view of their comparative rarity (the next transits of Venus being 2004 June 8, 2012 June 6: these transits occur at intervals of $121\frac{1}{2}$,8, $105\frac{1}{2}$,8 years). Transits of Mercury are more frequent (1973 Nov. 10, 1986, Nov. 13, 1993 Nov. 6, 1999 Nov. 15) and because of its larger shadow cone at the Earth's distance, the approximations made make a lesser contribution to the error than is the case for Venus.

The geometrical conditions for a transit of Mercury are illustrated in Fig. 8.16 which is drawn for a geocentric interior contact and so corresponds to the annular phase of a solar eclipse. The motion of the planet across the limb of the sun is

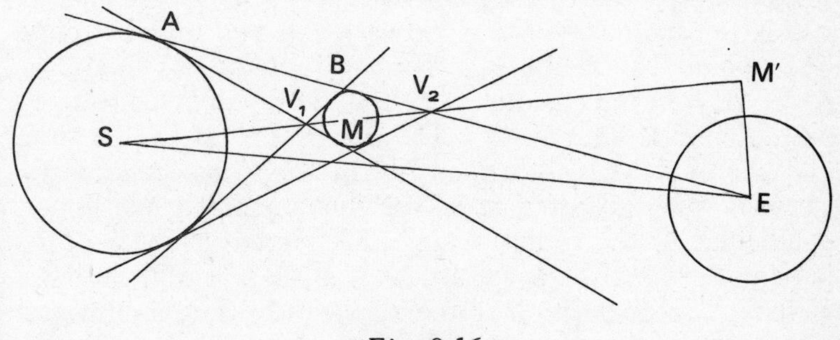

Fig. 8.16

called *ingress* at the beginning of the transit and *egress* at the end of transit. In the diagram S, E, M are the centres of the Sun, Earth and Mercury respectively. The umbral shadow cone of Mercury has vertex V_2 and the penumbral shadow cone has vertex V_1. It is convenient to consider *geocentric contact* in the case of transits of Mercury. A geocentric contact is such that a generator of the shadow cones passes through the centre of the Earth, E. Exterior ingress or egress occurs when a generator of the penumbral shadow cone passes through E and interior ingress or egress occurs when a generator of the umbral shadow cone passes through E. Fig. 8.16 is drawn for the latter situation.

As for eclipses, a fundamental plane is defined. The fundamental plane is that plane, perpendicular to the axis of the shadow cones, containing the centre of the Earth. In Fig. 8.16 the line M′E lies in the fundamental plane. Unlike eclipses, however, the origin of coordinates on the fundamental plane is not the centre of the Earth, but is that point on the fundamental plane intersected by the axis of the shadow cones. The x-axis of the fundamental plane is chosen to lie in a plane parallel to the ecliptic and is oppositely directed to the motion of Mercury. The y-axis is at right angles in a northerly direction. The Earth is regarded as occupying a position diametrically opposite that of the Sun and may therefore be displaced through a small angle in ecliptic latitude from the ecliptic of date. The position of Mercury must be calculated allowing for the light travel time. For an instant of observation T, the light observed was emitted by the Sun at time $T - \Delta t_1$ where Δt_1 is the light travel time between the Sun and the Earth, and passed by Mercury at time $T - \Delta t_1 + \Delta t_2$ where Δt_2 is the light travel time between the Sun and Mercury. The position of Mercury is therefore obtained at time $T - \Delta t_1 + \Delta t_2$. In the algebra which follows it will be convenient to use ecliptic coordinates. The disposition of the Earth E and Mercury M is illustrated in Fig. 8.17 for a heliocentric celestial sphere centred at S. The fundamental plane is perpendicular to the radius vector SM. Mercury moves on its orbit in the direction of the arrow. K is the pole of the ecliptic of date. The disposition of the x, y-axes in the fundamental plane then follow. Let the Earth have ecliptic coordinates

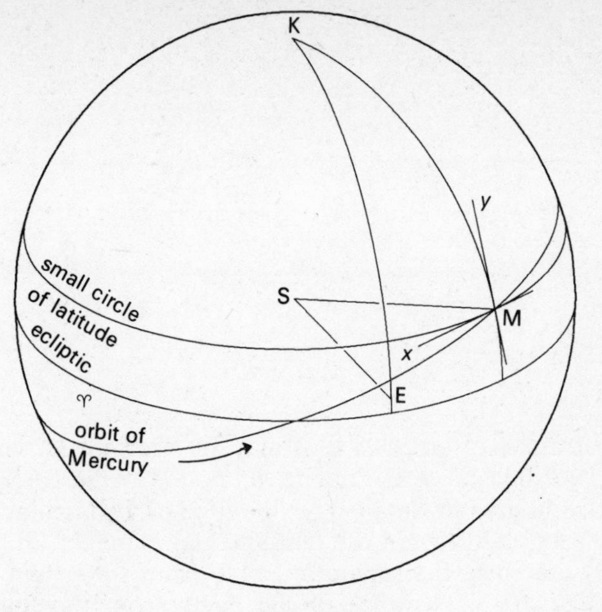

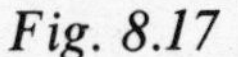

Fig. 8.17

$\lambda_{\oplus}, \beta_{\oplus}$ such that

$$\begin{aligned} \lambda_{\oplus} &= \lambda_{\odot} \pm 180°, \\ \beta_{\oplus} &= -\beta_{\odot}, \end{aligned} \tag{8.131}$$

where $(\lambda_{\odot}, \beta_{\odot})$ are the ecliptic coordinates of the Sun with respect to the ecliptic of date and let Mercury have ecliptic coordinates (λ_{M}, β_{M}). Let the heliocentric separation of the Earth and Mercury be $m = \mathrm{M\hat{S}E} = \mathrm{ME}$ and let the angle $\mathrm{K\hat{M}E} = \mathscr{M}$ where K is the pole of the ecliptic of date. In the spherical triangle KEM

$$\begin{aligned} \mathrm{KE} &= 90° - \beta_{\oplus}, \qquad \mathrm{KM} = 90° - \beta_{M}, \\ \mathrm{EM} &= m, \qquad \mathrm{E\hat{K}M} = \lambda_{M} - \lambda_{\oplus}. \end{aligned} \tag{8.132}$$

Application of the Sine and Transposed Cosine Rules to the spherical triangle KEM then gives

$$\begin{aligned} \sin m \sin \mathscr{M} &= \cos \beta_{\oplus} \sin (\lambda_{M} - \lambda_{\oplus}), \\ \sin m \cos \mathscr{M} &= \cos \beta_{M} \sin \beta_{\oplus} \\ &\quad - \sin \beta_{M} \cos \beta_{\oplus} \cos (\lambda_{M} - \lambda_{\oplus}). \end{aligned} \tag{8.133}$$

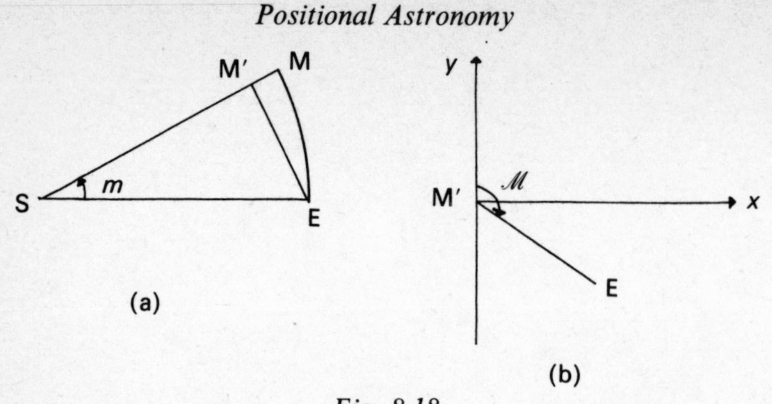

Fig. 8.18

The formulae of eqns. (8.133) may be used in deriving the (x, y) coordinates of E. In Fig. 8.18a the plane SEM is illustrated in greater detail. The line M′E perpendicular to the shadow axis SM lies in the fundamental plane (cf. Fig. 8.16). M′E is therefore the separation of E from the origin of coordinates. If the distance of the Earth from the Sun is R (= SE) then M′E = $R \sin m$. From the definition of spherical angles it is clear that the angle $K\hat{M}E$ of Fig. 8.17 is equal to the angle $y\hat{M}E$ of Fig. 8.18b. Hence the coordinates (X, Y) of E are

$$\begin{aligned} X &= R \sin m \cos (\mathscr{M} - 90°) \\ &= R \sin m \sin \mathscr{M}, \\ &= R \cos \beta_{\oplus} \sin (\lambda_{M} - \lambda_{\oplus}), \text{(using eqn. (8.133))} \end{aligned}$$

or

$$\begin{aligned} x = \frac{X}{R} &= \cos \beta_{\oplus} \sin (\lambda_{M} - \lambda_{\oplus}) \\ &\simeq \lambda_{M} - \lambda_{\oplus} \simeq m \sin \mathscr{M}, \end{aligned} \tag{8.134a}$$

since $m, \beta_{\oplus}, \lambda_{M} - \lambda_{\oplus}$ are small angles. Similarly

$$\begin{aligned} Y &= -R \sin m \cos (180° - \mathscr{M}) \\ &= R \sin m \cos \mathscr{M}, \\ &= R\{\cos \beta_{M} \sin \beta_{\oplus} \\ &\quad - \sin \beta_{M} \cos \beta_{\oplus} \cos (\lambda_{M} - \lambda_{\oplus})\}, \text{(using eqn. (8.133))} \end{aligned}$$

or

$$y = \frac{Y}{R} = \cos \beta_M \sin \beta_\oplus - \sin \beta_M \cos \beta_\oplus \cos (\lambda_M - \lambda_\oplus)$$

$$\simeq \beta_\oplus - \beta_M \simeq m \cos \mathcal{M}. \tag{8.134b}$$

The units of x, y are usually taken to be seconds of arc. As before quantities u, v are introduced, here defined by

$$u = m \cos \mathcal{M} = \beta_\oplus - \beta_M = y; \; v = m \sin \mathcal{M} = \lambda_M - \lambda_\oplus = x;$$

$$m^2 = u^2 + v^2; \tag{8.135}$$

where m is chosen to be positive. Denoting the hourly variations of $u, v, \lambda_M, \lambda_\oplus, \beta_M, \beta_\oplus$ by a prime then n, M may be defined by

$$u' = \beta'_\oplus - \beta'_M = n \cos N; \; v' = \lambda'_M - \lambda'_\oplus = n \sin N;$$

$$n^2 = u'^2 + v'^2; \tag{8.136}$$

where n is chosen to be positive.

In Fig. 8.16 V_1, V_2 are the vertices of the penumbral and umbral cones respectively of Mercury. Let f_1, f_2 be the semi-angles of the respective shadow cones. Then

$$\sin f_1 = \frac{SA}{SV_1} = \frac{MB}{V_1M}; \qquad \sin f_2 = \frac{SA}{SV_2} = \frac{MB}{MV_2};$$

$$= \frac{SA + MB}{SV_1 + V_1M}; \qquad = \frac{SA - MB}{SV_2 - MV_2};$$

$$= \frac{k_\odot + k_M}{r}; \qquad = \frac{k_\odot - k_M}{r}; \tag{8.137}$$

where $k_\odot, k_M$ are the linear radii of the Sun and Mercury respectively and r is the heliocentric distance of Mercury. Expressing r in A.U. f_1, f_2 in seconds of arc

$$f_1 \simeq \frac{S_\odot + S_M}{r}, \qquad f_2 \simeq \frac{S_\odot - S_M}{r}, \tag{8.137a}$$

where $S_\odot, S_M$ are the angular radii of the Sun and Mercury at unit distance (i.e. 1 A.U.) respectively. The radii of the shadow cones at the fundamental plane may now be determined.

Let R_p, R_u denote the radii of the penumbral and umbral shadow cones respectively on the fundamental plane where (see Fig. 8.16)

$$R_p = \mathrm{V_1M'} \tan f_1 = \left\{R \cos m - \frac{k_\odot}{\sin f_1}\right\} \tan f_1,$$

$$= R \sec f_1 \left\{\cos m \sin f_1 - \frac{k_\odot}{R}\right\};$$

$$R_u = \mathrm{V_2M'} \tan f_2 = \left\{R \cos m - \frac{k_\odot}{\sin f_2}\right\} \tan f_2,$$

$$= R \sec f_2 \left\{\cos m \sin f_2 - \frac{k_\odot}{R}\right\}.$$

Setting $\sin L_p = R_p/R$ and $\sin L_u = R_u/R$,

$$\begin{aligned} \sin L_p &= \sec f_1 \left\{\cos m \sin f_1 - \frac{\sin S_\odot}{R}\right\}, \\ \sin L_u &= \sec f_2 \left\{\cos m \sin f_2 - \frac{\sin S_\odot}{R}\right\}, \end{aligned} \tag{8.138a}$$

or

$$\begin{aligned} L_p &\simeq \sec f_1 \left\{f_1 \cos m - \frac{S_\odot}{R}\right\}, \\ L_u &\simeq \sec f_2 \left\{f_2 \cos m - \frac{S_\odot}{R}\right\}, \end{aligned} \tag{8.138b}$$

where R is now in A.U. and in eqn. (8.138b) $L_p, L_u, f_1, f_2, S_\odot$ are in seconds of arc.

For a transit of Mercury exterior contact occurs for $m = L_p$ and interior contact occurs for $m = L_u$ where L_p, L_u are given by eqns. (8.138b). Let a subscript zero denote values of u, v, u', v', n, m at an ephemeris time T_0 near a time of contact. Then, from eqns. (8.135, 8.136, 8.138b) if contact occurs at time $T_0 + t$,

$$(u_0 + u'_0 t)^2 + (v_0 + v'_0 t)^2 - L_{p,u}^2 = 0, \tag{8.139a}$$

or

$$n_0^2 t^2 + 2D_0 t + (m_0^2 - L_{p,u}^2) = 0, \tag{8.139b}$$

where

$$D_0 = u_0 u_0' + v_0 v_0'. \tag{8.139c}$$

The time of contact may therefore be sought by finding the zeros of eqn. (8.139a) in the manner described for eclipses or from solution of the quadratic equation (8.139b) whence

$$t = -\frac{D_0}{n_0^2} \mp \frac{1}{n_0^2}\{D_0^2 - n_0^2(m_0^2 - L_{p,u}^2)\}^{\frac{1}{2}} \tag{8.140}$$

The calculation is iterated until the check $m \equiv L_{p,u}$ is satisfied at the time of contact.

A geocentric calculation does not give accurate timings of contact for an observer on the surface of the Earth. Allowance must be made for the effects of the displacement of the observer from the centre of the Earth. Since the displacement of the observer from the centre of the Earth ($\ngtr 6{\cdot}4 \,.\, 10^3$ km) is small by comparison with the radius of the shadow cone ($R_u \sim 10^6$ km) approximate values for the displacement will suffice. The method of calculation requires the determination of the rectangular coordinates P, Q, W of the observer's geocentric zenith with respect to the centre of the Earth. The fundamental plane for this purpose will be regarded as passing through the centre of the Earth and the pole of the ecliptic of date. The actual fundamental plane is inclined at angle m to the approximate fundamental plane. The value of cos m is unity to four significant figures (which is the accuracy to which the appropriate parameters are given in the Astronomical Ephemeris) and the corrections made necessary by the displacement of the observer from the centre of the Earth amount to no more than 3% of the geocentric parameters. Hence the approximation is of sufficient accuracy. In Fig. 8.19 a sphere of radius ρ equal to the geocentric distance of the observer is drawn. The observer's geocentric zenith lies in the direction EO where the centre of the Earth is at E. The approximate fundamental plane is perpendicular to the line E′E at E′ where E′E defines the solar direction (for the purpose of this approximation the displacement of E′ in latitude with respect to the ecliptic of date is ignored). The longitude of E′ is therefore $\lambda_{\oplus}$. Let (λ, β) be the ecliptic coordinates of the observer's geocentric zenith. Then rectangular co-ordinates of O with

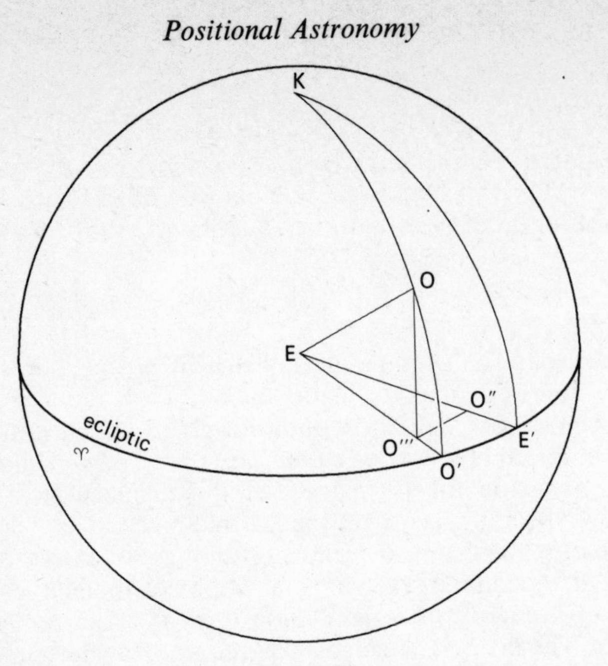

Fig. 8.19

respect to E are

$$\begin{aligned} P &= \mathrm{O''O'''} = \rho \cos \beta \sin (\lambda_\oplus - \lambda), \\ Q &= \mathrm{O'''O} = \rho \sin \beta, \\ W &= \mathrm{EO''} = \rho \cos \beta \cos (\lambda_\oplus - \lambda). \end{aligned} \tag{8.141}$$

It is customary to treat P, Q, W as positive quantities, the signs being taken care of by the trigonometric functions. It is usual to scale P, Q in terms of R giving p, q where,

$$\begin{aligned} p &= \frac{P}{R} = \frac{\rho}{\rho_0} \frac{\rho_0}{R} \cos \beta \sin (\lambda_\oplus - \lambda) \\ &= \frac{\rho}{\rho_0} \sin \pi \cos \beta \sin (\lambda_\oplus - \lambda), \\ q &= \frac{Q}{R} = \frac{\rho}{\rho_0} \frac{\rho_0}{R} \sin \beta = \frac{\rho}{\rho_0} \sin \pi \sin \beta, \end{aligned} \tag{8.142}$$

where π is the horizontal parallax of the Sun at the instant of contact. Since π is small, $\sin \pi$ may be replaced by π expressed in seconds of arc so giving p, q in the same units as x, y, and u, v. For an observer on the surface of the Earth u, v are replaced by u_1, v_1 where

$$u_1 = u + q, \qquad v_1 = v + p \tag{8.143}$$

The radii of the shadow cones for the observer will differ from the geocentric values. The observer is distant $R + W$ from the Sun in the direction of SE. The alteration that must be made to the radius of the shadow cone will be illustrated for the case of the umbral shadow cone. In eqn. (8.138), R must be replaced by $R + W$ so that

$$R_{u,1} = \left\{(R + W)\cos m - \frac{k_\odot}{\sin f_2}\right\}\tan f_2$$

so that the heliocentric angular radius $L_{u,1}$ of the shadow cone is

$$\sin L_{u,1} = \frac{R_{u,1}}{R + W} = \frac{1}{R}\left(1 - \frac{W}{R}\right)\left\{(R + W)\cos m - \frac{k_\odot}{\sin f_2}\right\}\tan f_2, \tag{8.144}$$

since $W \ll R$. Expanding eqn. (8.144)

$$\begin{aligned}\sin L_{u,1} &= \sec f_2 \left\{\cos m \sin f_2 - \frac{k_\odot}{R} - \frac{W}{R}\cos m \sin f_2 - \frac{W^2}{R^2}\cos m \sin f_2 + \frac{k_\odot}{R}.\frac{W}{R} + \frac{W}{R}\cos m \sin f_2\right\},\\ &= \sec f_2\left\{\cos m \sin f_2 - \frac{k_\odot}{R}\right\} + \frac{k_\odot}{R}\frac{W}{R}\sec f_2,\\ &= \sin L_u + \frac{k_\odot}{R}.\frac{W}{R}\sec f_2.\end{aligned} \tag{8.145}$$

But to this order of approximation the following substitutions may be made in the correction terms in eqn. (8.145)

$$\sec f_2 = 1, \frac{k_\odot}{R} = \sin S_\odot,$$

$$\frac{W}{R} = \frac{\rho}{\rho_0}\frac{\rho_0}{R}\cos\beta\cos(\lambda_\oplus - \lambda) = \frac{\rho}{\rho_0}\sin\pi\cos\beta\cos(\lambda_\oplus - \lambda), \tag{8.146}$$

so that

$$\sin L_{u,1} = \sin L_u + w \text{ or } L_{u,1} \simeq L_u + w, \tag{8.147}$$

where

$$\begin{aligned} w &= \frac{\rho}{\rho_0}\sin S_\odot \sin\pi\cos\beta\cos(\lambda_\oplus - \lambda), \\ &= \frac{\rho}{\rho_0}\pi\sin S_\odot\cos\beta\cos(\lambda_\oplus - \lambda), \end{aligned} \tag{8.148}$$

where w, π are in seconds of arc in the second of eqns. (8.148). Eqns. (8.139) can now be evaluated with $u_1, v_1, u'_1, v'_1, m_1, n_1, L_{p,u,1}$ replacing $u, v, u', v', m, n, L_{p,u}$. The parameters are evaluated at the time of geocentric contact and a correction to this time found for the site of observation.

Finally the ecliptic coordinates (λ, β) of the observer's geocentric zenith must be evaluated in terms of more readily evaluated parameters. The equatorial coordinates of the observer's geocentric zenith are Hour Angle $H = 0$ and declination $\delta = \phi'$ where ϕ' is the geocentric latitude of the observer, while the right ascension $\alpha = S_t$ the local Sidereal Time. Then from eqns. (3.8)

$$\begin{aligned} \sin\beta &= \cos\varepsilon\sin\phi' - \sin\varepsilon\cos\phi'\sin S_t, \\ \cos\beta\sin\lambda &= \sin\varepsilon\sin\phi' + \cos\varepsilon\cos\phi'\sin S_t, \\ \cos\beta\cos\lambda &= \cos\phi'\cos S_t. \end{aligned} \tag{8.149}$$

Substituting from eqns. (8.149) in eqns. (8.142, 8.148) for λ, β gives

$$\begin{aligned} p &= \pi\{-\rho \sin\phi' \sin\varepsilon \cos\lambda_{\oplus} + \rho\cos\phi'(-\cos\varepsilon\cos\lambda_{\oplus} \\ &\quad \times \sin S_t + \sin\lambda_{\oplus}\cos S_t)\}, \\ q &= \pi\{\rho\sin\phi'\cos\varepsilon - \rho\cos\phi'\sin\varepsilon\sin S_t\}, \\ w &= \pi\sin S_{\odot}\{\rho\sin\phi'\sin\varepsilon\sin\lambda_{\oplus} + \rho\cos\phi'(\cos\varepsilon \\ &\quad \times \sin\lambda_{\oplus}\sin S_t + \cos\lambda_{\oplus}\cos S_t)\}, \end{aligned} \tag{8.150}$$

where ρ is now expressed in units of ρ_0. u_1, v_1, L_1 all have the same form namely,

$$A + B\rho\sin\phi' + \rho\cos\phi'(C\sin S_t + D\cos S_t) \tag{8.151a}$$

where,

	u_1	v_1	L_1
A	u	v	L
B	$\pi\cos\varepsilon$	$-\pi\sin\varepsilon\cos\lambda_{\oplus}$	$\pi\sin S_{\odot}\sin\varepsilon\sin\lambda_{\oplus}$
C	$-\pi\sin\varepsilon$	$-\pi\cos\varepsilon\cos\lambda_{\oplus}$	$\pi\sin S_{\odot}\cos\varepsilon\sin\lambda_{\oplus}$
D	0	$\pi\sin\lambda_{\oplus}$	$\pi\sin S_{\odot}\cos\lambda_{\oplus}$

(8.151b)

The local sidereal time is given by $S_t = \mu - \lambda_E$ where μ is the ephemeris sidereal time of geocentric contact and λ_E is the ephemeris longitude of the observer. Clearly the time variations of u_1, v_1, namely $u'_1 v'_1$, can be computed. The forms of u'_1, v'_1 will be similar to those of u_1 and v_1 since only $u, v, \lambda_{\oplus}$ and S_t vary with time. In the Astronomical Ephemeris A, B, C, D are listed for the transit. The values are computed at the times of exterior and interior ingress and egress. Interior contacts are found using the angle f_2 as outlined above. Exterior contacts are obtained for the penumbral cone of Mercury replacing f_2 by f_1.

The above treatment of the transits of Mercury is approximate. The terms in B, C, D are usually not more than 3 per cent of the term in A. The correction to the time of geocentric contact is small. The Astronomical Ephemeris, as in the case of eclipses, gives the Elements of the Transit and the Geocentric Phases. A map is provided illustrating those regions of the Earth's surface from which the transit may be observed. The map gives curves, at intervals of 10 s, of times of interior ingress and egress.

Chapter IX

Plate Measurements

The aim of this chapter is not to discuss the detailed procedures involved in the measurement of photographic plates for astrometric purposes, since these techniques are highly specialised. The aim of this chapter is a more restricted one, being the establishment of a group of formulae which permit the rapid determination from plates of the rectangular equatorial coordinates of any celestial object. It is sometimes essential to determine the rectangular equatorial coordinates of a comet rapidly so that a preliminary orbit for the comet may be computed. A preliminary orbit has value in predicting where the comet may be observed in the future and as a basis for constructing a more accurate orbit once further observations are available.

In order to realise the above aim, standard coordinates must be considered again. The formal definition of standard coordinates (ξ, η) was given in Chapter IV. Standard coordinates were defined for a given epoch and in terms of a refracting telescope of specified focal length. While such formal definitions are quite precise, the realisation of standard coordinates in practice would be a very difficult, and probably unrewarding, problem were it not for the fact that most of the errors of measurement are linear in the coordinates. If (ξ, η) are the standard coordinates of an object and (p, q) are the best measurements of the standard coordinates then the errors are linear in the coordinates if

$$\begin{aligned} \xi - p &= a\xi + b\eta + c, \\ \eta - q &= d\xi + e\eta + f, \end{aligned} \tag{9.1}$$

where a, b, c, d, e, f are constants known as the *plate constants*. Clearly by measuring a number of stars whose standard coordinates are known, the plate constants may be derived by a least squares analysis. Eqn. (9.1) in fact represents the first terms of a series but for many purposes a representation to the first order is adequate. In this case ξ, η on the right hand side of eqns. (9.1) may be replaced by p, q since the error introduced will be of an order higher than the first and so may be discounted; i.e.,

$$\begin{aligned} \xi - p &= ap + bq + c, \\ \eta - q &= dp + eq + f. \end{aligned} \tag{9.2}$$

Either eqns. (9.1) or (9.2) may be used, depending on which form is the more suitable for the problem under investigation.

The determination of the plate constants by the method of least squares is laborious since a large number of stars must be measured before a satisfactory analysis can be attempted. In principle the existence of fast automatic plate measuring machines such as GALAXY have removed the labour, but such machines are by no means universal, and a simple accurate means of determining the positions of celestial objects is still required. The *method of dependences* gives a simple means whereby star positions may be determined directly.

9.1 The Method of Dependences

The dependences must first be derived from standard coordinates and the associated plate constants. Suppose a photographic plate has been used to record the position of a comet. The standard coordinates (ξ_*, η_*) of the comet could be determined from its measured position (p_*, q_*) using eqn. (9.2) if the plate constants were known. To avoid the use of plate constants, select three stars whose standard coordinates (ξ_i, η_i) $(i = 1, 2, 3)$ are known. The stars $S_i (i = 1, 2, 3)$ should be chosen so that the comet C lies within the triangle formed by them as illustrated in Fig. 9.1. Let (p_*, q_*) be the measured position of the comet C and (p_i, q_i) $(i = 1, 2, 3)$ be the measured positions of the reference stars S_i.

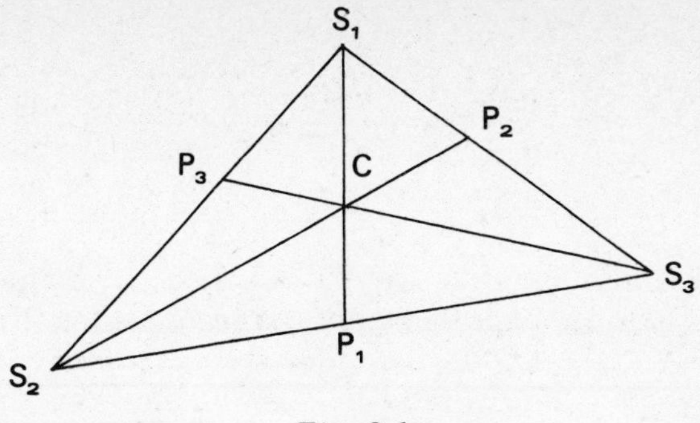

Fig. 9.1

Then, from eqns. (9.1),

$$\begin{aligned} \xi_* - p_* &= a\xi_* + b\eta_* + c, \\ \xi_i - p_i &= a\xi_i + b\eta_i + c, \quad i = 1, 2, 3. \end{aligned} \tag{9.3}$$

Multipliers D_i, $i = 1, 2, 3$ called *dependences* are chosen so that

$$\begin{aligned} D_1(\xi_1 - p_1) &+ D_2(\xi_2 - p_2) + D_3(\xi_3 - p_3) - (\xi_* - p_*) \\ &= a\{D_1\xi_1 + D_2\xi_2 + D_3\xi_3 - \xi_*\} \\ &\quad + b\{D_1\eta_1 + D_2\eta_2 + D_3\eta_3 - \eta_*\} \\ &\quad + c\{D_1 + D_2 + D_3 - 1\} \\ &= 0. \end{aligned} \tag{9.4}$$

It follows from eqn. (9.4) that since a, b, c are constants

$$\begin{aligned} D_1\xi_1 + D_2\xi_2 + D_3\xi_3 - \xi_* &= 0, \\ D_1\eta_1 + D_2\eta_2 + D_3\eta_3 - \eta_* &= 0, \\ D_1 + D_2 + D_3 - 1 &= 0. \end{aligned} \tag{9.5}$$

If eqns. (9.5) are used in eqn. (9.4) it also follows that

$$D_1p_1 + D_2p_2 + D_3p_3 - p_* = 0,$$

and similarly that, (9.6)

$$D_1q_1 + D_2q_2 + D_3q_3 - q_* = 0.$$

The standard coordinates ξ_*, η_* of the comet may then be written

$$\xi_* = p_* + \sum_{i=1}^{3} D_i\xi_i - \sum_{i=1}^{3} D_i p_i,$$
$$\eta_* = q_* + \sum_{i=1}^{3} D_i\eta_i - \sum_{i=1}^{3} D_i q_i. \tag{9.7}$$

In Chapter IV standard coordinates were derived with respect to the equatorial coordinates of the plate centre. Tabulated values of standard coordinates for stars are usually referred to specific plate centres. It is obviously unlikely that the centre of the photographic plate on which the comet was recorded would correspond to one of these specified plate centres. Suppose that with respect to a given specified plate centre the coordinates of the comet were (ξ'_*, η'_*) such that

$$\xi_* = \xi'_* - k_\xi$$
$$\eta_* = \eta'_* - k_\eta \tag{9.8}$$

then,

$$\xi'_* = \xi_* + k_\xi = p_* + \sum_{i=1}^{3} D_i(\xi_i + k_\xi) - \sum_{i=1}^{3} D_i p_i,$$

(using the third of eqns. (9.5))

$$= p_* + \sum_{i=1}^{3} D_i\xi'_i - \sum_{i=1}^{3} D_i p_i, \tag{9.9a}$$

and similarly,

$$\eta'_* = q_* + \sum_{i=1}^{3} D_i\eta'_i - \sum_{i=1}^{3} D_i q_i, \tag{9.9b}$$

where (ξ'_i, η'_i) $(i = 1, 2, 3)$ are the standard coordinates of the reference stars S_i with respect to the same specified plate centre. Therefore the standard coordinates can be referred to any suitable plate centre, not necessarily the centre of the plate actually measured. The above analysis is an indication of the value of using dependences.

The units of the standard coordinates (ξ, η) and by implication those of the measured coordinates (p, q) are such that the focal length of the standard refractor is unity. However, by re-arranging eqns. (9.5) and (9.6), the following group of equations may be derived:

$$
\begin{aligned}
D_1 p_1 + D_2 p_2 + D_3 p_3 - p_* &= 0, \\
D_1 q_1 + D_2 q_2 + D_3 q_3 - q_* &= 0, \\
D_1 + D_2 + D_3 - 1 &= 0.
\end{aligned} \tag{9.10}
$$

Eqns. (9.10) contain only the dependences D_i and the measured coordinates (p, q). There is no explicit reference in eqns. (9.10) to standard coordinates. Clearly, any convenient units may be used for (p, q).

Eqns. (9.10) allow a simple geometrical interpretation of the dependences. Suppose that the measured coordinates (p, q) are available for the reference stars S_i and comet C. Then eqns. (9.10) are three equations for the three unknowns D_1, D_2, D_3. The usual solution of such a system of equations gives

$$
\frac{D_1}{\begin{vmatrix} p_* & p_2 & p_3 \\ q_* & q_2 & q_3 \\ 1 & 1 & 1 \end{vmatrix}} = \frac{D_2}{\begin{vmatrix} p_* & p_1 & p_3 \\ q_* & q_1 & q_3 \\ 1 & 1 & 1 \end{vmatrix}} = \frac{D_3}{\begin{vmatrix} p_* & p_1 & p_2 \\ q_* & q_1 & q_2 \\ 1 & 1 & 1 \end{vmatrix}} = \frac{1}{\begin{vmatrix} p_1 & p_2 & p_3 \\ q_1 & q_2 & q_3 \\ 1 & 1 & 1 \end{vmatrix}}. \tag{9.11}
$$

The determinant

$$
\begin{vmatrix} p_1 & p_2 & p_3 \\ q_1 & q_2 & q_3 \\ 1 & 1 & 1 \end{vmatrix}
$$

is twice the area of the triangle formed by the reference stars S_i. The other determinants in the denominators of eqns. (9.11) may be interpreted as twice the areas of the triangles formed by the comet C and two of the reference stars. Hence

$$
\begin{aligned}
D_1 &= \frac{\text{area } CS_2S_3}{\text{area } S_1S_2S_3} = \frac{\mathrm{CP}_1}{\mathrm{S}_1\mathrm{P}_1}, \\
D_2 &= \frac{\text{area } CS_3S_1}{\text{area } S_1S_2S_3} = \frac{\mathrm{CP}_2}{\mathrm{S}_2\mathrm{P}_2}, \\
D_3 &= \frac{\text{area } CS_1S_2}{\text{area } S_1S_2S_3} = \frac{\mathrm{CP}_3}{\mathrm{S}_3\mathrm{P}_3}.
\end{aligned} \tag{9.12}
$$

Therefore, by a simple graphical method, the dependences D_1, D_2, D_3 can be determined. The relative configurations of reference stars and comet are marked on a sheet of tracing paper and the dependences determined from measurement with a ruler. Less easily, the measured coordinates (p, q) of the reference stars and comet can be obtained and the dependences calculated from eqns. (9.11). From the form of the dependences, the most accurate results can be achieved if C is near the centroid of the triangle $S_1S_2S_3$.

Once the dependences have been determined, the rectangular equatorial coordinates of the comet may be determined directly given the rectangular equatorial coordinates of the reference stars. There is no need to determine standard coordinates. However, the formulae for standard coordinates in terms of rectangular equatorial coordinates are required. Making use of eqns. (4.17) of Chapter IV, the first of eqns. (9.5) can be written in the form

$$\begin{aligned} D_1\xi_1 + D_2\xi_2 + D_3\xi_3 - \xi_* &= 0 \\ &= \frac{1}{R}\left\{D_1\frac{(-x_1y_0 + y_1x_0)}{(x_1x_0 + y_1y_0 + z_1z_0)} + D_2\frac{(-x_2y_0 + y_2x_0)}{(x_2x_0 + y_2y_0 + z_2z_0)}\right. \\ &\quad \left. + D_3\frac{(-x_3y_0 + y_3x_0)}{(x_3x_0 + y_3y_0 + z_3z_0)} - \frac{(-x_*y_0 + y_*x_0)}{(x_*x_0 + y_*y_0 + z_*z_0)}\right\}, \end{aligned} \quad (9.13)$$

where (x_0, y_0, z_0) are the rectangular coordinates of the plate centre. A first approximation to eqn. (9.13) may be obtained by assuming

$$I_i = x_ix_0 + y_iy_0 + z_iz_0 = 1, \qquad i = 1, 2, 3, *. \quad (9.14)$$

Such an approximation is accurate to six significant figures if the reference stars lie within a circle of radius 2′ centred on the comet. Accuracy to four significant figures is obtained for a circle of 15′ radius and to three significant figures for a circle of 30′ radius centred on the comet. If the approximation of eqn. (9.14) is made, eqn. (9.13) reduces to

$$\{-y_0(D_1x_1 + D_2x_2 + D_3x_3 - x_*) \\ + x_0(D_1y_1 + D_2y_2 + D_3y_3 - y_*)\} = 0, \qquad (9.15)$$

from which it may be deduced that

$$D_1x_1 + D_2x_2 + D_3x_3 - x_* = 0, \qquad (9.16a)$$

$$D_1y_1 + D_2y_2 + D_3y_3 - y_* = 0, \qquad (9.16b)$$

since x_0, y_0 are constants. From the second of eqns. (9.5) it may be deduced similarly that

$$D_1z_1 + D_2z_2 + D_3z_3 - z_* = 0. \qquad (9.16c)$$

A first approximation for x_*, y_*, z_* may be determined from eqns. (9.16). Such an approximation may have sufficient accuracy. Should improved accuracy be required a further approximation to eqn. (9.13) may be obtained as follows. Using the definition of I_i contained in eqn. (9.14), eqn. (9.13) may be written in the form

$$\left[-y_0\left\{\sum_{i=1}^{3} D_i\left(\frac{x_i}{I_i}\right) - \left(\frac{x_*}{I_*}\right)\right\} \\ + x_0\left\{\sum_{i=1}^{3} D_i\left(\frac{y_i}{I_i}\right) - \left(\frac{y_*}{I_*}\right)\right\}\right] = 0 \qquad (9.17)$$

from which it may be deduced that

$$\sum_{i=1}^{3} D_i\left(\frac{x_i}{I_i}\right) - \left(\frac{x_*}{I_*}\right) = 0, \qquad (9.18a)$$

$$\sum_{i=1}^{3} D_i\left(\frac{y_i}{I_i}\right) - \left(\frac{y_*}{I_*}\right) = 0, \qquad (9.18b)$$

and from the corresponding equations for η it may be deduced that

$$\sum_{i=1}^{3} D_i\left(\frac{z_i}{I_i}\right) - \left(\frac{z_*}{I_*}\right) = 0. \qquad (9.18c)$$

For a plate centred on a point (presuming that the comet will be nearly at the plate centre) whose rectangular equatorial coordinates are (x_0, y_0, z_0), the rectangular equatorial coordinates (x_i, y_i, z_i) of any other point on the plate may be written in the form

$$x_i = x_0 + \Delta x_i, \qquad y_i = y_0 + \Delta y_i, \qquad z_i = z_0 + \Delta z_i \quad (9.19)$$

where $\Delta x_i, \Delta y_i, \Delta z_i$ are the increments necessary to convert from the plate centre to any other point on the plate. It will be assumed in the following analysis that the increments are small by comparison with x_0, y_0, z_0. Eqns. (9.18) may be expanded in terms of $\Delta x, \Delta y, \Delta z$. The algebra will be given only for eqn. (9.18a) since the treatment is exactly similar for eqns. (9.18b,c). Eqn. (9.18a) becomes

$$\sum_{i=1}^{3} D_i\{x_i(1 - x_0\,\Delta x_i - y_0\,\Delta y_i - z_0\,\Delta z_i)\} \\ - x_*(1 - x_0\,\Delta x_* - y_0\,\Delta y_* - z_0\,\Delta z_*) = 0 \quad (9.20)$$

to the first order in the increments. Eqn. (9.20) may be re-expressed in the form

$$\left\{\sum_{i=1}^{3} D_i x_i - x_*\right\} \\ - x_0\left\{\sum_{i=1}^{3} D_i x_i(x_i - x_0) - x_*(x_* - x_0)\right\} \\ - y_0\left\{\sum_{i=1}^{3} D_i x_i(y_i - y_0) - x_*(y_* - y_0)\right\} \\ - z_0\left\{\sum_{i=1}^{3} D_i x_i(z_i - z_0) - x_*(z_* - z_0)\right\} = 0. \quad (9.21)$$

i.e.,

$$x_* = \sum_{i=1}^{3} D_i x_i - \frac{1}{2}\left[x_0\left\{\sum_{i=1}^{3} D_i x_i^2 - x_*^2\right\} \right. \\ + y_0\left\{\sum_{i=1}^{3} D_i x_i y_i - x_* y_*\right\} \\ \left. + z_0\left\{\sum_{i=1}^{3} D_i x_i z_i - x_* z_*\right\}\right] \quad (9.22a)$$

Similarly, from eqns. (9.18b,c),

$$y_* = \sum_{i=1}^{3} D_i y_i - \frac{1}{2}\left[x_0\left\{\sum_{i=1}^{3} D_i x_i y_i - x_* y_*\right\}\right.$$

$$+ y_0\left\{\sum_{i=1}^{3} D_i y_i^2 - y_*^2\right\}$$

$$\left. + z_0\left\{\sum_{i=1}^{3} D_i y_i z_i - y_* z_*\right\}\right], \quad (9.22b)$$

$$z_* = \sum_{i=1}^{3} D_i z_i - \frac{1}{2}\left[x_0\left\{\sum_{i=1}^{3} D_i x_i z_i - x_* z_*\right\}\right.$$

$$+ y_0\left\{\sum_{i=1}^{3} D_i y_i z_i - y_* z_*\right\}$$

$$\left. + z_0\left\{\sum_{x=1}^{3} D_i z_i^2 - z_*^2\right\}\right]. \quad (9.22c)$$

Since the leading term on the right hand side of eqn. (9.22) represents the zero order solution of eqn. (9.16), then terms in [ ] in eqn. (9.22) may be regarded as a correction term. Therefore in [ ] the values of x_*, y_*, z_* given by eqn. (9.16) are acceptable. Again for the terms in [ ], x_0, y_0, z_0 are not known. One way of proceeding is to set x_0, y_0, z_0 equal to the values for a reference star. Alternatively x_0, y_0, z_0 could represent a mean for the three reference stars. If possible, the plate might be exposed centred on one of the reference stars. Clearly eqns. (9.22) can be solved iteratively until the solution for x_*, y_*, z_* converges. Sufficient accuracy is usually obtained after two iterations. The values x_*, y_*, z_* obtained in the previous iteration are used to evaluate the terms in [ ] on the right hand side of eqns. (9.22) in obtaining the next iteration. A similar scheme has been given by L. J. Comrie (J. Brit. Astr. Ass., **39**, 203, 1929) for the direct determination of equatorial coordinates (α_*, δ_*) using dependences.

In this case the equations are

$$\begin{aligned}\alpha_* = \alpha_1 + D_2(\alpha_2 - \alpha_1) + D_3(\alpha_3 - \alpha_1) - \tan\delta_1\{D_2(\alpha_2 \\ - \alpha_1)(\delta_2 - \delta_1) + D_3(\alpha_3 - \alpha_1)(\delta_3 - \delta_1) \\ - (\alpha_* - \alpha_1)(\delta_* - \delta_1)\}, \\ \delta_* = \delta_1 + D_2(\delta_2 - \delta_1) + D_3(\delta_3 - \delta_1) \\ + \tfrac{1}{4}\sin 2\delta_1\{D_2(\alpha_2 - \alpha_1)^2 + D_3(\alpha_3 - \alpha_1)^2 \\ - (\alpha_* - \alpha_1)^2\},\end{aligned} \tag{9.23}$$

where $(\alpha_i, \delta_i)(i = 1, 2, 3)$ are the equatorial coordinates of the reference stars.

The value of dependences lies in their ease of measurement and wide domain of applicability. Although based on the concept of standard coordinates, they may be used with equatorial and rectangular equatorial coordinates. Because of this, they form a useful laboratory method of rapidly determining the positions of celestial objects. However, the method outlined in this section can be no more accurate than the measurement of the dependences. Clearly least squares methods could be employed though the advantage of rapid evaluation would be lost. The method of dependences would have no application in accurate astrometric work and other methods would be used. Indeed it is clear from the above analysis that the need for standard coordinates in astrometry is becoming minimal.

9.2 Standard Coordinates with Respect to Errors of Measurement

In Chapter IV a formal definition of standard coordinates was given. The formal definition presumed a perfect standard refractor both in regard to optics and mounting and that the conditions imposed on the choice of (ξ, η) coordinate axes could be realised in practice. However, it is clear that neither telescope nor measuring machine can be constructed with perfect accuracy. In the introduction to this chapter it was stated (see eqns. (9.1 or 2)) that the errors could be made linear in the coordinates. Hence by a simple analysis the principal part of the errors of measurement and telescope construction could be removed. Indeed, if non-linear (i.e., higher order) terms were introduced, an extension of the same analysis could be used to remove them as well. It will be shown in this section that the major contribution to the errors is linear. The true standard coordinates will be denoted by (ξ, η) and the measured coordinates by (p, q). In each case the units will be that of the focal length of the standard refractor.

9.2.1 Errors of measurement. A conventional type of astronomical plate measuring machine has a plate holder which can be moved by means of screws in two directions at right angles. The measured coordinates (p, q) may not correspond with standard coordinates (ξ, η). However, in this situation the errors are exactly linear in the coordinates. There are three ways in which the measured coordinates may not correspond with the standard coordinates—the origin of each system may be different, the axes of one system may be rotated with respect to the ones of the other, the measured axes may not be at right angles. These situations are illustrated in Fig. 9.2. Fig. 9.2(i) illustrates the situation in which the measured and standard coordinates have a different origin, Fig. 9.2(ii) illustrates rotation of the axes of the measured coordinates with respect to the axes of the standard coordinates while Fig. 9.2(iii) illustrates non-perpendicularity of the axes of the measured coordinates. In any given situation all three such errors may be present but it is useful to separate them in order to more easily determine their effects. Let S be a star whose standard coordinates are ξ, η and whose measured coordinates are p, q. Clearly in the situation of Fig. 9.2(i)

$$\begin{aligned} p &= \xi - c \\ q &= \eta - f \end{aligned} \tag{9.24}$$

where the standard coordinates of O′ are (c, f). The errors are then

$$\begin{aligned} \xi - p &= c, \\ \eta - q &= f, \end{aligned} \tag{9.25}$$

which is a special case of linearity in that the coordinates

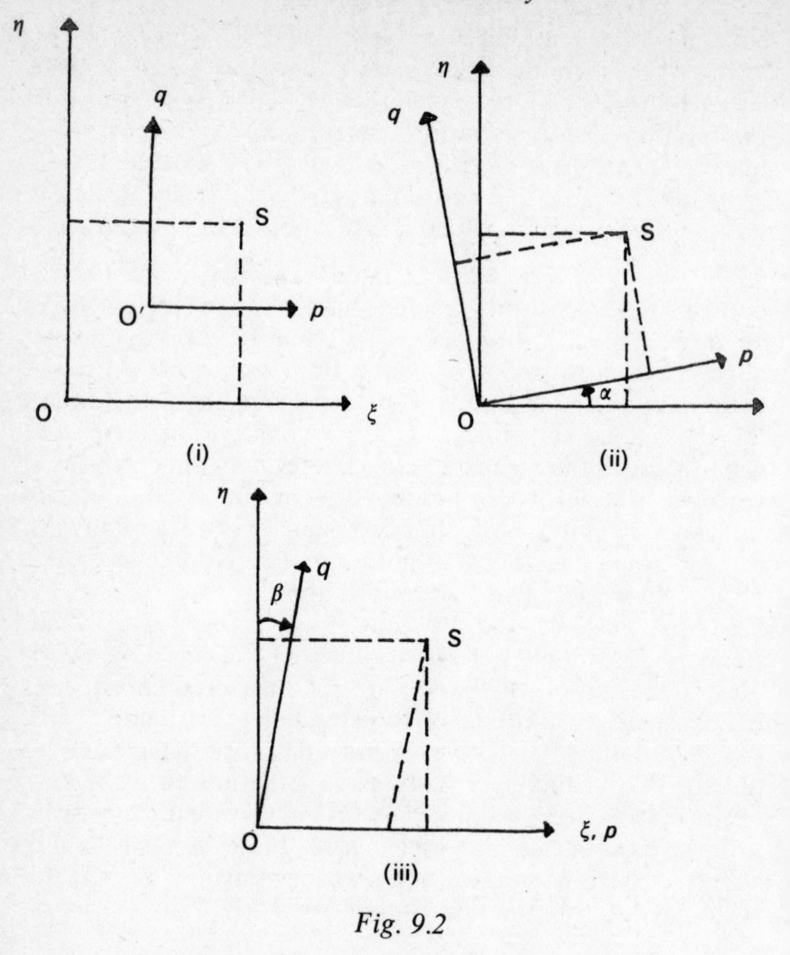

Fig. 9.2

differ by a constant. In the situation of Fig. 9.2(ii)

$$p = \xi \cos \alpha + \eta \sin \alpha$$
$$q = -\xi \sin \alpha + \eta \cos \alpha \tag{9.26}$$

where α is the angle through which the ξ, η axes must be rotated to bring them in alignment with the p, q axes. The errors are then

$$\xi - p = \xi(1 - \cos \alpha) - \eta \sin \alpha$$
$$\eta - q = \xi \sin \alpha + \eta(1 - \cos \alpha) \tag{9.27}$$

It is clear from the form of eqns. (9.27) that the errors are linear in the standard coordinates since $\sin\alpha$ and $(1 - \cos\alpha)$ are constants. In the situation of Fig. 9.2(iii)

$$\begin{aligned} p &= \xi - q\sin\beta = \xi - \eta\tan\beta, \\ q &= \eta\sec\beta, \end{aligned} \tag{9.28}$$

so that the errors are

$$\begin{aligned} \xi - p &= \eta\tan\beta \\ \eta - q &= \eta(1 - \sec\beta) \end{aligned} \tag{9.29}$$

from which it is seen that the errors are again linear in the standard coordinates since $\tan\beta$ and $(1 - \sec\beta)$ are constants. It is usual to construct measuring equipment so that α, β, are as small as possible, since from the form of eqns. (9.27, 29) interchangeability of ξ, η, with p, q, is only possible for α, β small. If α, β are made very small either standard or measured coordinates can be used to express the error between measured and standard coordinates.

9.2.2 Errors of the telescope. It is unlikely that the construction of a real telescope will be ideal and therefore its focal length may differ from that of the standard refractor, the optical axis may not pass through the origin of the measured coordinates, and the plate holder may not hold the photographic plate exactly perpendicular to the optical axis. The error in focal length gives rise to an error which is linear in the coordinates but the remaining errors are not linear, though of course, the principal contribution to the error comes from the linear term.

Suppose that the standard refractor has focal length f, whereas the telescope used for observation has focal length f'; $f - f'$ may be regarded as the error in the focal length. Standard coordinates referred to f' as the unit of length will therefore differ from standard coordinates referred to f as the unit of length. Suppose u is some measurement on a photographic plate. In units of f, f' the scaled measurements are u/f, u/f'. The error in the scaled measurement is therefore

$$\frac{u}{f} - \frac{u}{f'} = \frac{u}{f}\left(1 - \frac{f}{f'}\right) \tag{9.30}$$

i.e. the error is a simple scaling of the true value u/f. The errors of construction affecting focal length are therefore linear in the coordinates. In point of fact the focal length of the telescope used does not have to be identical with that of the standard refractor since determination of the plate constants will automatically allow for the scaling.

The photographic plate may not be perpendicular to the optical axis of the telescope. If the construction of the telescope has been carefully executed such errors may be second order in the sense used in the case of centering error (see below) and can be usually neglected. In Fig. 9.3 the centre of the object glass of the telescope is at C, the origin of standard coordinates at O. The plate should be perpendicular to OC, i.e. containing the line OS, but has been tilted through an angle τ from the perpendicular so that it contains the line OS′. A star which

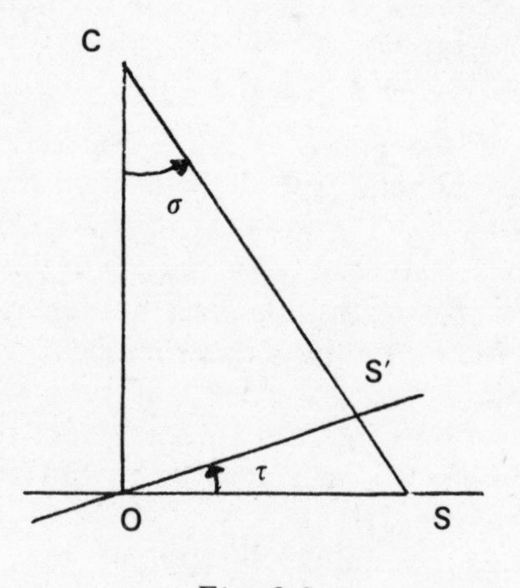

Fig. 9.3

should have been recorded at S is now recorded at S′. Let l = OS and l' = OS′. l' is the actual measured position of the star while l would be the true measured position. From the geometry of the triangle OSS′

$$\frac{l}{\sin\{180° - (90° - \sigma + \tau)\}} = \frac{l'}{\sin(90° - \sigma)}$$

i.e.

$$\frac{l}{l'} = \frac{\cos(\sigma - \tau)}{\cos\sigma} = \cos\tau + \sin\tau\tan\sigma \simeq 1 + \tau\sigma, \qquad (9.31)$$

if τ and σ are small angles. Hence to the same order

$$\frac{l'}{l} = 1 - \tau\sigma. \qquad (9.32)$$

Since OC is the focal length of the telescope and is treated as being of unit radius

$$l = \text{OC}\tan\sigma = \tan\sigma = \sigma. \qquad (9.33)$$

Using eqn. (9.33) in eqn. (9.32)

$$l' - l = -\tau\sigma^2. \qquad (9.34)$$

Clearly if τ is small and the angular coverage of the plate is reasonably small the error caused by a tilt of the plate to the optical axis is second order and may be neglected.

Because of imperfections of construction, the optical axis of a telescope may not pass through that point on the sky whose equatorial coordinates are (α_0, δ_0) on which the telescope is nominally centred. As a consequence the optical axis may not pass through the origin of standard coordinates for the nominal plate centre. Suppose that the centering error is small and of amount $d\alpha_0$ in α_0 and $d\delta_0$ in δ_0. For the purposes of this chapter it will be shown that such a centering error gives rise to a principal contribution which is linear in the coordinates. Only the η coordinate will be considered, the treatment of the ξ coordinate being similar. From Chapter IV, eqn. (4.19) the expression for η in terms of rectangular equatorial coordinates is

$$\begin{aligned}\eta &= \frac{z}{(x_0^2 + y_0^2)^{\frac{1}{2}}(xx_0 + yy_0 + zz_0)} - \frac{z_0}{(x_0^2 + y_0^2)^{\frac{1}{2}}} \\ &= f(x_0, y_0, z_0),\end{aligned} \qquad (4.19)$$

where x, y, z are the rectangular coordinates of the object whose standard coordinates are (ξ, η) and (x_0, y_0, z_0) are the rectangular equivalents of (α_0, δ_0). If errors $\mathrm{d}x_0, \mathrm{d}y_0, \mathrm{d}z_0$ are made in the coordinates (x_0, y_0, z_0) then the corresponding error in η is given to the first order by

$$\begin{aligned}
\mathrm{d}\eta = \eta - q &= \frac{\partial f}{\partial x_0}\mathrm{d}x_0 + \frac{\partial f}{\partial y_0}\mathrm{d}y_0 + \frac{\partial f}{\partial z_0}\mathrm{d}z_0 \\
&= -z\left\{\frac{x_0\,\mathrm{d}x_0 + y_0\,\mathrm{d}y_0}{(x_0^2 + y_0^2)^{\frac{3}{2}}(xx_0 + yy_0 + zz_0)}\right. \\
&\qquad \left. + \frac{x\,\mathrm{d}x_0 + y\,\mathrm{d}y_0 + z\,\mathrm{d}z_0}{(x_0^2 + y_0^2)^{\frac{1}{2}}(xx_0 + yy_0 + zz_0)^2}\right\} \\
&\qquad - \frac{\mathrm{d}z_0}{(x_0^2 + y_0^2)^{\frac{1}{2}}} + \frac{z_0(x_0\,\mathrm{d}x_0 + y_0\,\mathrm{d}y_0)}{(x_0^2 + y_0^2)^{\frac{3}{2}}} \\
&= A\eta + B, \qquad (9.35a)
\end{aligned}$$

where

$$\begin{aligned}
A &= -\left\{\frac{x_0\,\mathrm{d}x_0 + y_0\,\mathrm{d}y_0}{(x_0^2 + y_0^2)}\right. \\
&\qquad \left. + \frac{x\,\mathrm{d}x_0 + y\,\mathrm{d}y_0 + z\,\mathrm{d}z_0}{(xx_0 + yy_0 + zz_0)}\right\}, \qquad (9.35b) \\
B &= -\frac{1}{(x_0^2 + y_0^2)^{\frac{1}{2}}} \\
&\qquad \times \left\{\frac{z_0(x\,\mathrm{d}x_0 + y\,\mathrm{d}y_0 + z\,\mathrm{d}z_0)}{(xx_0 + yy_0 + zz_0)} + \mathrm{d}z_0\right\}. \qquad (9.35c)
\end{aligned}$$

From the form of eqns. (9.35b,c) it is clear that A,B depend on the rectangular equatorial coordinates (x, y, z) of the object measured. The rectangular equatorial coordinates of any object on a plate may be expressed in terms of the rectangular equatorial coordinates of the nominal plate centre, i.e.

$$x = x_0 + \Delta x, \qquad y = y_0 + \Delta y, \qquad z = z_0 + \Delta z. \qquad (9.36)$$

Then

$$\frac{x\,\mathrm{d}x_0 + y\,\mathrm{d}y_0 + z\,\mathrm{d}z_0}{(xx_0 + yy_0 + zz_0)}$$

$$= \frac{(x_0\,\mathrm{d}x_0 + y_0\,\mathrm{d}y_0 + z_0\,\mathrm{d}z_0) + (\Delta x\,\mathrm{d}x_0 + \Delta y\,\mathrm{d}y_0 + \Delta z\,\mathrm{d}z_0)}{1 + x_0\,\Delta x + y_0\,\Delta y + z_0\,\Delta z}$$

$$\simeq x_0\,\mathrm{d}x_0 + y_0\,\mathrm{d}y_0 + z_0\,\mathrm{d}z_0, \tag{9.37}$$

provided that products $\Delta x\,\mathrm{d}x_0$, etc., may be regarded as small in comparison with other quantities. To this order of approximation only can A, B be regarded as constants. If the angular extent of the plate is large an approximation to higher order must be employed. In this respect centering error differs from errors of misalignment in the measuring machine and it is therefore important that the telescope be adjusted to make the centering error as small as possible. The neglected terms in the centering error could, in principle, be larger than the error of tilt.

9.2.3 The effect of refraction and annual aberration. The effects of refraction and annual aberration may be treated in a similar fashion as with centering error. Again illustrating the argument by consideration of the standard coordinate η, eqn. (4.19) may be expressed in the form

$$\eta = f(x, y, z, x_0, y_0, z_0). \tag{4.19}$$

Then, assuming that the displacements caused by refraction or aberration may be represented by a first order Taylor series,

$$\begin{aligned} \mathrm{d}\eta = \eta - q &= \frac{\partial f}{\partial x}\mathrm{d}x + \frac{\partial f}{\partial y}\mathrm{d}y + \frac{\partial f}{\partial z}\mathrm{d}z + \frac{\partial f}{\partial x_0}\mathrm{d}x_0 \\ &\quad + \frac{\partial f}{\partial y_0}\mathrm{d}y_0 + \frac{\partial f}{\partial z_0}\mathrm{d}z_0 \\ &= \frac{\mathrm{d}z}{(x_0^2 + y_0^2)^{\frac{1}{2}}(xx_0 + yy_0 + zz_0)} \\ &\quad - \frac{z(x_0\,\mathrm{d}x + y_0\,\mathrm{d}y + z_0\,\mathrm{d}z)}{(x_0^2 + y_0^2)^{\frac{1}{2}}(x_0x + yy_0 + zz_0)^2} \\ &\quad + A\eta + B, \end{aligned} \tag{9.38}$$

using eqn. (9.35). Re-arrangement of eqn. (9.38) gives

$$d\eta = -\frac{2z_0}{(x_0^2 + y_0^2)^{\frac{1}{2}}} \times (x_0\, dx_0 + y_0\, dy_0 + z_0\, dz_0) - C\eta, \qquad (9.39a)$$

where

$$C = Dx_0\, dx_0 + Dy_0\, dy_0 + 2z_0\, dz_0 \qquad (9.39b)$$

and

$$D = \{2 + (x_0^2 + y_0^2)^{-1}\},$$

provided that products of the type $\Delta x\, dx$ may be neglected and where setting $(dx + dx_0) = 2dx_0$ and $(dz - dz_0) = 0$ introduces an error of order higher than the first. Again the error is linear in the coordinates. dx_0, dy_0, dz_0 may be evaluated using eqns. (6.84, 6.85, 6.86) in the case of refraction and eqns. (6.96, 6.97, 6.98) in the case of aberration. A similar result may be derived in the case of ξ.

Since the errors arising from misalignment of the plate in the measuring machine, misadjustment of the telescope and the displacements caused by refraction and annual aberration can all be expressed in linear form to the first order the total error may be written in the form

$$\begin{aligned} \xi - p &= a\xi + b\eta + c = ap + bq + c \\ \eta - q &= d\xi + e\eta + f = dp + eq + f \end{aligned} \qquad (9.1)$$

since the second pair of equations on the right of eqn. (9.1) only introduces errors of the second order.

By deriving the plate constants a, b, c, d, e, f it is clear that mechanical errors and the effects of refraction and annual aberration may be removed in one operation and the standard coordinates for any object on the plate can then be derived. In practice, standard coordinates are not now essential since direct solutions for rectangular equatorial coordinates can be carried out. The introduction of dependences is simplified through the use of standard coordinates and dependences give a useful rapid means of determining the rectangular equatorial coordinates from direct measurement when the rectangular equatorial coordinates of three reference stars are known.

Chapter X

Orbital Motion

The gravitational interaction of two or more celestial objects is of great importance in astronomy. The gravitational interaction of a planet, the Sun and the rest of the solar system determines the path in space, or orbit, taken by the planet. Two or more stars may form a stable configuration under the influence of their mutual gravitational attractions to give a binary or multiple star. From a study of planetary motions or the motions of multiple stars it is possible to derive masses for the interacting objects. It is only from the study of such orbital motion that direct determinations of mass can be made. Such measurements are of fundamental importance in calibrating other astrophysical indicators of mass.

Unfortunately the gravitational interaction of more than two bodies cannot be treated simply, and for the purposes of this book the gravitational interaction between two bodies only will be discussed. Fortunately in many practical situations the major interaction is between two bodies and interactions with other bodies can be treated as a perturbation. The methods used in treating more complex situations can be found in standard texts on Celestial Mechanics. The aim of the present chapter is to describe the nature of the bound orbital motions of two bodies under the influence of their mutual gravitational attraction. This theory will be applied in Chapter XI to the planetary system from the point of view of the determination of the position of the planets on the sky at any instant. The determination of the constants specifying the orbit will also be discussed. In Chapter XII, the theory of this chapter will be applied to the measurement of the orbits of binary stars with a view to the determination of stellar masses.

In this and the two subsequent chapters it will be assumed that celestial objects have spherical symmetry and are composed of concentric spherical shells in which the density is constant. In such circumstances the objects may be replaced by point masses at their geometrical centres.

10.1 Newton's Law of Gravitation

Newton proposed that the gravitational interaction between two bodies of masses m_1, m_2 separated by a distance d was proportional to the product of the masses and inversely proportional to the square of their separation. He showed that such a law of force could explain the empirical laws determined by Kepler from observation of the planets. However, Newton could not give an account of the nature of gravitation (which is still not known) and the idea of action at a distance was considered obscurantist by many of his contemporaries. The prediction, on the basis of Newton's law, by Halley of the return of the comet now named after him established Newton's description of gravitational interaction as a working hypothesis. Newtonian theory has since been superseded by relativistic theory but for most astronomical purposes Newtonian theory is sufficiently accurate. In mathematical terms the gravitation force $\boldsymbol{F}$ is given by

$$\boldsymbol{F} = -\frac{Gm_1m_2}{d^2}\hat{\boldsymbol{d}} \tag{10.1}$$

where $\hat{\boldsymbol{d}}$ is the unit vector defining the line of centres between the interacting bodies and G is the gravitational constant.

In order to consider the motion of two bodies under their self gravitation, suppose that a body of mass M_* has rectangular coordinates (X, Y, Z) and a body of mass m has rectangular coordinates (x, y, z) where the rectangular coordinates are referred to an inertial frame of reference $\mathrm{C}(x\ y\ z)$ as illustrated in Fig. 10.1. Let the position vector of the mass M_* be $\boldsymbol{R}$ and the position vector of the mass m be $\boldsymbol{r}$ with respect to C. The position vector of m with respect to M_* is $\boldsymbol{d} = \boldsymbol{r} - \boldsymbol{R}$. If the only force acting on the bodies is their

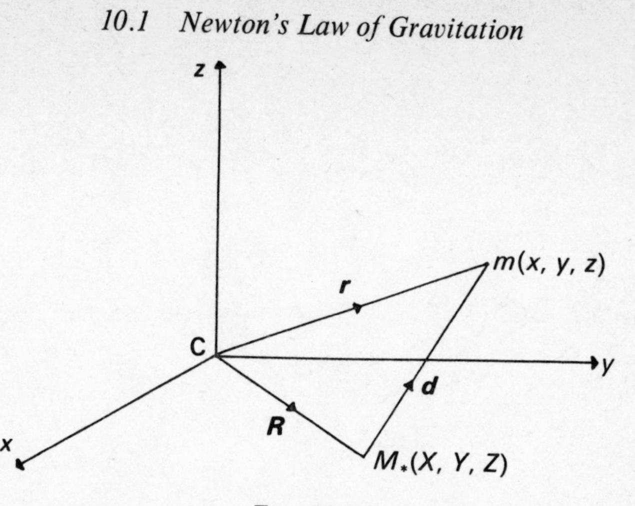

Fig. 10.1

mutual gravitational interaction then using Newton's law, their equations of motion are

$$M_* \frac{\mathrm{d}^2\boldsymbol{R}}{\mathrm{d}t^2} = \frac{GmM_*}{d^3}\boldsymbol{d}, \qquad m\frac{\mathrm{d}^2\boldsymbol{r}}{\mathrm{d}t^2} = -\frac{GM_*m}{d^3}\boldsymbol{d}. \qquad (10.2)$$

Expressing eqn. (10.2) in terms of rectangular components

$$\begin{aligned} M_*\ddot{X} &= \frac{GmM_*}{d^3}(x - X), & m\ddot{x} &= -\frac{GM_*m}{d^3}(x - X), \\ M_*\ddot{Y} &= \frac{GmM_*}{d^3}(y - Y), & m\ddot{y} &= -\frac{GM_*m}{d^3}(y - Y), \qquad (10.3) \\ M_*\ddot{Z} &= \frac{GmM_*}{d^3}(z - Z), & m\ddot{z} &= -\frac{GM_*m}{d^3}(z - Z), \end{aligned}$$

where a dot placed above a parameter denotes its derivative with respect to time. Subtracting the corresponding components of acceleration to find the motion of the mass m with respect to the mass M_*

$$\begin{aligned} \ddot{x} - \ddot{X} &= -G(M_* + m)(x - X), \\ \ddot{y} - \ddot{Y} &= -G(M_* + m)(y - Y), \qquad (10.4) \\ \ddot{z} - \ddot{Z} &= -G(M_* + m)(z - Z). \end{aligned}$$

If (ξ, η, ζ) are the rectangular coordinates of the mass m with respect to the mass M_* then

$$\xi = x - X, \qquad \eta = y - Y, \qquad \zeta = z - Z$$

and $\hfill (10.5)$

$$d^2 = (x - X)^2 + (y - Y)^2 + (z - Z)^2 = \xi^2 + \eta^2 + \zeta^2.$$

Eqns. (10.4) then become

$$\ddot{\xi} + \frac{\mu}{d^3}\xi = 0, \qquad \ddot{\eta} + \frac{\mu}{d^3}\eta = 0, \qquad \ddot{\zeta} + \frac{\mu}{d^3}\zeta = 0, \tag{10.6}$$

where

$$\mu = G(M_* + m).$$

Since

$$\zeta\ddot{\eta} - \eta\ddot{\zeta} = -\frac{\mu}{d^3}(\zeta\eta - \eta\zeta) = 0 = \frac{d}{dt}(\zeta\dot{\eta} - \eta\dot{\zeta}),$$

$$\zeta\dot{\eta} - \eta\dot{\zeta} = \text{constant} = A. \tag{10.7a}$$

Similarly it may be shown that

$$\xi\dot{\zeta} - \zeta\dot{\xi} = \text{constant} = B, \tag{10.7b}$$

$$\eta\dot{\xi} - \xi\dot{\eta} = \text{constant} = C. \tag{10.7c}$$

From eqns. (10.7) it follows that

$$A\xi + B\eta + C\zeta = 0. \tag{10.8}$$

Eqn. (10.8) means that the motion of the mass m relative to the mass M_* takes place in a plane. A full three dimensional treatment of motion under gravity may be avoided and the motion considered only in the plane defined by eqn. (10.8). When the motion of a planet on the sky is considered in Chapter XI, the orientation of this plane in space will need to be considered. However, the mathematical formulation of motion under gravity can be simplified by considering only motion in a single plane. If coordinate axes (x, y) are established with origin at the mass M_* in the plane defined by the motion of the mass m relative to the mass M_* the two dimensional equations equivalent to eqns. (10.4) are

$$\ddot{x} + \frac{\mu}{d^3}x = 0, \qquad \ddot{y} + \frac{\mu}{d^3}y = 0,$$

where (10.4a)

$$d^2 = x^2 + y^2.$$

Eqns. (10.4a) must be integrated simultaneously with respect to time in order to find the path in space.

10.2 Orbits for Motion Under the Influence of Gravity

The gravitational force between two mass points M_* and m is directed along the line joining the points. The treatment of gravitational forces is therefore a special case of the more general theory of central forces. It is convenient to transform from cartesian to polar coordinates. In Fig. 10.2 an initial frame of reference $M_*(x, y)$ is centred at a mass point M_*.

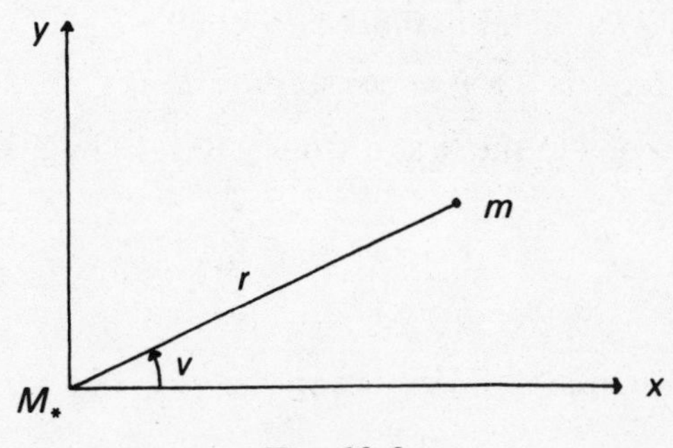

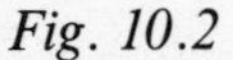
Fig. 10.2

The position of a mass point m may be specified either by its cartesian coordinates (x, y) or its polar coordinates (r, v), where

$$x = r\cos v, \qquad y = r\sin v. \tag{10.9}$$

v is called the *true anomaly* in case of astronomical orbits. Differentiation of eqns. (10.9) with respect to time gives

$$\begin{aligned} \dot{x} &= \dot{r}\cos v - r\sin v\dot{v}, \\ \dot{y} &= \dot{r}\sin v + r\cos v\dot{v} \end{aligned} \tag{10.10a}$$

and differentiation of eqns. (10.10a) with respect to time gives

$$\begin{aligned} \ddot{x} &= \ddot{r}\cos v - 2\dot{r}\sin v\dot{v} - r\cos v\dot{v}^2 - r\sin v\ddot{v}, \\ \ddot{y} &= \ddot{r}\sin v + 2\dot{r}\cos v\dot{v} - r\sin v\dot{v}^2 + r\cos v\ddot{v}. \end{aligned} \tag{10.10b}$$

From eqns. (10.10a) $\dot{r}$ and $r\dot{v}$ may be determined giving

$$\dot{r} = \dot{x}\cos v + y\sin v, \qquad r\dot{v} = -\dot{x}\sin v + y\cos v, \tag{10.11a}$$

where $\dot{r}$ is the magnitude of that component of velocity parallel to the position vector $\boldsymbol{r}$ of the mass point m with respect to the mass point M_* and $r\dot{v}$ is the magnitude of that component of velocity perpendicular to $\boldsymbol{r}$. Similarly the accelerations parallel and perpendicular to $\boldsymbol{r}$ are

$$\begin{aligned} \ddot{r} - r\dot{v}^2 &= \ddot{x}\cos v + \ddot{y}\sin v = -\frac{\mu}{r^2} \\ 2\dot{r}\dot{v} + r\ddot{v} &= -\ddot{x}\sin v + \ddot{y}\cos v = 0 \end{aligned} \tag{10.11b}$$

where use has been made of eqns. (10.4a). The second of eqns. (10.11b) may be integrated directly to give,

$$r^2\dot{v} = \text{constant} = h. \tag{10.12}$$

Elimination of $\dot{v}$ in the first of eqns. (10.11b) by means of eqn. (10.12) gives

$$\ddot{r} - \frac{h^2}{r^3} = -\frac{\mu}{r^2} \tag{10.13}$$

which is the differential equation defining the variation of r. Setting

$$u = 1/r, \tag{10.14}$$

and replacing $\mathrm{d}/\mathrm{d}t$ by $\dot{v}\,\mathrm{d}/dv$ eqn. (10.13) becomes

$$\frac{\mathrm{d}^2u}{\mathrm{d}v^2} + u = \frac{\mu}{h^2}. \tag{10.15}$$

Eqn. (10.15) has solution

$$u = A + B\cos v, \tag{10.16}$$

where A, B are the constants of integration. The constant A

may be found from eqns. (10.15, 10.16) and r is then given by

$$r = \frac{h^2/\mu}{1 + (h^2B/\mu)\cos v} = \frac{h^2/\mu}{1 + e\cos v}. \tag{10.17}$$

Eqn. (10.17) is the general equation of a conic with respect to an origin at a focus and h^2B/μ is identified with the eccentricity e (i.e. $B = \mu e/h^2$). If $e > 1$, the orbit is a hyperbola. Such motion is obtained if the mass point m is projected from infinity with a finite (non-zero) velocity. If $e = 1$, the orbit would be a parabola corresponding to motion from rest from infinity under the mutual attraction. If $e < 1$, the orbit is an ellipse and corresponds to bound motion of the mass point m about the mass point M_*. It is motion in closed elliptical orbits which is the concern of this chapter.

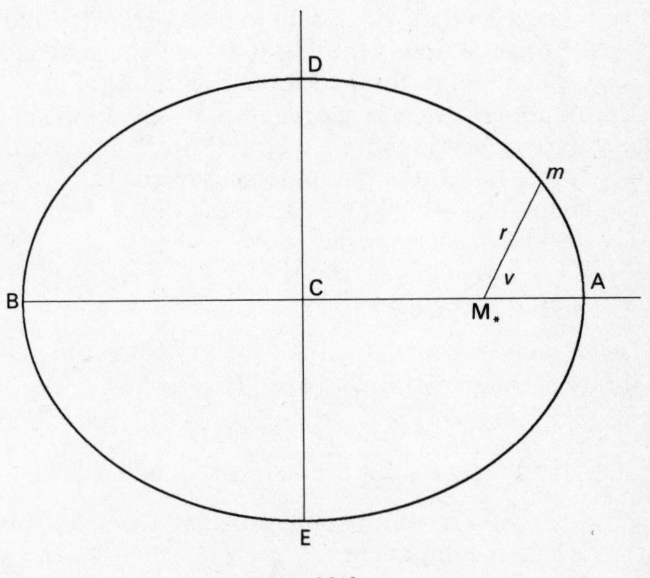

Fig. 10.3

An elliptical orbit is illustrated in Fig. 10.3. The mass point M_* is at the origin of coordinates. The line ACB is the major axis of the ellipse and the line DCE is the minor axis. C is the centre of the ellipse and M_* is a focus. If the semi-major axis of an ellipse has length a, the semi-minor axis length b and the eccentricity is e then

$$b^2 = a^2(1 - e^2),$$

$$CM_* = ae, \qquad M_*A = a(1 - e), \qquad M_*B = a(1 + e), \tag{10.18}$$

and

$$h^2 = \mu a(1 - e^2),$$

making use of eqn. (10.17).

Then r may be written in the form,

$$r = \frac{h^2/\mu}{1 + e \cos v} = \frac{a(1 - e^2)}{1 + e \cos v} \tag{10.19}$$

In planetary studies, M_* is identified with the centre of the Sun and m with the centre of a planet. v is the angular separation of m from A at the focus of the orbit. When the planet is nearest the Sun, i.e. at A, it is said to be at *perihelion* and when furthest from the Sun, i.e. at B, it is said to be at *aphelion*. The equivalent points for the Moon in its elliptical orbit about the Earth are *perigee* and *apogee* respectively. In the case of a binary star one star is chosen to be at the origin of coordinates and the orbit of the other about it considered. The equivalents of perihelion and aphelion in the case of a binary star are *periastron* and *apastron* respectively.

10.3 Orbital Kinematics

The components of velocity of the mass point m may now be derived. Differentiation of eqn. (10.19) gives

$$\dot{r} = \frac{h^2/\mu}{(1 + e \cos v)^2} e \sin v \frac{dv}{dt} = \frac{\mu e}{h} \sin v, \tag{10.20}$$

using eqns. (10.12, 10.19). Suppose that the mass point m has a velocity component V_1 perpendicular to the radius vector M_*m and a component V_2 perpendicular to the major axis AB. The component of V_2 parallel to the radius vector is then $V_2 \sin v$. This component must clearly be the velocity $\dot{r}$. Hence

$$\dot{r} = V_2 \sin v = \frac{\mu e}{h} \sin v,$$

i.e.

$$V_2 = e\frac{\mu}{h} = eh/p, \tag{10.21}$$

if use is made of eqn. (10.20) and $p = h^2/\mu = a(1 - e^2)$. Again the component of velocity perpendicular to the radius vector is V_1 where

$$V_1 = r\dot{v} - V_2 \cos v = \frac{h}{r} - \frac{\mu e}{h} \cos v = \frac{\mu}{h} = h/p, \tag{10.22}$$

if use is made of eqn. (10.19). The motion of m about M_* can therefore be resolved into components μ/h perpendicular to the radius vector $M_* m$ and $e\mu/h$ perpendicular to the major axis AB. The results expressed in eqns. (10.21, 10.22) were used when considering annual aberration in Chapter VI, section (6.3.1).

The total velocity W of the mass point m in an elliptical orbit is given by,

$$\begin{aligned} W^2 = \dot{r}^2 + r^2\dot{v}^2 &= \frac{\mu^2 e^2}{h^2} \sin^2 v + \frac{\mu^2}{h^2}(1 + 2e \cos v \\ &\quad + e^2 \cos v) \\ &= \frac{\mu^2}{h^2}(1 + 2e \cos v + e^2) \\ &= \mu\left(\frac{2}{r} - \frac{1}{a}\right), \end{aligned} \tag{10.23}$$

using eqn. (10.19). It is clear from eqn. (10.23) that

$$\text{if } v = 0; \qquad |W| = W_1 = \frac{\mu}{h}(1 + e)$$

$$\text{if } v = 180; \qquad |W| = W_2 = \frac{\mu}{h}(1 - e) \tag{10.24}$$

and

$$W_1 W_2 = \frac{\mu^2}{h^2}(1 - e^2) = \frac{\mu}{a}.$$

10.4 Kepler's Laws of Planetary Motion

From the analysis of the observations of Tycho Brahe, Kepler derived three empirical laws describing planetary motion.

1. The orbit of a planet about the Sun is an ellipse with the Sun at one focus.
2. The radius vector connecting the Sun and planet sweeps out equal areas in equal times.
3. The square of the orbital period is proportional to the cube of the length of the semi-major axis.

From Newton's law of gravitation it was shown (eqn. (10.17)) that the path of a mass point m moving under the gravitational attraction of a mass M_* was in general a conic with M_* at its focus—the only closed orbit of this type is an ellipse so agreeing with the first of Kepler's Laws. It was also derived that $r^2\dot{v}$ was a constant (eqn. (10.12)). If the mass point m moves through an angle dv in time dt, it sweeps out an area $\frac{1}{2}r^2\,dv$ in time dt. Hence Kepler's second law may also be derived from Newton's law of gravitation. If the mass point m takes a time T to move once around its elliptical orbit the constant h may be evaluated from

$$h = \frac{2\pi ab}{T} = \frac{2\pi a^2(1 - e^2)^{\frac{1}{2}}}{T}, \tag{10.25}$$

since $2\pi ab$ is the area swept out in time T. Defining n by

$$n = \frac{2\pi}{T}, \tag{10.26}$$

then

$$n = \frac{h}{a^2(1 - e^2)^{\frac{1}{2}}} = \frac{\mu^{\frac{1}{2}}}{a^{\frac{3}{2}}},$$

or

$$n^2a^3 = \mu. \tag{10.27}$$

Since μ is a constant, eqn. (10.27) expresses Kepler's Third Law. Kepler's Laws can be applied to any orbital motion under gravitational forces and are not restricted to planetary orbits. Their mathematical expression is given by eqns. (10.17, 12, 27).

10.5 Kepler's Equation

Although the formulae derived in section 10.3 are those which are usually quoted for the position of a mass point m in an elliptical orbit, these formulae are not useful in deriving values for r and v. The problem of determining r, v is not approached through these formulae but in a rather different way. Three parameters, called the *dynamical elements*, are used to specify the orbit. The dynamical elements are the period T, the eccentricity e and τ the time of perihelion (or perigee or periastron) passage. The period T—and hence n (see eqn. (10.26))—for the orbit can be determined with great accuracy. A simple geometrical construction may then be used to relate the mean and actual motions. In Fig. 10.4 the

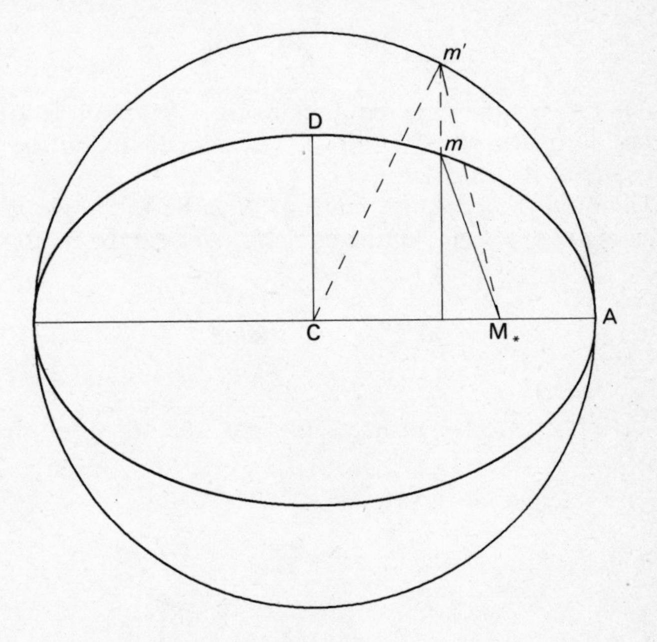

Fig. 10.4

elliptical orbit of a mass point m about the mass point M_* is AmD. The centre of the elliptical orbit is at C. A circle of centre C and radius CA $= a$ is drawn so that the plane of the circle is inclined at an angle i to the plane of the ellipse.

The angle i is chosen so that if the circle were projected onto the plane of the ellipse, the projection of the circle would be identical with the orbital ellipse, i.e.

$$\cos i = \frac{b}{a} = (1 - e^2)^{\frac{1}{2}}. \qquad (10.28)$$

The point m' on the circle, if projected onto the plane of the ellipse, would coincide with the mass point m. The area of the sector of the ellipse M_*Am is given by

$$\Delta M_* Am = \frac{\pi ab(t - \tau)}{T} = \frac{1}{2}nab(t - \tau) = \frac{1}{2}Mab,$$

i.e.

$$M = \frac{2\pi}{T}(t - \tau), \qquad (10.29)$$

where $t - \tau$ is the time required for the mass point m to move from A to m and $M = n(t - \tau)$ is defined to be the *mean anomaly* of the mass point m.

Defining the *eccentric anomaly* E to be the angle $m'\hat{C}M_*$ the area of the sector of the circle M_*Am' can be written

$$\begin{aligned}\Delta M_* Am' &= \Delta CAm' - \Delta CM_* m' \\ &= \tfrac{1}{2}a^2 E - \tfrac{1}{2}ae \, . \, a \sin E \qquad (10.30) \\ &= \tfrac{1}{2}a^2(E - e \sin E).\end{aligned}$$

But the area M_*Am' projects into the area M_*Am so that

$$\frac{b}{a}\Delta M_* Am' = \Delta M_* Am,$$

or

$$\tfrac{1}{2}Mab = \tfrac{1}{2}ab(E - e \sin E),$$

i.e.

$$M = E - e \sin E. \qquad (10.31)$$

Eqn. (10.31) is *Kepler's Equation*. Since the period T is known, n and so M can be calculated. If the eccentricity e is also known eqn. (10.31) can be solved for E. For many cases of orbital motion e is small. It is therefore usual to solve eqn. (10.31) iteratively for E. A zero order approximation for E may be found by setting

$$E_0 = M \tag{10.32a}$$

and seeking further approximations of the form

$$E_{i+1} = E_i + \Delta E_{i+1}. \tag{10.32b}$$

Then eqn. (10.31) has the form

$$\Delta E_{i+1} = \frac{(M - E_i) + e \sin E_i}{1 - e \cos E_i}, \tag{10.33}$$

where ΔE_{i+1} has been assumed to be a small angle. If e is large, eqn. (10.31) may be used to tabulate M for a range of values of E. A value of E can then be interpolated, using the tabular values of M so obtained, for the required value of M and eqn. (10.33) used to give a correction. The calculation required for the solution of Kepler's equation is readily performed by an automatic digital computer. Convergence is rapid for small values of e and two iterations are normally sufficient. When e is large, the usual numerical procedures for improving convergences can be used since it is inconvenient to use tabulation when automatic computers are used.

Having determined the eccentric anomaly E, the position r, v of the mass m in its orbit must be determined. Again using Fig. 10.4 it is clear that,

$$\frac{b}{a}(a \sin E) = b \sin E = r \sin (180° - v)$$

$$= r \sin v, \tag{10.34}$$

and

$$ae = a \cos E + r \cos (180° - v),$$

or

$$r \cos v = a (\cos E - e). \tag{10.35}$$

Combination of eqns. (10.34), (10.35) gives

$$r = (r^2 \sin^2 v + r^2 \cos^2 v)^{\frac{1}{2}} = a(1 - e \cos E), \tag{10.36a}$$

$$\cos v = \frac{\cos E - e}{1 - e \cos E}, \tag{10.36b}$$

from which both r and v may be determined. Therefore, knowing the period T, the time of perihelion passage τ and the eccentricity e, the mean anomaly M can be calculated, the eccentric anomaly can be derived from Kepler's equation and finally eqns. (10.36) used to determine r and v.

Alternatively quantities x, y may be calculated such that

$$\begin{aligned} x &= \cos E - e, \\ y &= (1 - e^2)^{\frac{1}{2}} \sin E, \end{aligned} \tag{10.37}$$

allowing the re-expression of eqns. (10.35, 10.36) in the form

$$\begin{aligned} r \cos v &= ax, \\ r \sin v &= ay. \end{aligned} \tag{10.38}$$

Tabulations of x, y as functions of e, M have been prepared (e.g. Tables of x and y rectangular coordinates, *Astronomical Papers prepared for the use of the American Ephemeris and Nautical Almanac*, Vol. 19, Part 1, 1964 by O. G. Franz and B. F. Mintz). Therefore interpolation in the Table gives x, y for the known values of e and M. r/a and v can then be derived directly. Such tables are of value for desk calculations since they save considerable computation. However, if access to a computing facility is available, a simple program can be devised to handle the entire calculation.

Chapter XI

Planetary Orbits

To a high degree of approximation, the planets comprising the solar system move in elliptical orbits about the Sun. The same is true for satellites in relation to their parent planets. Again, cometary orbits in the solar system are elliptical apart from perturbations which may, of course, be very large if the comet makes close approaches to, for example, Jupiter. It will be assumed for the purposes of this chapter that the planets and other bodies in the solar system move in closed elliptical orbits.

In Chapter X, it was shown how the dynamical elements T, e, τ could be used, through Kepler's Equation, to find the position r, v of an object in an elliptical orbit. In terms of the planetary system, the position of any planet is completely determined with respect to the Sun in terms of r and v. However, the planet is observed from the Earth and the problem is therefore to predict the position of the planet on the sky with respect to an observer at the centre of the Earth. Again, there is the converse problem of determining the elements of the orbit from observed positions of a comet or minor planet. This latter problem is more difficult. In this chapter both these problems will be discussed. The determination of the position of a planet on the sky is a fairly straightforward application of the techniques of spherical astronomy. The determination of the elements of an orbit is a detailed problem in celestial mechanics and the method discussed gives a straightforward derivation of preliminary elements.

In order to visualise the geocentric view of orbital motion about the Sun a simple model of planetary motion will be discussed in order to establish such phenomena as retrograde motion and phases of the planets. The simple model of planetary motion will be that the planets move in circular orbits and that all the orbits lie in the same plane. This discussion will be given in the next section.

11.1 Geocentric Phenomena

Since observation is carried out from the Earth, it is the combined effect of the motion of the Earth and the planet that is observed. While the planets all move around the Sun in the same sense, observation from the Earth can cause reversal of the apparent motion of the planet across the sky. Again, it is the *synodic* period of the planet in its orbit which is readily determined by observation and not its orbital period. Associated with these phenomena is the question of the intensity of the planets' illumination. Although not a positional phenomena some idea of the expected variation in brightness can be obtained from a simple consideration of relative planetary motions.

11.1.1 The determination of the period of a planet. The approximation is made that the planets move around the Sun in circular orbits. Let the radius of the Earth's orbit be a_E and the radius of the planetary orbit be a_P. The situation is illustrated in Fig. 11.1. The Sun is at S. Suppose that at some instant t_1 the Earth is at E_1 and the planet is at conjunction at P_1. Suppose that conjunction next occurs at the instant t_2 when the Earth is at E_2 and the planet is at P_2. Let the Earth's period be T_E and the planet's period be T_P. In the time interval $t_2 - t_1$ the Earth has moved through an angle $2\pi + \phi$ where $\phi = E_2\widehat{S}E_1$. The planet has moved through the angle ϕ in the same time. Then

$$2\pi + \phi = \frac{2\pi}{T_E}(t_2 - t_1) \qquad (11.1a)$$

and

$$\phi = \frac{2\pi}{T_P}(t_2 - t_1). \qquad (11.1b)$$

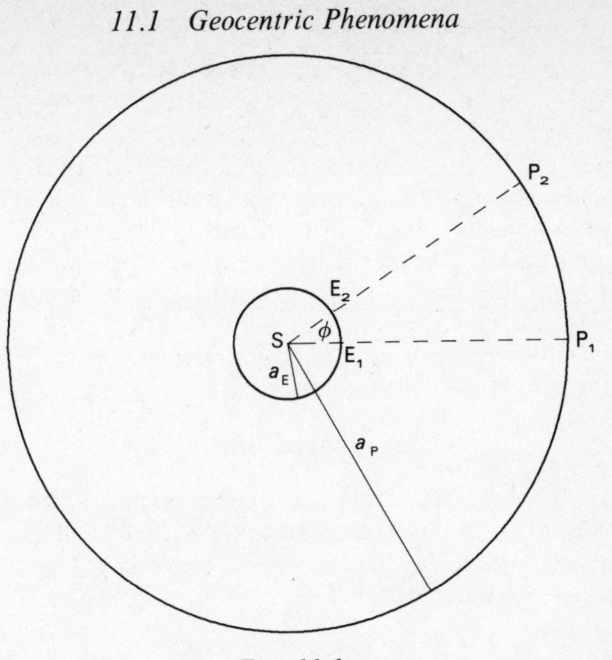

Fig. 11.1

Hence, eliminating ϕ between eqns. (11.1a, b),

$$1 = (t_2 - t_1)\left\{\frac{1}{T_E} - \frac{1}{T_P}\right\}.$$

But $t_2 - t_1 = S$ where S is the *synodic period*, i.e. the time interval between successive conjunctions of the planet with the Earth. Therefore,

$$\frac{1}{S} = \frac{1}{T_E} - \frac{1}{T_P}. \qquad (11.2)$$

Since S can be measured and T_E is known, T_P can be determined from eqn. (11.2). The above discussion forms the basis of the measurement of planetary periods. It should be noted that were the orbit of the planet to lie inside the orbit of the Earth the minus sign on the right hand side of eqn. (11.2) would be replaced by a plus sign.

11.1.2 Direct and retrograde motion. At some instant the relative dispositions of the Sun S, Earth E and planet P are illustrated in Fig. 11.2. The directions EP″, SE′P′ are towards the Vernal Equinox ♈. Then ♈$\widehat{S}$E = l_E, is the heliocentric ecliptic longitude of the Earth and ♈$\widehat{S}$P = l_P is the heliocentric ecliptic longitude of the planet since the orbits of the planets are being assumed to lie in the plane of the ecliptic. The geocentric ecliptic longitude of the planet is P″$\widehat{E}$P = λ. It should be noted that in Chapter X the position of a planet was given in terms of the true anomaly v, i.e. the angular separation of the planet from its position at perihelion. In order to determine v from the ecliptic longitude, the heliocentric ecliptic longitude ω of perihelion must be known. Then

$$\text{ecliptic longitude} = v + \omega \tag{11.3}$$

when the planetary orbit lies in the plane of the ecliptic. Let the radii of the orbits (assumed circular) of the Earth and planet be a_E (=SE), a_P (=SP) respectively and let the geocentric distance of the planet be ρ (=EP). At any instant

$$\begin{aligned} \rho \sin \lambda &= a_P \sin l_P - a_E \sin l_E, \\ \rho \cos \lambda &= a_P \cos l_P - a_E \cos l_E. \end{aligned} \tag{11.4}$$

Differentiation of eqns. (11.4) with respect to time gives

$$\begin{aligned} &\rho \cos \lambda \frac{d\lambda}{dt} + \sin \lambda \frac{d\rho}{dt} \\ &\qquad = a_P \cos l_P \frac{dl_P}{dt} - a_E \cos l_E \frac{dl_E}{dt}, \\ &\rho \sin \lambda \frac{d\lambda}{dt} - \cos \lambda \frac{d\rho}{dt} \\ &\qquad = a_P \sin l_P \frac{dl_P}{dt} - a_E \sin l_E \frac{dl_E}{dt}. \end{aligned} \tag{11.5}$$

From eqns. (11.5) it may be derived that

$$\rho \frac{d\lambda}{dt} = a_P \cos (l_P - \lambda) \frac{dl_P}{dt} - a_E \cos (l_E - \lambda) \frac{dl_E}{dt}, \tag{11.6}$$

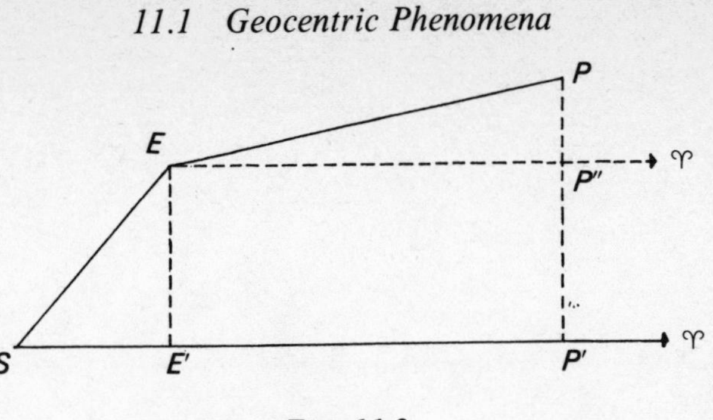

Fig. 11.2

while from eqns. (11.4) it may be derived that

$$\begin{aligned} \rho \cos(l_{\mathrm{P}} - \lambda) &= a_{\mathrm{P}} - a_{\mathrm{E}} \cos(l_{\mathrm{E}} - l_{\mathrm{P}}), \\ \rho \cos(l_{\mathrm{E}} - \lambda) &= a_{\mathrm{P}} \cos(l_{\mathrm{E}} - l_{\mathrm{P}}) - a_{\mathrm{E}}. \end{aligned} \tag{11.7}$$

Using eqns. (11.7) in eqn. (11.6) it follows that

$$\begin{aligned} \rho^2 \frac{\mathrm{d}\lambda}{\mathrm{d}t} = &-\{a_{\mathrm{P}} a_{\mathrm{E}} \cos(l_{\mathrm{E}} - l_{\mathrm{P}}) - a_{\mathrm{P}}^2\} \frac{\mathrm{d}l_{\mathrm{P}}}{\mathrm{d}t} \\ &-\{a_{\mathrm{P}} a_{\mathrm{E}} \cos(l_{\mathrm{E}} - l_{\mathrm{P}}) - a_{\mathrm{E}}^2\} \frac{\mathrm{d}l_{\mathrm{E}}}{\mathrm{d}t}. \end{aligned} \tag{11.8}$$

Since the Earth and planet are being assumed to move uniformly in circular orbits, Kepler's third law (eqn. (10.27)) may be used giving

$$\begin{aligned} \frac{\mathrm{d}l_{\mathrm{E}}}{\mathrm{d}t} &= n_{\mathrm{E}} = \frac{\mu_{\mathrm{E}}^{\frac{1}{2}}}{a_{\mathrm{E}}^{\frac{3}{2}}} \simeq \frac{(GM)^{\frac{1}{2}}}{a_{\mathrm{E}}^{\frac{3}{2}}}; \\ \frac{\mathrm{d}l_{\mathrm{P}}}{\mathrm{d}t} &= n_{\mathrm{P}} = \frac{\mu_{\mathrm{P}}^{\frac{1}{2}}}{a_{\mathrm{P}}^{\frac{3}{2}}} \simeq \frac{(GM)^{\frac{1}{2}}}{a_{\mathrm{P}}^{\frac{3}{2}}}, \end{aligned} \tag{11.9}$$

where it has been assumed that the masses of the Earth and planet may be neglected by comparison with the solar mass M. Since ρ^2 can be determined in the form

$$\rho^2 = a_{\mathrm{P}}^2 + a_{\mathrm{E}}^2 - 2a_{\mathrm{P}} a_{\mathrm{E}} \cos(l_{\mathrm{E}} - l_{\mathrm{P}}) \tag{11.10}$$

using the Cosine Rule for plane triangles, use of eqns. (11.9) in eqn. (11.8) gives

$$\rho^2 \frac{d\lambda}{dt} = \kappa\{\cos\psi - \cos(l_E - l_P)\}, \qquad (11.11a)$$

where,

$$\cos\psi = \frac{(a_P^{\frac{1}{2}} + a_E^{\frac{1}{2}})(a_P a_E)^{\frac{1}{2}}}{a_P^{\frac{3}{2}} + a_E^{\frac{3}{2}}}, \qquad (11.11b)$$

and

$$\kappa = (GM)^{\frac{1}{2}}(a_P^{-\frac{1}{2}} a_E + a_P a_E^{-\frac{1}{2}}). \qquad (11.11c)$$

Since ρ^2 is necessarily positive and κ is a constant whose sign may be assumed positive without loss of generality, the sign of $d\lambda/dt$ depends only on the sign of $\{\cos\psi - \cos(l_E - l_P)\}$. If $d\lambda/dt$ is positive the geocentric longitude of the planet increases with time and the motion is said to be *direct*. If $d\lambda/dt$ is negative the geocentric longitude of the planet decreases with time and the motion is said to be *retrograde*. If $d\lambda/dt$ is zero the planet is said to be *stationary*. The synodic period S of the planet, being the time interval between successive conjunctions may now be formally defined as the interval of time required for the geocentric longitude of the planet to increase by 2π.

Retrograde motion will only occur when $l_E - l_P$ is in the angular range $360° - \psi \rightarrow 0° \rightarrow \psi$. The interval of time spent in direct and retrograde motion is therefore $\{(180° - \psi)/180°\}S$ and $(\psi/180°)S$ respectively. It should be noted that $\cos\psi < 1$ for all values of a_P.

In discussing planetary motion the term *elongation* is occasionally encountered. The elongation of the planet P from the Sun S as viewed from the Earth is the angle $\hat{SEP} = \mathscr{E}$ (see Fig. 11.2). If $\odot$ is the geocentric longitude of the Sun then

$$\mathscr{E} = (360° - \odot) + \lambda = \lambda - \odot, \qquad (11.12)$$

provided the plane of the planetary orbit lies in the ecliptic. The elongation may be used to save tabulation but such a policy is not usually adopted.

11.1.3 The phases and brightness of planets. Since any planet is illuminated from one direction and observed from a different direction only a fraction of the illuminated surface of the planet can be seen at any time. The effect is most clearly demonstrated by the phases of the Moon, but similar effects are exhibited by the planets. In Fig. 11.3 the planet P is illustrated as a spherical body centred at P. The radius vector from the Sun to the planet is in the direction SP and the line of centres for the Earth and Planet is EP. The illuminated

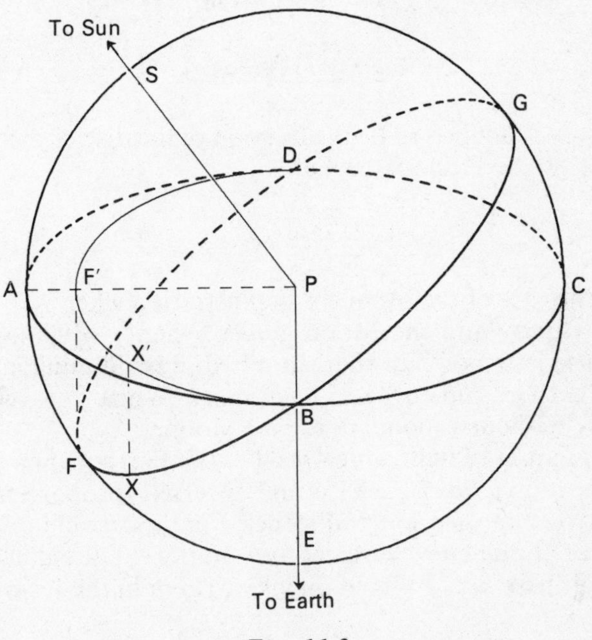

Fig. 11.3

hemisphere of the planet is that lying on the sunward side of the great circle FBGD. The planet as seen from the Earth would be the disc defined by the great circle ABCD. However, not all of the planetary disc is illuminated—only that part bounded by the arcs DAB and DFB can be seen. The section of the disc that is observed to be illuminated may be found by projecting the arc of great circle DFB onto the plane of the great circle ABCD. Projecting the point F parallel to the line of sight EP gives the point F′ on the diameter AP, while any other point X on DFB projects parallel to the line of sight to X′. The curve DF′X′B is part of an ellipse. The illuminated area A is therefore

$$A = \tfrac{1}{2}\pi r^2 - \tfrac{1}{2}\pi r \,.\, \mathrm{PF}', \tag{11.13a}$$

where r is the radius of the planet. Denoting the angle $\mathrm{S\hat{P}E}$ by χ (where χ may be interpreted as the elongation of the Earth from the Sun as viewed from the planet and $\chi = l_P - \lambda$) then $\mathrm{PF}' = r \cos(180° - \chi)$ so that eqn. (11.13a) becomes

$$A = \frac{\pi r^2}{2}(1 + \cos\chi). \tag{11.13b}$$

The phase is defined to be the fraction of the disc of the planet observed to be illuminated, i.e.

$$\text{phase} = \frac{A}{\pi r^2} = \tfrac{1}{2}(1 + \cos\chi). \tag{11.14}$$

Using the case of the Moon as an illustration when $\chi = 0°$ the phase is unity and the Moon is *full*. When $\chi = 90°$ or $270°$ the phase is 0·5 or half the Moon's disc is illuminated. The Moon is then said to be at *quadrature*. When $\chi = 180°$ the phase is zero corresponding to *new* Moon.

The amount of light which reaches the Earth from a planet is proportional to the phase and inversely proportional to the square of the geocentric distance. For the case of uniformly illuminated planetary surfaces and motion in circular orbits the brightness B of a planet may be written in the form

$$B = \frac{I(1 + \cos\chi)}{\rho^2} \tag{11.15}$$

where I is a constant. I is determined by the distance of the planet from the Sun and by the reflectivity of the planetary surface. (Since the distance of a planet from the Sun varies I will vary slightly.) Referring to Fig. 11.2, denoting the angle $\mathrm{E\hat{S}P}$ by ϕ $(= l_E - l_P)$ and applying the Sine Rule for plane triangle to the triangle SEP,

$$\frac{\sin(180° - \chi - \phi)}{a_P} = \frac{\sin\chi}{a_E}\left(= \frac{\sin\mathscr{E}}{a_P}\right), \tag{11.16a}$$

from which it may be deduced that

$$\tan\chi = \frac{2a_E a_P \sin\phi}{\rho^2 + a_P^2 - a_E^2}, \quad \cos\chi = \frac{a_P - a_E\cos\phi}{(a_E^2 - 2a_E a_P\cos\phi + a_P^2)^{\frac{1}{2}}}. \tag{11.16b}$$

Then, using eqn. (11.10) in (11.16b) and using eqn. (11.16b) in (11.15) the brightness B of the planet may be written in the form

$$B = \frac{I\{1 + (a_P - a_E\cos\phi)/\rho\}}{\rho^2} = I\left\{\frac{1}{\rho^2} + \frac{a_P - a_E\cos\phi}{\rho^3}\right\}. \tag{11.17}$$

It is of some interest to determine those occasions when the brightness B of the planet is a maximum or minimum. Since ρ is a function of ϕ only, B is also a function of ϕ only. The condition of maximum or minimum brightness is therefore given by

$$\frac{dB}{d\phi} = I\left\{-\frac{2}{\rho^3}\frac{d\rho}{d\phi} - \frac{3(a_P - a_E\cos\phi)}{\rho^4}\frac{d\rho}{d\phi} + \frac{a_E\sin\phi}{\rho^3}\right\} = 0. \tag{11.18}$$

$d\rho/d\phi$ may be determined from eqn. (11.10) to be

$$\rho\frac{d\rho}{d\phi} = a_E a_P \sin\phi. \tag{11.19}$$

Hence using eqn. (11.19) in (11.18) the condition for maximum or minimum brightness becomes

$$\rho^{-5} a_E I \sin\phi\,\{\rho^2 + 4a_P\rho - 3(a_E^2 - a_P^2)\} = 0. \tag{11.20}$$

The five solutions of eqn. (11.20) are

$$\phi = 0, \quad \phi = 180°, \quad \rho = \pm(3a_E^2 + a_P^2)^{\frac{1}{2}} - 2a_P, \quad \rho = \infty, \tag{11.21}$$

though clearly the solutions $\rho = -(3a_E^2 + a_P^2)^{\frac{1}{2}} - 2a_P$, $\rho = \infty$ must be excluded since ρ must be positive and a planet at infinity would have zero brightness. Three real solutions are therefore possible for $a_P/a_E < 1$, while the solution $\rho = (3a_E^2 + a_P^2)^{\frac{1}{2}} - 2a_P$ becomes unphysical ($\rho < 0$) for $a_P/a_E > 1$, so that only two solutions then remain. The solution for

$a_P/a_E = \frac{1}{4}$ separates two classes of solutions; namely, those solutions for which there is a maximum of brightness for $\phi = 180°$, $a_P/a_E \leqslant \frac{1}{4}$ and those solutions for which there is a relative minimum of brightness for $\phi = 180°$, $a_P/a_E > \frac{1}{4}$. The variation of B (in units of I/a_E^2) with ϕ is illustrated in Fig. 11.4 for $a_P < a_E$. None of the curves in Fig. 11.4 represent any particular planetary situation—the diagram illustrates trends. Since $a_P/a_E > \frac{1}{4}$ for both Mercury and Venus their brightnesses will vary in a manner similar to that for the curve $a_P/a_E = 0{\cdot}45$. However, were Venus and Mercury observed from Jupiter

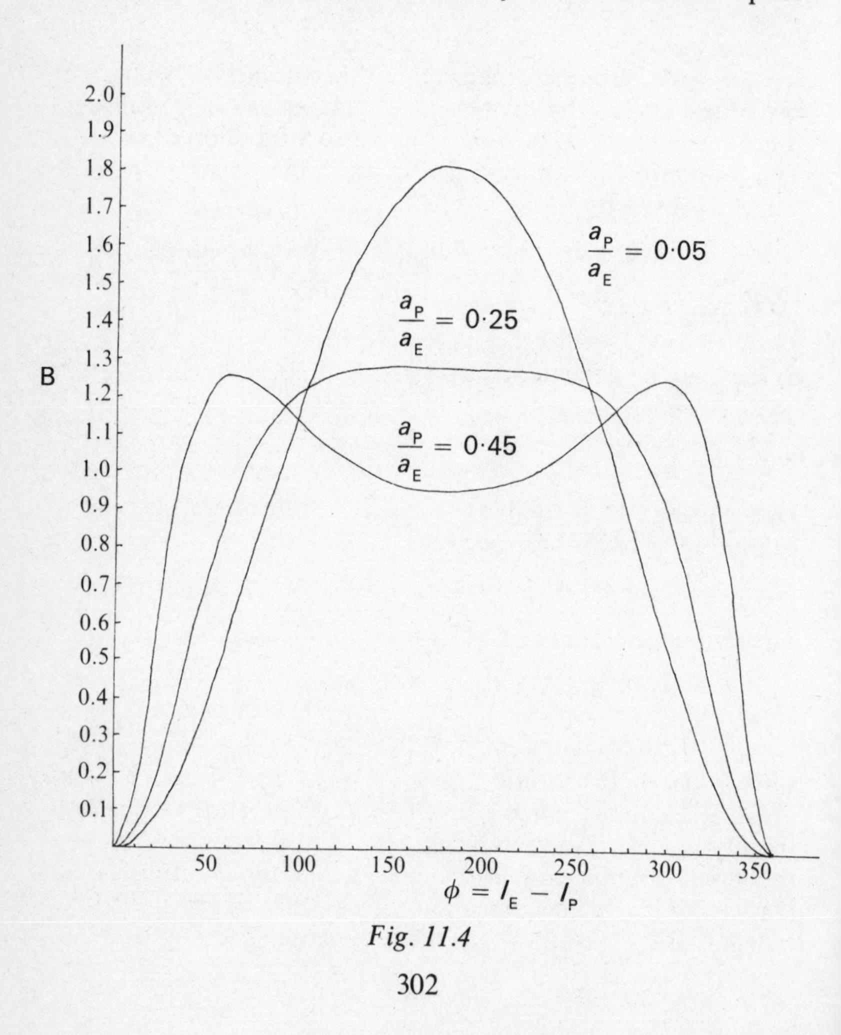

Fig. 11.4

(i.e. identifying a_E as the heliocentric distance of Jupiter) their brightnesses would vary in a manner similar to the curve for $a_P/a_E = 0{\cdot}05$. In the cases where $a_P/a_E > 1$, the only physically realistic solutions are those for $\phi = 0°$, $\phi = 180°$. The brightness B then fluctuates between its values at $\phi = 0°$ and $\phi = 180°$ as is the case for those planets at a greater distance from the Sun than the Earth.

The considerations of this section, based as they are on the concept of uniform planetary motion in circular orbits in the plane of the ecliptic, can only be an approximation to actual geocentric observations. However, discussion of these phenomena on the basis of such an approximation gives a useful means of visualising the effects. The gross picture so obtained can then be modified for more accurate work.

11.2 The Determination of Planetary Geocentric Coordinates

In this section the means of deriving the actual geocentric position of a planet will be discussed assuming only the properties of its orbit about the Sun. The transformation from orbit to position on the sky can be made in three parts. The position of the planet in its orbit and the orientation of the orbit with respect to rectangular equatorial axes must first be established and then its position on the sky with respect to heliocentric rectangular equatorial coordinates is then established. Finally the heliocentric position is transformed to the geocentric position.

11.2.1 The elements of the orbit. In Chapter X the dynamical elements T, e, τ of an orbit were introduced. From these elements it was possible to locate the position of the planet in its orbit with respect to the Sun. In order to specify the orientation in space of the orbit with respect to the frame of reference defining rectangular ecliptic coordinates three further coordinates or elements are required. In Fig. 11.5, the great circle FNG defines the plane of the ecliptic, the great circle NA′P′B defines the plane of the planetary orbit. The celestial sphere is assumed centred on the Sun S and heliocentric rectangular ecliptic coordinates are defined by S(x', y', z') where the x'-axis is in the direction of the Vernal Equinox ♈, the z'-axis is in the direction of the pole of the ecliptic and the y'-axis completes the right handed set. K is the pole of the ecliptic in the northern (celestial) hemisphere. The point P is the position of a planet in its elliptical orbit about the Sun S and A is the position of perihelion. The projections of

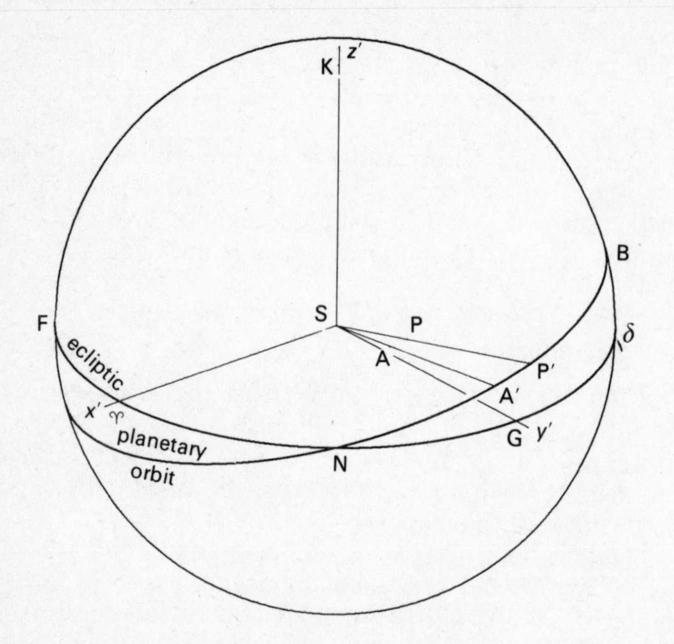

Fig. 11.5

A and P on the heliocentric celestial sphere are A′, P′ respectively. The orientation of the planetary orbit is specified uniquely by its inclination i to the ecliptic, the longitude ♈N $\equiv \theta$ of the ascending node on the ecliptic and the separation NA′ $= \omega$ of perihelion from the ascending node of the orbit. The orbit is then specified completely if $T, e, \tau, \theta, \omega, i$ are known. The length a of the semi-major axis is usually included in the elements to establish the scale of the orbit (clearly knowledge of a is unnecessary to fix the position of the planet since r may be expressed in units of a). The true anomaly v in terms of Fig. 11.4 can then be identified with the angle

$A'\hat{S}P' = A\hat{S}P$. Collecting the identifications together

$$\theta = \Upsilon N, \qquad \omega = NA', \qquad i = B\hat{N}G,$$
$$v = A\hat{S}P = A'P'. \tag{11.22}$$

11.2.2 Heliocentric rectangular equatorial coordinates of a planet. A system of rectangular ecliptic coordinates $S(x', y', z')$ (see Fig. 11.5) has been established. The actual position of the planet in space can be established with respect to these coordinates. To establish the position of the planet in space, the value of r must be taken into account. Denoting the rectangular ecliptic coordinates of the planet by (x', y', z')

$$\begin{aligned} x' &= r\cos P\hat{S}\Upsilon = r\cos P'\Upsilon, \\ y' &= r\cos P\hat{S}G = r\cos P'G, \\ z' &= r\cos P\hat{S}K = r\cos P'K. \end{aligned} \tag{11.23}$$

Applications of the Cosine Rule to the spherical triangles $P'N\Upsilon$, $P'NG$, $P'NK$ and use of the definitions (11.22) gives

$$\begin{aligned} x' &= r\{\cos\theta\cos(\omega + v) - \sin\theta\sin(\omega + v)\cos i\}, \\ y' &= r\{\sin\theta\cos(\omega + v) + \cos\theta\sin(\omega + v)\cos i\}, \\ z' &= r\sin(\omega + v)\sin i. \end{aligned} \tag{11.24}$$

As was shown in Chapter III eqn. (3.6) rectangular ecliptic coordinates (x', y', z') are related to rectangular equatorial coordinates (x, y, z) by

$$\begin{aligned} x &= x', \\ y &= y'\cos\varepsilon - z'\sin\varepsilon, \\ z &= y'\sin\varepsilon + z'\cos\varepsilon. \end{aligned} \tag{3.6}$$

Substitution for x', y', z' from eqns. (11.24) in eqns. (3.6) gives

$$x = r\{\cos\theta\cos(\omega + v) - \sin\theta\sin(\omega + v)\cos i\},$$
$$= r\{P_x\cos v + Q_x\sin v\}, \tag{11.25a}$$
$$= aP_x(\cos E - e) + bQ_x\sin E; \tag{11.25b}$$
$$y = r[\{\sin\theta\cos(\omega + v) + \cos\theta\sin(\omega + v)\cos i\}\cos\varepsilon - \sin(\omega + v)\sin i\sin\varepsilon],$$

$$= r\{P_y \cos v + Q_y \sin v\}, \tag{11.26a}$$

$$= aP_y(\cos E - e) + bQ_y \sin E; \tag{11.26b}$$

$$z = r[\{\sin\theta\cos(\omega + v) + \cos\theta\sin(\omega + v)\cos i\}\sin\varepsilon + \sin(\omega + v)\sin i\cos\varepsilon],$$

$$= r\{P_z \cos v + Q_z \sin v\}, \tag{11.27a}$$

$$= aP_z(\cos E - e) + bQ_z \sin E; \tag{11.27b}$$

where use has been made of eqns. (10.34, 10.35) and

$$\begin{aligned}
P_x &= \cos\theta\cos\omega - \sin\theta\sin\omega\cos i,\\
P_y &= (\sin\theta\cos\omega + \cos\theta\sin\omega\cos i)\\
&\quad \times \cos\varepsilon - \sin\omega\sin i\sin\varepsilon,\\
P_z &= (\sin\theta\cos\omega + \cos\theta\sin\omega\cos i)\\
&\quad \times \sin\varepsilon + \sin\omega\sin i\cos\varepsilon,\\
Q_x &= -\cos\theta\sin\omega - \sin\theta\cos\omega\cos i,\\
Q_y &= (-\sin\theta\sin\omega + \cos\theta\cos\omega\cos i)\\
&\quad \times \cos\varepsilon - \cos\omega\sin i\sin\varepsilon.\\
Q_z &= (-\sin\theta\sin\omega + \cos\theta\cos\omega\cos i)\\
&\quad \times \sin\varepsilon + \cos\omega\sin i\cos\varepsilon.
\end{aligned} \tag{11.28}$$

It is clear that it is not necessary to determine v in order to find the rectangular equatorial coordinates of a planet. If the length a of the semi major axis is known, then solution of Kepler's equation for E is sufficient to allow calculation of the heliocentric rectangular equatorial coordinates. Because of precession the elements θ, ω, i undergo changes and must therefore be defined for some epoch. However, the Ps and Qs need be determined once only for any epoch and thereafter solution of Kepler's equation is all that is required to determine E for any proposed observation of the planet. The equatorial coordinates determined in this way refer to the epoch for which the elements θ, ω, i were defined. It would be usual to take the epoch as the instant for which the planetary position was required. Formulae for the elements of planetary orbits are given in *Astrophysical Quantities*.

11.2.3 Geocentric rectangular equatorial coordinates of a planet. The heliocentric rectangular coordinates (x_E, y_E, z_E) of the Earth can be readily found. Eqns. (11.28) simplify since $\theta = i \equiv 0$. The longitude of perihelion of the Earth may be denoted by ω_E so that

$$\begin{aligned} P_x &= \cos \omega_E, & Q_x &= -\sin \omega_E, \\ P_y &= \sin \omega_E \cos \varepsilon, & Q_y &= \cos \omega_E \cos \varepsilon, \\ P_z &= \sin \omega_E \sin \varepsilon, & Q_z &= \cos \omega_E \sin \varepsilon, \end{aligned} \tag{11.29}$$

from which the heliocentric rectangular equatorial coordinates of the Earth may be found when used in eqns. (11.25b, 11.26b, 11.27b). If (ξ_P, η_P, ζ_P) are the geocentric rectangular equatorial coordinates of the planet then

$$\begin{aligned} \xi_P &= x_P - x_E = x_P + X, \\ \eta_P &= y_P - y_E = y_P + Y, \\ \zeta_P &= z_P - z_E = z_P + Z, \end{aligned} \tag{11.30}$$

where (x_P, y_P, z_P) are the heliocentric rectangular coordinates of a planet and $(X = -x_E, Y = -y_E, Z = -z_E)$ are the geocentric rectangular coordinates of the Sun. It would be usual to express all parameters in eqns. (11.30) in units of a_E. Since the geocentric rectangular coordinates of the Sun are given for each day of the year eqns. (11.25b, 11.26b, 11.27b, 11.28) can be used to determine (x_P, y_P, z_P) and eqns. (11.30) to determine the geocentric rectangular coordinates (ξ_P, η_P, ζ_P). The geocentric rectangular coordinates give the actual position of the planet in space. However, in the sense of Chapter III, rectangular equatorial coordinates refer to a unit sphere. The coordinates are scaled by the geocentric distance of the planet to give

$$\frac{\xi_P}{\rho} = \cos \alpha_P \cos \delta_P, \qquad \frac{\eta_P}{\rho} = \sin \alpha_P \cos \delta_P, \qquad \frac{\zeta_P}{\rho} = \sin \delta_P, \tag{11.31a}$$

where

$$\rho = +(\xi_P^2 + \eta_P^2 + \zeta_P^2)^{\frac{1}{2}}, \tag{11.31b}$$

and (α_P, δ_P) are the geocentric equatorial coordinates of the planet in the sense of Chapter III. The geocentric equatorial coordinates of the planet are finally determined using eqns. (11.31a). Clearly the rectangular geocentric equatorial co-ordinates, scaled by the factor ρ, of the planet may also be determined from eqns. (11.31a) if the geocentric equatorial coordinates of the planet have been determined from observation.

In order to determine the position of any planet the seven quantities $a, e, T, \tau, \theta, i, \omega$ must be known. a, e, T determine the characteristics of the orbit while τ gives the time of latest perihelion passage. θ, i, ω determine the orientation of the orbit in space. The quantities are known as the elements of the orbit. The determination of the position of a planet on the sky requires solution of Kepler's equation to determine the eccentric anomaly E and then the heliocentric coordinates of the planet can be determined. Knowing the geocentric position of the Sun allows the determination of the geocentric coordinates of the planet. Strictly, a correction for planetary aberration should be applied to the geocentric positions determined by the method outlined above before the computed position can be used.

11.3 The Determination of the Elements of a Planetary Orbit

The determination of the elements of a planetary orbit is strictly a topic in celestial mechanics. The problem cannot be treated as simply as the geometrical problem of converting a position in an orbit to position on the sky. The problem of determination of the elements is one which is central to many texts on celestial mechanics and which for ultimate accuracy requires mathematical and numerical sophistication.

However, it is sometimes necessary to derive the elements of a preliminary orbit for a comet so that predictions of the position of the comet can be made with a view to aiding identification of the comet for future observations. A method suitable for this purpose was devised by Merton (Mon. Not. R. Astr. Soc. **85**, 693, 1925) based on three observations of the position of the comet (or other object within the solar system). A similar method had been discussed by Veithen (Uber die Verwendung der Rechenmachine bei der Bahnbestimmung von Planeten, Leipzig, 1912). A simplified treatment of Merton's discussion will be given here since it gives a useful first approximation to the elements of a cometary orbit before the more accurate methods of celestial mechanics are employed. In what follows it will be assumed that the observed positions of the comet or similar object have been corrected for precession and the contribution to aberration resulting from the motion of the Earth as already discussed in Chapters VI, VII.

The mean coordinates, referred to the beginning of the Besselian Year nearest the date of observation, are obtained. The coordinates of the Sun are corrected for parallax and for the terrestrial position of the observer (see eqns. (6.121, 6.122, 6.123)). With these preliminary corrections, the elements of the orbit can then be determined from three independent observations of the position of the comet.

11.3.1 Fundamental formulae. The rectangular heliocentric coordinates of the comet cannot be determined since the geocentric distance ρ of the comet is unknown. Suppose the comet is observed at three separate times t_i, $i = 1, 2, 3$. The mean equatorial coordinates of the comet are (α_i, δ_i) at time t_i. At these times the geocentric rectangular equatorial coordinates of the Sun are (X_i, Y_i, Z_i) and may be determined by the procedures outlined in section (11.2) or from the Astronomical Ephemeris. In Fig. 11.6 the relative positions of the Sun (centred at S), the Earth (centred at E) and the comet (centred at C)

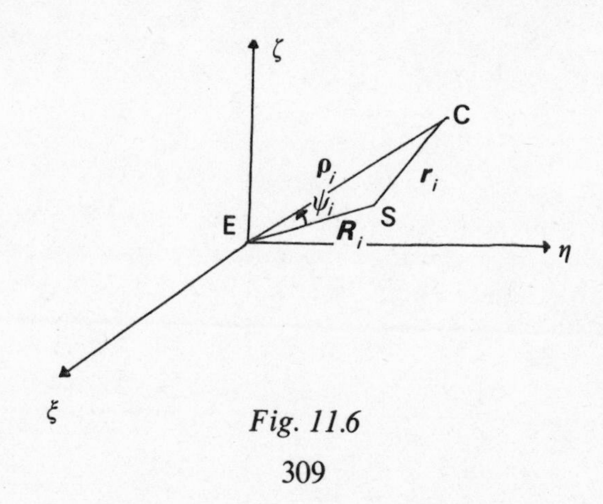

Fig. 11.6

are illustrated for any instant t_i with respect to the frame of reference $E(\xi, \eta, \zeta)$ defining geocentric rectangular equatorial coordinates. With respect to E at time t_i the position vector of the comet is $\boldsymbol{\rho}_i$ and the position vector of the Sun is $\boldsymbol{R}_i$. The position vector of the comet with respect to the Sun is $\boldsymbol{r}_i$ at time t_i. The geocentric rectangular equatorial coordinates of the comet are then given by eqn. (11.31a) namely

$$\begin{aligned} \xi_i &= \rho_i \cos \alpha_i \cos \delta_i, & \eta_i &= \rho_i \sin \alpha_i \cos \delta_i, & \zeta_i &= \rho_i \sin \delta_i \\ &= \rho_i l_i, & &= \rho_i m_i, & &= \rho_i n_i. \end{aligned} \tag{11.32}$$

l_i, m_i, n_i can be determined from the observed positions (α_i, δ_i). From Fig. 11.6 it is also clear that

$$R_i^2 = X_i^2 + Y_i^2 + Z_i^2,$$

and (11.33)

$$R_i \cos \psi_i = l_i X_i + m_i Y_i + n_i Z_i,$$

where ψ_i is the (internal) angle C$\hat{\text{E}}$S between the vectors $\boldsymbol{\rho}_i$, $\boldsymbol{R}_i$.

In terms of heliocentric rectangular equatorial coordinates the position vectors of the comet are $\boldsymbol{r}_1, \boldsymbol{r}_2, \boldsymbol{r}_3$ at the three instants of observation. Since the interval between observation will be quite small (with respect to the period of the comet) $\boldsymbol{r}_1, \boldsymbol{r}_3$ may be expressed in terms of $\boldsymbol{r}_2$ by means of Taylor's Theorem, namely,

$$\begin{aligned} \boldsymbol{r}_1 &= \boldsymbol{r}_2 - t_{12}\dot{\boldsymbol{r}}_2 + \tfrac{1}{2}t_{12}^2\ddot{\boldsymbol{r}}_2 - \tfrac{1}{6}t_{12}^3\dddot{\boldsymbol{r}}_2 + \dots, \\ \boldsymbol{r}_3 &= \boldsymbol{r}_2 + t_{23}\dot{\boldsymbol{r}}_2 + \tfrac{1}{2}t_{23}^2\ddot{\boldsymbol{r}}_2 + \tfrac{1}{6}t_{23}^3\dddot{\boldsymbol{r}}_2 + \dots, \end{aligned} \tag{11.34}$$

where

$$t_{12} = t_2 - t_1, \qquad t_{23} = t_3 - t_2. \tag{11.35}$$

The equation of motion for the comet, moving under the influence of its gravitational interaction with the Sun, may be used to evaluate $\ddot{\boldsymbol{r}}$ and $\dddot{\boldsymbol{r}}$. Eqns. (10.6) may be written in the vector form

$$\ddot{\boldsymbol{r}} + \frac{\mu}{r^3}\boldsymbol{r} = 0 \tag{10.6a}$$

where $\mu = GM$, M being the mass of the Sun. The mass of the comet m_c is neglected in the definition of μ since cometary masses are negligible by comparison with the mass of the Sun. By differentiation of eqn. (10.6a) with respect to time,

$$\dddot{\boldsymbol{r}} = -\frac{\mu}{r^3}\dot{\boldsymbol{r}} + \frac{3\mu \boldsymbol{r}}{r^4}\dot{r} = -k\dot{\boldsymbol{r}} + 3k\boldsymbol{r}\frac{\dot{r}}{r} \simeq -k\dot{\boldsymbol{r}}, \qquad (11.36)$$

where

$$k = \mu/r^3.$$

The Taylor series expansions for $\boldsymbol{r}_1$, $\boldsymbol{r}_3$ will be assumed to be of sufficient accuracy if terms in t^4 and higher powers are neglected. Such an assumption is usually sufficiently accurate for preliminary purposes. Substituting for $\ddot{\boldsymbol{r}}$ and $\dddot{\boldsymbol{r}}$ from eqns. (10.6a), (11.36) respectively, eqns. (11.34) become

$$\boldsymbol{r}_1 = \boldsymbol{r}_2\left(1 - \tfrac{1}{2}t_{12}^2 k_2 - \tfrac{1}{2}t_{12}^3 k_2 \frac{\dot{r}_2}{r_2}\right) - t_{12}\dot{\boldsymbol{r}}_2(1 - \tfrac{1}{6}t_{12}^2 k_2),$$

$$\boldsymbol{r}_3 = \boldsymbol{r}_2\left(1 - \tfrac{1}{2}t_{23}^2 k_2 + \tfrac{1}{2}t_{23}^3 k_2 \frac{r_2}{\dot{r}_2}\right) + t_{23}\dot{\boldsymbol{r}}_2(1 - \tfrac{1}{6}t_{23}^2 k_2). \qquad (11.37)$$

Defining f_1, g_1, f_2, g_2 and p_2 by

$$\begin{aligned} f_1 &= (1 - \tfrac{1}{2}t_{12}^2 k_2 - \tfrac{1}{2}t_{12}^3 k_2 p_2),\\ g_1 &= (1 - \tfrac{1}{2}t_{23}^2 k_2 + \tfrac{1}{2}t_{23}^3 k_2 p_2),\\ f_2 &= (1 - \tfrac{1}{6}t_{12}^2 k_2),\\ g_2 &= (1 - \tfrac{1}{6}t_{23}^2 k_2), \end{aligned} \qquad (11.38)$$

$$p_2 = \dot{r}_2/r_2.$$

$\dot{\boldsymbol{r}}_2$ may be written in the form

$$\dot{\boldsymbol{r}}_2 = \frac{f_1\boldsymbol{r}_2 - \boldsymbol{r}_1}{t_{12}f_2} = \frac{\boldsymbol{r}_3 - g_1\boldsymbol{r}_2}{t_{23}g_2}, \qquad (11.39a)$$

or

$$c_1\boldsymbol{r}_1 - \boldsymbol{r}_2 + c_3\boldsymbol{r}_3 = 0, \qquad (11.39b)$$

where

$$c_1 = \frac{t_{23}g_2}{t_{23}f_1g_2 + t_{12}f_2g_1}, \quad c_3 = \frac{t_{12}f_2}{t_{23}f_1g_2 + t_{12}f_2g_1}. \qquad (11.39c)$$

Eqn. (11.39b) may be expressed in component form namely,

$$\begin{aligned} c_1x_1 - x_2 + c_3x_3 &= 0, \\ c_1y_1 - y_2 + c_3y_3 &= 0, \\ c_1z_1 - z_2 + c_3z_3 &= 0. \end{aligned} \qquad (11.40)$$

(x_i, y_i, z_i) $(i = 1, 2, 3)$ may be replaced through use of eqns. (11.30, 11.33), so that eqns. (11.40) may be written in the form

$$\begin{aligned} c_1\rho_1 l_1 - \rho_2 l_2 + c_3\rho_3 l_3 &= c_1X_1 - X_2 + c_3X_3, \\ c_1\rho_1 m_1 - \rho_2 m_2 + c_3\rho_3 m_3 &= c_1Y_1 - Y_2 + c_3Y_3, \\ c_1\rho_1 n_1 - \rho_2 n_2 + c_3\rho_3 n_3 &= c_1Z_1 - Z_2 + c_3Z_3. \end{aligned} \qquad (11.41)$$

Eqns. (11.41) are the three fundamental equations from which the five quantities c_1, c_3, $c_1\rho_1$, ρ_2, $c_3\rho_3$ must be determined. Clearly since five quantities cannot be obtained from three equations some procedure which permits estimates of at least two of the quantities must be sought.

11.3.2 Approximate solutions. Since the right hand side of eqns. (11.41) is determinate apart from c_1, c_3 it would seem desirable to seek approximate values for c_1, c_3. Using the definitions (11.38)

$$\begin{aligned} t_{23}f_1g_2 + t_{12}f_2g_1 &= t_{23}(1 - \tfrac{1}{2}t_{12}^2k_2 - \tfrac{1}{2}t_{12}^3k_2p_2) \\ &\quad \times (1 - \tfrac{1}{6}t_{23}^2k_2) + t_{12}(1 - \tfrac{1}{2}t_{23}^2k_2 \\ &\quad + \tfrac{1}{2}t_{23}^3k_2p_2)(1 - \tfrac{1}{6}t_{12}^2k_2), \\ &= (t_{12} + t_{23})\{(1 - \tfrac{1}{6}k_2(t_{12} + t_{23})^2 \\ &\quad - \tfrac{1}{2}k_2p_2t_{12}t_{23}(t_{12} - t_{23})\}; \end{aligned} \qquad (11.42a)$$

$$\begin{aligned} c_1 &= \frac{t_{23}}{t_{12} + t_{23}}(1 - \tfrac{1}{6}t_{23}^2k_2)\{1 + \tfrac{1}{6}k_2(t_{12} + t_{23})^2 \\ &\quad + \tfrac{1}{2}k_2p_2t_{12}t_{23}(t_{12} - t_{23})\}, \\ &= \frac{t_{23}}{t_{12} + t_{23}} + \frac{1}{6}k_2t_{12}t_{23}\left(1 + \frac{t_{23}}{t_{12} + t_{23}}\right) \\ &\quad + \frac{1}{2}k_2p_2\frac{t_{12}t_{23}^2(t_{12} - t_{23})}{t_{12} + t_{23}}; \end{aligned} \qquad (11.42b)$$

$$c_3 = \frac{t_{12}}{t_{12} + t_{23}} + \frac{1}{6}k_2 t_{12} t_{23}\left(1 + \frac{t_{12}}{t_{12} + t_{23}}\right) + \frac{1}{2}k_2 p_2 \frac{t_{12}^2 t_{23}(t_{12} - t_{23})}{t_{12} + t_{23}}; \qquad (11.42c)$$

where terms of order k_2^2 and higher have been neglected since k_2 is a small quantity. From eqns. (11.42b, c) it is clear that a zero order approximation c_1^0, c_3^0 for c_1, c_3 is

$$c_1^0 = \frac{t_{23}}{t_{12} + t_{23}}, \qquad c_3^0 = \frac{t_{12}}{t_{12} + t_{23}}, \qquad (11.43a)$$

while the next order of approximation is

$$\begin{aligned} c_1^1 &= c_1^0\{1 + \tfrac{1}{6}k_2 t_{12} t_{13}(1 + c_1^0)\}, \\ c_3^1 &= c_3^0\{1 + \tfrac{1}{6}k_2 t_{12} t_{23}(1 + c_3^0)\}, \end{aligned} \qquad (11.43b)$$

where k_2 is evaluated on the basis of a solution of eqns. (11.41) for the zero order approximation. Since p_2 contains a term in $1/r_2$, the third term on the right hand sides of eqns. (11.42b, c) is smaller than the other terms and is further minimized in that a quantity $(t_{12} - t_{23})$ is involved. It is therefore desirable that observations are made for almost equal intervals of time. Frequently such a condition cannot be met and it is sufficient to approximate $\dot{r}_2$ by

$$\dot{r}_2 = \frac{r_3 - r_1}{t_{12} + t_{23}}. \qquad (11.44)$$

$\dot{r}_2$ cannot be evaluated until some estimate of r_3, r_1 has been obtained.

If approximations c_1^0, c_3^0 to c_1, c_2 respectively are obtained from eqns. (11.43a), eqns. (11.41) may then be solved using standard techniques for $c_1\rho_1$, ρ_2, $c_3\rho_3$ and hence for ρ_1, ρ_2, ρ_3. From Fig. 11.6 it is clear that

$$r_2^2 = \rho_2^2 + R_2^2 - 2\rho_2 R_2 \cos\psi_2 = \rho_2(\rho_2 - 2R_2\cos\psi_2) + R_2^2, \qquad (11.45)$$

using the Cosine Rule for plane triangles. Since R_2 and $R_2 \cos \psi_2$ may be determined from eqns. (11.33) a first estimate r_2^0 for r_2 may be found from eqn. (11.45). In the same way first estimates r_1^0, r_3^0 may be obtained for r_1, r_3 respectively. k_2 may then be found and an estimate made of the term in $k_2 p_2$. If this latter term is significant, it can be retained in the determination of c_1^1, c_3^1. Eqns. (11.41) are solved again using the approximate c_1^1, c_3^1 on the right hand side and $c_1\rho_1$, ρ_2, $c_3\rho_3$ once more obtained. Correction, if appropriate, is made for planetary aberration by adjusting the times of observation according to

$$t_i^* = t_i - \rho_i/c, \qquad i = 1, 2, 3 \tag{11.46}$$

where t_i is the time of observation and t_i^* is the time at which the light left the comet. Again new values r_1^1, r_2^1, r_3^1 are computed. A further set of values c_1^2, c_3^2 are computed using eqns. (11.42b, c) replacing the t_i by t_i^* and evaluating $k_2 p_2$ on the basis of the last approximation to r_1, r_2, r_3. Eqns. (11.41) are again solved for the final values of $c_1\rho_1$, ρ_2, $c_3\rho_3$ and values r_1^2, r_2^2, r_3^2 are obtained.

The values obtained may be checked since the heliocentric rectangular equatorial coordinates (x_i, y_i, z_i) $(i = 1, 3)$ of the comet may be found using eqns. (11.30) once values for the ρ_i $(i = 1, 3)$ have been obtained. Then, using the values c_1^2, c_3^2 for c_1, c_3, eqns. (11.40) may be used to calculate x_2, y_2, z_2 and hence α_2, δ_2. The computed and observed values of α_2, δ_2 can then be compared and errors detected. Clearly the iteration of the solution should be stopped once an accuracy concomitant with the accuracy of the observations has been attained.

11.3.3 Determination of the elements of the orbit. In the previous section r_1, r_2, r_3 have been determined. Since ρ_1, ρ_2, ρ_3 have also been determined in the course of the calculation it is a simple matter to compute (x_i, y_i, z_i) $(i = 1, 2, 3)$ from eqns. (11.30). Knowing the lengths of the vectors $\boldsymbol{r}_1, \boldsymbol{r}_2, \boldsymbol{r}_3$ and the heliocentric rectangular equatorial coordinates (x_i, y_i, z_i) of the comet corresponding to these vectors, the elements defining the orbit may be found. However, the determination involves the true anomaly v. The scalar products of the position vectors $\boldsymbol{r}_i$ give

$$\begin{aligned} \boldsymbol{r}_3 \cdot \boldsymbol{r}_1 &= r_3 r_1 \cos(v_3 - v_1) = x_3 x_1 + y_3 y_1 + z_3 z_1 \\ \boldsymbol{r}_2 \cdot \boldsymbol{r}_1 &= r_2 r_1 \cos(v_2 - v_1) = x_2 x_1 + y_2 y_1 + z_2 z_1 \end{aligned} \tag{11.47}$$

from which

$$v_3 - v_1 = p_{31}, \qquad v_2 - v_1 = p_{21} \tag{11.48}$$

may be computed. Again the equation of the orbit of the comet gives

$$\begin{aligned} r_1(1 + e\cos v_1) &= r_2(1 + e\cos v_2) = r_3(1 + e\cos v_3) \\ &= a(1 - e^2). \end{aligned} \tag{11.49}$$

Ratios of the lengths of the radius vectors can be established, namely,

$$\frac{r_1}{r_3} = \frac{1 + e\cos v_3}{1 + e\cos v_1} = q_{31}, \qquad \frac{r_1}{r_2} = \frac{1 + e\cos v_2}{1 + e\cos v_1} = q_{21}, \tag{11.50}$$

enabling q_{31}, q_{21} to be computed since r_1, r_2, r_3 are known. Then from eqns. (11.50, 11.48)

$$\begin{aligned} e\cos v_1(\cos p_{21} - q_{21}) - e\sin v_1 \sin p_{21} &= q_{21} - 1, \\ e\cos v_1(\cos p_{31} - q_{31}) - e\sin v_1 \sin p_{31} &= q_{31} - 1. \end{aligned} \tag{11.51}$$

Eqns. (11.51) may be solved for $e\cos v_1$, $e\sin v_1$. Cancellation may be encountered if $r_1 \simeq r_2 \simeq r_3$ and p_{21}, p_{31} are small. Loss of accuracy may then ensue. However, in the case of comets seen from the Earth v, at least, is usually changing with some rapidity as the comet approaches the Sun and the effects of cancellation are not as severe as for a more slowly moving planetary object. Having found $e\cos v_1$, $e\sin v_1$, $\tan v_1$ and so v_1 can be calculated. v_2, v_3 can then be calculated from eqn. (11.48). Finally e is determined and a can be calculated from eqn. (11.49).

From eqn. (10.36a)

$$r_i = a(1 - e\cos E_i), \qquad i = 1, 2, 3 \tag{10.36a}$$

from which a value of the eccentric anomaly E_i can be computed for each observation. Kepler's equation eqn. (10.31) allows the computation of the mean anomaly M_i corresponding to E_i, namely

$$M_i = E_i - e\sin E_i = \frac{2\pi}{T}(t_i^* - \tau), \tag{11.52}$$

on making use of the definition of the mean anomaly M in

terms of the period T and time of perihelium passage τ (see also eqn. (10.29)). Three equations are available from which to evaluate the two unknowns—the period T and time of perihelion passage τ—and check for consistency. The elements a, e, T, τ defining the orbit are then determined.

The elements defining the orientation of the orbit may be found using eqns. (11.25a, 11.26a, 11.27a) since

$$\begin{aligned} x_i &= P_x r_i \cos v_i + Q_x r_i \sin v_i, \\ y_i &= P_y r_i \cos v_i + Q_y r_i \sin v_i, \qquad i = 1, 2, 3. \\ z_i &= P_z r_i \cos v_i + Q_z r_i \sin v_i, \end{aligned} \tag{11.53}$$

giving 9 equations in all, three equations involving each pair of unknowns (P_x, Q_x), (P_y, Q_y), (P_z, Q_z). Hence the Ps and Qs may be determined and checked for consistency. Additional checks are provided by

$$\begin{aligned} P_x^2 + P_y^2 + P_z^2 &= Q_x^2 + Q_y^2 + Q_z^2 = 1, \\ P_x Q_x + P_y Q_y + P_z Q_z &= 0. \end{aligned} \tag{11.54}$$

The elements θ, ω, i may be derived using eqns. (11.28) from which it may be deduced that

$$\begin{aligned} \sin i \sin \omega &= P_z \cos \varepsilon - P_y \sin \varepsilon, \\ \sin i \cos \omega &= Q_z \cos \varepsilon - Q_y \sin \varepsilon, \\ \cos \theta &= P_x \cos \omega - Q_x \sin \omega. \end{aligned} \tag{11.55}$$

The orbital elements are now completely determined from three observations of the comet. The accuracy of the calculation as outlined above is limited by the accuracy of the observations and the degree of cancellation involved in eqn. (11.51). It may be sufficient for the purposes of a preliminary orbit to use only the first approximation to c_1, c_3 in which case the calculation becomes rapid. The effects of cancellation in eqn. (11.51) should only prove serious in special circumstances and their influence on the elements of the preliminary orbit can be assessed.

The above method is a modification of Gauss's method of determination of the elements of a planetary orbit. It is clear that if a calculation of anything other than a preliminary orbit is required, it would be desirable to use the full Gaussian, or other appropriate, procedure.

Chapter XII

Binary Stars

The occurrence of two stars in almost the same line of sight may be the result of chance or a real physical association may exist between the pair of stars. If the association is a result of chance, i.e. the two stars are at quite different distances from the observer, the pair is called an *optical double.* If the two stars are moving under their mutual gravitational interaction the pair is a *binary system.* A *multiple system* is one in which a number (>2) of stars are in motion under their mutual gravitational attraction. In this chapter only binary systems will be considered.

Binary stars may be divided into three broad classes—*visual binaries*, *spectroscopic binaries* and *eclipsing binaries. Visual binaries* are two stars which can be resolved telescopically: spectroscopic binaries and eclipsing binaries cannot be resolved telescopically. *Spectroscopic binaries* reveal their nature through periodicity in their radial velocity. The radial velocity of a star is determined from the Doppler shift of the lines of its spectrum. The orbital motion of a star in a binary system gives a periodic component to the radial velocity. An *eclipsing binary*, as its name suggests, reveals its presence by variability of its light output. One star of the system is alternately obscuring the other star and passing behind it, so giving the pair a varying light output as seen by an observer.

Binary Stars are important because they offer a direct way of determining stellar masses (visual and spectroscopic binaries) and, before a recent development in interferometric techniques introduced a direct method of measuring stellar radii, the only way of obtaining information on relative stellar radii (eclipsing binaries). However, the establishment of good orbits for binary stars requires considerable painstaking work and data on stellar masses and radii accumulates slowly.

The principles on which the determinations of the elements of binary star orbits are based will be discussed in this chapter. There are many practical details which must be considered before a satisfactory solution in any given case can be attained and such details are beyond the scope of this book. For more detailed treatments, reference should be made to specialist monographs.

In the case of visual binaries, the brighter (though not necessarily the more massive) star is called the *primary* and the fainter is called the *secondary*. The orbit of the secondary about the primary is established by direct observation. The motion of the secondary about the primary is exactly parallel with the case of planetary motion about the Sun. When the secondary is nearest the primary in its orbit about the primary, it is said to be at *periastron*: when the secondary is at the opposite end of the major axis of its orbit, it is said to be at *apastron*. In the case of spectroscopic and eclipsing binaries the motion of the components about the centre of gravity of the binary is determined. This difference does not alter the fundamental equations governing the relative motion of two bodies under the influence of their mutual self-gravitation. In particular, the period T of a binary star is the same whether the motion is relative to one of the components or relative to the centre of mass of the system. The equations developed in Chapter X apply equally to the motion of binary stars as to planetary motion, where μ is re-interpreted as $G(m_1 + m_2)$, m_1, m_2 being the respective masses of the component stars of the binary.

12.1 Stellar Masses

In Fig. 12.1 it is supposed that P is the primary and S the secondary of a binary system. C is the centre of mass of the system. Let P have mass m_1 and S have mass m_2. With respect to the primary P the position vector of the secondary S is $\boldsymbol{R}$;

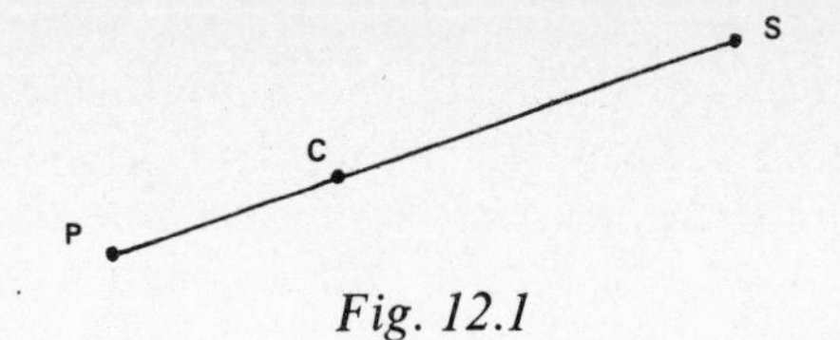

Fig. 12.1

with respect to the centre of mass C the position vector of P is $\boldsymbol{r}_1$ and the position vector of S is $\boldsymbol{r}_2$. Then

$$\boldsymbol{r}_1 = -\frac{m_2}{m_1 + m_2}\boldsymbol{R}, \qquad \boldsymbol{r}_2 = \frac{m_1}{m_1 + m_2}\boldsymbol{R}, \tag{12.1}$$
$$\frac{r_1}{r_2} = \frac{a_1}{a_2} = \frac{m_2}{m_1},$$

if the minus sign defining the direction of r_1 is neglected when magnitudes only are being considered and a_1, a_2 are the semi-major axes of the orbits of primary and secondary respectively about the centre of mass C. From Kepler's Third Law

$$n^2a^3 = G(m_1 + m_2) = n^2(a_1 + a_2)^3, \tag{12.2}$$

where $n = 2\pi/T$ is the mean angular motion and T is the period of the binary. If n, a_1, a_2 can be determined the mass of each component of the binary can be determined.

In the case of visual binaries, the relative orbit of the secondary about the primary is determined. The values of n, a are determined so that only the mass $(m_1 + m_2)$ of the system can be determined. If the spectra of both components of a spectroscopic binary are available, then m_2/m_1 and $(m_1 + m_2) \sin^3 i$ can be determined. Since i (where i is the inclination of the orbit of the binary to a plane perpendicular to the line of sight) cannot be determined for spectroscopic binaries, separation of the masses of the components is again not possible in individual cases. Eclipsing binaries give no information (with respect to astrometric measurements) on stellar masses.

The case of spectroscopic binaries may be considered further. If the spectrum of only one component of a binary can be obtained (the spectrum of the other component being too faint to be recorded—as is frequently the case) any knowledge of the ratio of masses is lost. However, in this case an astrophysically useful quantity—the *mass function*—can be derived. Suppose that the spectrum of P (see Fig. 12.1) is available. P, a star of mass m_1, moves about the centre of mass C in an elliptical orbit whose semi-major axis has length a_1. Then from eqns. (12.1)

$$\frac{a_1}{a_1 + a_2} = \frac{a_1}{a} = \frac{m_2}{m_1 + m_2}. \tag{12.3}$$

Using eqn. (12.2)

$$G(m_1 + m_2) = n^2 a_1^3 \left(1 + \frac{a_2}{a_1}\right)^3 = n^2 a_1^3 \left(1 + \frac{m_1}{m_2}\right)^3$$

i.e.

$$\frac{m_2^3}{(m_1 + m_2)} = \frac{n^2}{G} a_1^3. \tag{12.4}$$

Hence, multiplying eqn. (12.4) by $\sin^3 i$

$$\frac{(m_2 \sin i)^3}{(m_1 + m_3)^2} = \frac{n^2}{G}(a_1 \sin i)^3. \tag{12.4a}$$

Since $n^2(a_1 \sin i)^3/G$ can be determined from the orbit of one component of a spectroscopic binary, the mass function $(m_2 \sin i)^3/(m_1 + m_2)^2$ can be determined. The average mass of spectroscopic binaries of the same spectral type can also be found. Let the average mass of a spectroscopic binary of a specified spectral class be M and suppose a sample N spectroscopic binaries of the specified spectral class have had their orbits determined. Then multiplying eqn. (12.2) by $\sin^3 i$,

$$(m_1 + m_2) \sin^3 i = \frac{n^2}{G}(a_1 \sin i + a_2 \sin i)^3. \tag{12.5}$$

Assuming that the spectra of both components of the N spectroscopic binaries were available the right hand side of eqn. (12.5) can be evaluated for each binary. Summing eqn. (12.5) over all N binaries gives

$$\sum_{j=1}^{N} (m_1 + m_2)_j \sin^3 i_j = NMS \tag{12.6}$$

where $S = \frac{2}{3}$ is the average value of $\sin^3 i$. Hence M may be deduced.

In the case of the Earth, Kepler's Third Law takes the form

$$n_E^2 a_E^3 = GM_{\odot},$$

since the mass of the Earth is negligible by comparison with the mass, $M_{\odot}$, of the Sun. Then the ratio of the mass of a visual binary and the mass of the Sun is

$$\frac{m_1 + m_2}{M_{\odot}} = \frac{n^2 a^3}{n_E^2 a_E^2} \tag{12.7}$$

If the angular separation of primary and secondary is A when the true separation is a, then $A = a/d$ where d is the distance of the binary from the Earth. If the parallax of the binary is π, then $a = A/\pi$ where A, π are measured in consistent units. Then the ratio of mass may be written in the form

$$\frac{m_1 + m_2}{M_{\odot}} = \frac{n^2 A^3}{n_E^2 a_E^2 \pi^3}. \tag{12.8}$$

If $m_1 + m_2$ can be determined (or estimated in the case of a spectroscopic binary) eqn. (12.8) gives the possibility of determining the parallax of the binary.

12.2 The Elements of the Orbit of a Binary Star

As in the case of planetary orbits the length of the semi-major axis of the relative orbit a (or a_1, a_2), the eccentricity e, period T and time of periastron passage τ are required to establish the orbit of a binary star. The orbit, like a planetary orbit, has to be fixed in space and consequently three further elements are required to establish the spatial orientation of the orbit. The elements describing the spatial orientation of the orbit are defined on the basis that the binary system is a visual binary. Not all of the spatial coordinates so defined are relevant to the orbits of spectroscopic and eclipsing binaries.

The measurements which can be made for a visual binary are the separation of the pair ρ and the orientation θ of the line of centres of the primary and secondary with respect to some fixed direction. The fixed direction is the north direction which can be determined by stopping the drive of the telescope and noting the direction of drift of the primary. The direction of drift will be along a parallel of declination and the north direction will be perpendicular to this direction. The measurement of ρ, θ for visual binaries can be more easily discussed in terms of visual rather than photographic observations. Visual binaries are observed using a special eyepiece such that two parallel wires (wires A and B) are crossed by a third (wire C) at right angles. The separation of A and B can be measured. The entire system of wires can be rotated in such a way that the angular orientation of wire C can be determined. By allowing the primary to drift along C the orientation of the north direction can then be found. The relative orientation of the line of centres of primary and secondary can be found by rotating the eyepiece until wire C lies along the line of centres. The separation is measured by placing wire A through the primary and wire B through the secondary. The coordinates ρ, θ of the secondary may then be plotted using the convention that θ increases eastwards from the north direction. An equivalent system is used photographically. In Fig. 12.2 the primary is at P and the secondary is at S. PN is the north direction and is chosen to be the x-axis. The y-axis is in an easterly direction. A z-axis may be chosen in the line of sight

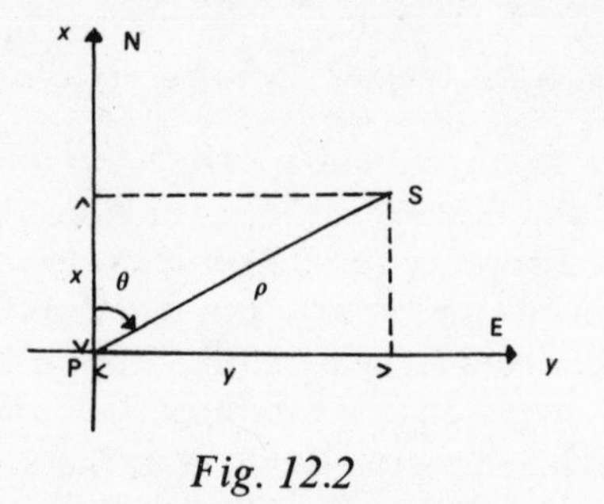

Fig. 12.2

for z increasing away from the observer, to complete the right handed set. Then ρ = PS, θ = N$\hat{\text{P}}$S and

$$x = \rho \cos \theta, \qquad y = \rho \sin \theta. \tag{12.9}$$

In Fig. 12.3 the true orbit of the binary is represented by the ellipse ASEBD. The primary P is at the focus of the ellipse which represents the relative orbit. A is the periastron point

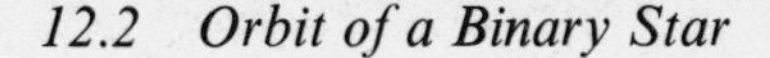

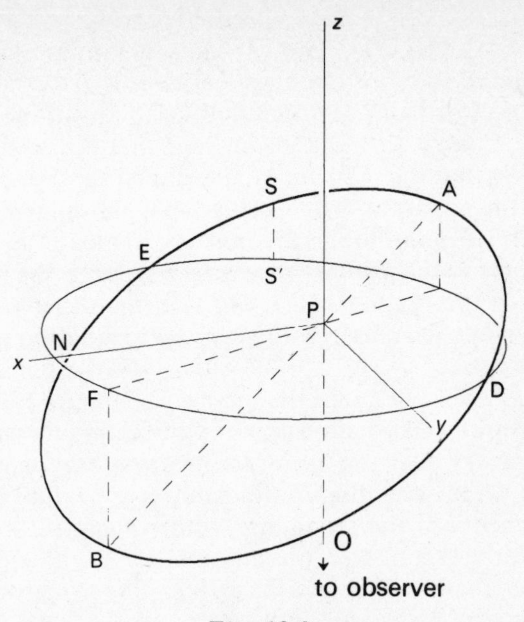

Fig. 12.3

and B is the apastron point. The line of sight to the observer O is PO. The projection of the true orbit onto a plane perpendicular to the line of sight gives the observed apparent orbit ENFD. D, E are the nodes of the true orbit on the apparent orbit. It will be assumed that D is the ascending node. PN is the north direction. The rectangular axes P(x, y, z) are defined by taking the z-axis in the line of sight away from the observer and the x-axis through N. The y-axis makes up the right handed set. This frame of reference is then identical with that of Fig. 12.2.

Let the true position of the secondary S project to S′ in the apparent orbit. Then $N\hat{P}S' = \theta$ and $PS' = \rho$. In order to define the orientation of the true orbit with respect to the apparent orbit three coordinates i, Ω, ω are required: $i = A\hat{D}S'$ is the inclination of the plane of the true orbit with respect to the apparent orbit; $\Omega = N\hat{P}D$ is the angular separation of the ascending node D from the north point N; $\omega = D\hat{P}A$ is the angular separation of the periastron point A from the ascending node D.

It is clear that there is some ambiguity inherent in the above definitions of i, Ω, ω. The inclination i does not define a precise plane for the true orbit with respect to the apparent orbit. It is clear from Fig. 12.3 that the plane of the true orbit could be rotated through an angle $2i$ about the line ED, in such a way that the plane of the true orbit passed through the plane of the apparent orbit during the rotation, without altering the above definition. If the position angle of the secondary with respect to the primary increases with time, the inclination is taken in the range $0 \leqslant i < 90°$. On the other hand, if the position angle decreases with time the inclination is taken to lie in the range $90° < i \leqslant 180°$. If $i = 90°$ the apparent orbit is a straight line through the primary. Direct observation of the separation and position angle of the secondary with respect to the primary does not allow a critical estimation of whether or not D is the ascending or descending node. Radial velocity measurements of the secondary with respect to the primary would settle the matter unambiguously. It is conventional to choose a value of Ω less than 180°. The longitude of periastron ω can have any value between 0° and 360° but is subject to the uncertainty of 180° in Ω.

For certain specialised circumstances conventions have been adopted. If $i = 0°$ or 180°, the longitude Ω of the node D is taken to be 0°. The longitude of periastron is taken to be ω for direct motion and $360° - \omega$ for retrograde motion. If the true orbit is a circle the longitude of periastron ω is taken to be 0°; τ is then the epoch of nodal passage and would require a change by a half period were the longitude of the ascending node found to be greater than 180°.

The observations of separation and orientation can be related to the parameters of the true orbit and the parameters determining its orientation in space. When the secondary is at S (see Fig. 12.3) in its true orbit, it appears at S′ in the apparent orbit. The vector PS has length r; $S\hat{P}A = v$, the true anomaly; $PS' = \rho$, the measured separation and $N\hat{P}S' = \theta$. The vector PD is common to both true and apparent orbit. Then

$$\rho \cos S'\hat{P}D = r \cos S\hat{P}D,$$

i.e.

$$\rho \cos (\theta - \Omega) = r \cos (\omega + v). \tag{12.10}$$

Similarly the component of ρ in the plane of the apparent orbit perpendicular to PD is

$$\rho \sin(\theta - \Omega) = r \sin(\omega + v) \cos i \tag{12.11}$$

and the z coordinate of S is

$$z = r \sin(\omega + v) \sin i. \tag{12.12}$$

Then,

$$\rho(\cos\theta \cos\Omega + \sin\theta \sin\Omega) = r(\cos\omega \cos v - \sin\omega \sin v), \tag{12.13a}$$

$$\rho(\sin\theta \cos\Omega - \cos\theta \sin\Omega) = r(\sin\omega \cos v + \cos\omega \sin v)\cos i. \tag{12.13b}$$

Subtraction of eqn. (12.13b) multiplied by $\sin\Omega$ from eqn. (12.13a) multiplied by $\cos\Omega$ gives,

$$\begin{aligned} x = \rho\cos\theta = {} & r\cos v(\cos\omega\cos\Omega - \sin\omega\sin\Omega\cos i) \\ & + r\sin v(-\sin\omega\cos\Omega - \cos\omega\sin\Omega\cos i). \end{aligned} \tag{12.14a}$$

Addition of eqn. (12.13a) multiplied by $\sin\Omega$ to eqn. (12.13b) multiplied by $\cos\Omega$ gives,

$$\begin{aligned} y = \rho\sin\theta = {} & r\cos v(\cos\omega\sin\Omega + \sin\omega\cos\Omega\cos i) \\ & + r\sin v(-\sin\omega\sin\Omega + \cos\omega\cos\Omega\cos i), \end{aligned} \tag{12.14b}$$

while

$$z = r\cos v(\sin\omega \sin i) + r\sin v(\cos\omega \sin i). \tag{12.14c}$$

Eqns. (12.14) are the only equations available from which the elements of the orbits of binary stars may be deduced. In the case of visual binaries the only measurements available are those of (ρ, θ). (x, y) can be determined from (ρ, θ) and the analysis is confined to eqns. (12.14a, b) i.e. to two equations in three unknowns. Additional information is required in order to solve for i, ω, Ω. In the case of spectroscopic binaries only measurements of the radial velocity with respect to the Sun are available. The motion of a component of the binary about its centre of mass only contributes that part of its velocity, dz/dt in the line of sight, to the total radial velocity. Since eqn. (12.14c) does not contain a term in Ω, it follows that Ω cannot be determined for spectroscopic binaries. Again $\sin i$ is a factor in eqn. (12.14c) and therefore i cannot be determined for spectroscopic binaries. In the case of eclipsing binaries eqns. (12.14) are replaced by a more relevant set of equations.

It will be assumed that the period of the binary system has been determined from observation and that observations are available over a time interval exceeding one period. Special methods exist for those situations where observations are available for a time interval less than the period.

12.3 The Determination of the Elements of the Orbit of a Visual Binary

The observations of a visual binary consist of a set of values (ρ, θ, t) where t is the time of observation. Such observations will contain errors of observation. However, the effects of such errors can be minimised since it is known that the apparent orbit must be an ellipse. Since Kepler's Second Law will apply to the apparent ellipse

$$\rho^2\dot{\theta} = x\frac{dy}{dt} - y\frac{dx}{dt} = k \qquad (12.15)$$

where k is a constant. Since $\dot{\theta}$ may be found by numerical differentiation a value of k can be found either by taking a simple mean or by a least squares analysis. The general equation of a conic may be written in the form

$$Ax^2 + 2Hxy + By^2 + 2Fx + 2Gy + 1 = 0. \qquad (12.16)$$

Clearly a more complex least squares analysis could be used to determine A, B, F, G, H. However, the solution for k must also apply and be taken into account in deriving the coefficients of eqn. (12.16). Differentiation of eqn. (12.16) gives

$$\frac{dx}{dt}(Ax + Hy + F) + \frac{dy}{dt}(By + Hx + G) = 0. \qquad (12.17)$$

dx/dt, dy/dt may be obtained by numerical differentiation of the values of x, y determined through eqn. (12.9) from ρ, θ. The value of k may be introduced into eqn. (12.17) by dividing dx/dt, dy/dt by the least squares value of k. A least squares solution is then obtained for A, B, F, G, H. Having minimised the errors of observation the elements of the true orbit must be derived from the values of A, B, F, G, H and k. Two methods are available—the first method is that discussed by Kowalsky and the second makes use of the Thiele–Innes constants. Kowalsky's method requires a least squares analysis of both eqns. (12.15), (12.16) but has the advantage of dealing directly with the algebraic quantities involved. The use of the Thiele–Innes method avoids the least squares solution of eqn. (12.16) and is the method most used in practice.

8.3.1 Kowalsky's method. Eqns. (12.14) may be written in the form

$$x = l_1 r \cos v + l_2 r \sin v, \qquad y = m_1 r \cos v + m_2 r \sin v,$$
$$z = n_1 r \cos v + n_2 r \sin v \tag{12.18}$$

where

$$\begin{aligned}
l_1 &= \cos\omega\cos\Omega - \sin\omega\sin\Omega\cos i,\\
m_1 &= \cos\omega\sin\Omega + \sin\omega\cos\Omega\cos i,\\
n_1 &= \sin\omega\sin i,\\
l_2 &= -\sin\omega\cos\Omega - \cos\omega\sin\Omega\cos i,\\
m_2 &= -\sin\omega\sin\Omega + \cos\omega\cos\Omega\cos i,\\
n_2 &= \cos\omega\sin i.
\end{aligned} \tag{12.19}$$

$r\cos v$, $r\sin v$ are the coordinates of the secondary with respect to a frame of reference in the plane of the true orbit such that the origin is at the focus (i.e. the primary) and one axis (ξ) taken along the major axis and the other (η) parallel to the minor axis. If (ξ, η) are the coordinates of the secondary at some point in its orbit then

$$\xi = r\cos v, \qquad \eta = r\sin v, \tag{12.20}$$

and the equation of the true orbit in

$$\frac{(\xi + ae)^2}{a^2} + \frac{\eta^2}{b^2} = 1, \tag{12.21}$$

where a is the length of the semi-major axis, b is the length of the semi-minor axis and e is the eccentricity of the true orbit. From eqns. (12.19)

$$x = \xi l_1 + \eta l_2, \tag{12.22a}$$

$$y = \xi m_1 + \eta m_2. \tag{12.22b}$$

Solution of eqns. (12.22) for ξ, η gives

$$\xi = \frac{m_2 x - l_2 y}{l_1 m_2 - l_2 m_1}, \qquad \eta = -\frac{m_1 x - l_1 y}{l_1 m_2 - l_2 m_1}. \tag{12.23}$$

Use of eqns. (12.23) in eqn. (12.21) gives,

$$\begin{aligned} b^2\{m_2^2 x^2 + l_2^2 y^2 - 2l_2 m_2 xy + a^2 e^2 (l_1 m_2 - l_2 m_1)^2 \\ + 2ae(m_2 x - l_2 y)(l_1 m_2 - l_2 m_1)\} \\ + a^2(m_1^2 x^2 + l_1^2 y^2 - 2l_1 m_1 xy) = a^2 b^2 (l_1 m_2 - l_2 m_1)^2. \end{aligned} \tag{12.24}$$

Equating coefficients between eqns. (12.16) and (12.24) gives

$$\begin{aligned} A &= f(m_2^2 b^2 + m_1^2 a^2), \\ B &= f(l_2^2 b^2 + l_1^2 a^2), \\ H &= -f(l_2 m_2 b^2 + l_1 m_1 a^2), \\ F &= faeb^2 m_2(l_1 m_2 - l_2 m_1), \\ G &= -faeb^2 l_2(l_1 m_2 - l_2 m_1), \\ 1 &= -fa^2 b^2 (l_1 m_2 - l_2 m_1)^2 (1 - e^2), \end{aligned} \tag{12.25}$$

where f is a constant multiplier. Hence,

$$A = -\frac{m_1^2 + m_2^2(1 - e^2)}{P^2}, \qquad B = -\frac{l_1^2 + l_2^2(1 - e^2)}{P^2}, \tag{12.26}$$

$$H = \frac{l_1 m_1 + l_2 m_2(1 - e^2)}{P^2}, \qquad F = -\frac{em_2}{P}, \qquad G = \frac{el_2}{P},$$

where

$$P^2 = [a(1 - e^2)(l_1 m_2 - l_2 m_1)]^2. \tag{12.27}$$

By forming suitable combinations of A, B, F, G, H the elements of the orbit can be determined.

$$\left.\begin{aligned}
-(F^2 - G^2) + (A - B) &= \frac{l_1^2 - m_1^2 + l_2^2 - m_2^2}{P^2} \\
&= \frac{\cos 2\Omega \tan^2 i}{a^2(1 - e^2)^2}, \qquad (12.28a) \\
FG - H &= -\frac{l_1 m_1 + l_2 m_2}{P^2}, \\
&= -\frac{\sin 2\Omega \tan^2 i}{2a^2(1 - e^2)^2}. \qquad (12.28b)
\end{aligned}\right\} \text{using eqns. (12.19).}$$

Hence, from eqns. (12.28)

$$\tan 2\Omega = -2\frac{FG - H}{-(F^2 - G^2) + (A - B)}, \tag{12.29}$$

from which Ω may be found. Then from either of eqns. (12.28) $\tan^2 i/a^2(1 - e^2)^2$ may be evaluated. Again,

$$\begin{aligned}
(F^2 + G^2) - (A + B) &= \frac{l_1^2 + l_2^2 + m_1^2 + m_2^2}{P^2}, \\
&= \frac{2}{a^2(1 - e^2)^2} + \frac{\tan^2 i}{a^2(1 - e^2)^2},
\end{aligned} \tag{12.30}$$

which allows the determination $a^2(1 - e^2)^2$ since $\tan^2 i/a(1 - e^2)^2$ is already known. Once $a^2(1 - e^2)^2$ is known, $\tan i$ and so i can be determined. Also,

$$\left.\begin{aligned}
l_2 &= \frac{PG}{e} = \frac{Ga(1 - e^2)\cos i}{e}, \\
m_2 &= -\frac{PF}{e} = -\frac{Fa(1 - e^2)\cos i}{e},
\end{aligned}\right\} \text{using eqns. (12.19)} \tag{12.31}$$

Evaluation of l_2 and m_2 from eqns. (12.19) gives,

$$\sin \omega = \frac{a(1 - e^2)}{e}\cos i\,(F \sin \Omega - G \cos \Omega), \tag{12.32a}$$

$$\cos\omega\cos i = -\frac{a(1-e^2)}{e}\cos i\,(F\cos\Omega + G\sin\Omega), \quad (12.32b)$$

from which

$$\tan\omega = \frac{(G\cos\Omega - F\sin\Omega)\cos i}{G\sin\Omega + F\cos\Omega}. \quad (12.32c)$$

ω may be determined from eqn. (12.32c). When ω has been obtained, e can be found from either of eqns. (12.32a, b) since $a(1 - e^2)$ is known. Finally a can be found. The period T is known from observation, a, e, Ω, i, ω are determined so that only the time τ of periastron passage remains to be determined. The true anomaly v may be calculated for those points in which the apparent ellipse cuts the x- or y-axis. These points will be $(x_1, 0)$, $(x_2, 0)$, $(0, y_3)$, $(0, y_4)$. Then, from eqn. (12.22b)

$$\begin{aligned} O &= m_1 r_{1,2}\cos v_{1,2} + m_2 r_{1,2}\sin v_{1,2} \\ O &= l_1 r_{3,4}\cos v_{3,4} + l_2 r_{3,4}\sin v_{3,4}. \end{aligned} \quad (12.33)$$

Hence,

$$\tan v_{1,2} = -\frac{m_1}{m_2}, \qquad \tan v_{3,4} = -\frac{l_1}{l_2}. \quad (12.34)$$

Two pairs of values of $v_{1,2}, v_{2,4}$ are obtained from eqns. (12.34). From these values, the corresponding eccentric anomalies $E_{1,2,3,4}$ may be computed. Kepler's Equation is then used to give $M_{1,2,3,4}$. Since,

$$M_{1,2,3,4} = \frac{360}{T}(t_{1,2,3,4} - \tau) \quad (12.35)$$

and since the times $t_{1,2,3,4}$ at which the secondary is on the x- or y-axis can be interpolated, T and τ may be obtained. It is clear that the period T need not be known *a priori* and indeed, if an adequate part of the apparent orbit is known, the above method can be applied.

Kowalsky's method is rather cumbersome, involving continued evaluation of trigonometric functions. Nevertheless it illustrates the principles of determination of the elements of the orbit in a direct manner which demonstrates the relationship with the algebraic representation of the apparent orbit.

12.3.2 Thiele–Innes method. Kowalsky's method of determining the elements of the orbit of a visual binary is rarely used in practice. This stems, in part, from the difficulties of applying the method of least squares to isolate five parameters when using a hand calculator, and the continued evaluation of trigonometric functions. Although such calculations can be handled readily by an automatic computer, it is very laborious to undertake by hand and in consequence a somewhat different method is more commonly used in practice.

An apparent orbit which satisfies Kepler's Second Law must be obtained. In principle the errors of measurement in ρ should be larger than those in θ. If, therefore, a smooth curve is drawn for the variation of θ with time (after the elimination of observational points clearly seen to be erroneous) such a curve should represent the variation of θ fairly accurately. Again, a least squares fit for $k = \rho^2\dot{\theta}$ may be obtained and the values of ρ adjusted to the smooth curve for θ, t and the least squares value for k. A smooth curve should then be available for ρ as function of time. From the curves of (ρ, t), (θ, t) curves of (x, t), (y, t) can be constructed. The correction procedure also contains an element of experience to ensure that no absurdities are introduced.

The curves of (x, t), (y, t) are used in deriving the orbital elements. The constants l_1, l_2, m_1, m_2, n_1, n_2 as defined by eqn. (12.18) in section 12.3.1 if multiplied by a, the length of the semi-major axis of the true orbit, are the Thiele–Innes Constants P, Q, R, S, T, U where,

$$\begin{aligned} P &= al_1, \qquad Q = am_1, \qquad R = an_1, \\ S &= al_2, \qquad T = am_2, \qquad V = an_2. \end{aligned} \tag{12.36}$$

Then, using eqn. (12.36) in eqns. (12.14),

$$x = PX + SY, \qquad y = QX + TY, \qquad z = RX + UY, \tag{12.37a}$$

where

$$X = \frac{r}{a}\cos v, \qquad Y = \frac{r}{a}\sin v. \tag{12.37b}$$

The Thiele–Innes constants can be used to derive the values of the elements, as follows:

$$\begin{aligned} P + T &= a\cos(\omega + \Omega)(1 + \cos i), \\ P - T &= a\cos(\omega - \Omega)(1 - \cos i), \\ Q - S &= a\sin(\omega + \Omega)(1 + \cos i), \\ -Q - S &= a\sin(\omega - \Omega)(1 - \cos i), \end{aligned} \tag{12.38}$$

if use is made of eqns. (12.36) and (12.19). Clearly,

$$\begin{aligned} \tan(\omega + \Omega) &= \frac{Q - S}{P + T}, \\ \tan(\omega - \Omega) &= -\frac{Q + S}{P - T}, \end{aligned} \tag{12.39}$$

from which ω, Ω can be found and hence

$$\begin{aligned} a(1 + \cos i) &= (P + T)\sec(\omega + \Omega) = (Q - S)\operatorname{cosec}(\omega + \Omega), \\ a(1 - \cos i) &= (P - T)\sec(\omega - \Omega) \\ &= -(Q + S)\operatorname{cosec}(\omega - \Omega), \end{aligned} \tag{12.40}$$

or

$$\frac{1 - \cos i}{1 + \cos i} = \frac{(P - T)\cos(\omega + \Omega)}{(P + T)\cos(\omega - \Omega)} = -\frac{(Q + S)\sin(\omega + \Omega)}{(Q - S)\sin(\omega - \Omega)} \tag{12.41}$$

from which i may be determined. a may be determined from either of eqns. (12.40).

The Thiele–Innes Constants may be found directly from the (x, t), (y, t) curves. It is clear that the coordinates of the centre of the apparent orbit (x_c, y_c) are given by

$$x_c = \tfrac{1}{2}(x_{\max} + x_{\min}), \qquad y_c = \tfrac{1}{2}(y_{\max} + y_{\min}) \tag{12.42}$$

where $x_{\max}$ is the maximum value of x and so on. The maximum and minimum values of x, y may be determined directly from the curves (x, t), (y, t). Let periastron have coordinates (x_p, y_p) and apastron have coordinates (x_a, y_a). At periastron the coordinates of the secondary in the true orbit are

$$X = 1 - e, \qquad Y = 0, \tag{12.43a}$$

so that,

$$x_p = P(1 - e), \qquad y_p = Q(1 - e). \tag{12.43b}$$

At apastron the coordinates of the secondary in the true orbit are

$$X = -1 - e, \qquad Y = 0, \tag{12.44a}$$

so that,

$$x_a = -P(1 + e), \qquad y_a = -Q(1 + e). \tag{12.44b}$$

The coordinates of the centre of the apparent orbit are therefore also given by

$$x_c = \tfrac{1}{2}(x_p + x_a) = -Pe, \qquad y_c = \tfrac{1}{2}(y_p + y_a) = -Qe. \tag{12.45}$$

Since (x_c, y_c) are known from eqn. (12.42) Pe, Qe can be found. To make further progress the projections of periastron and apastron on the apparent orbit must be located. Periastron and apastron are separated by the half period of the system. Also (see eqn. (12.45))

$$\begin{aligned} x_p - x_c &= x_c - x_a, \\ y_p - y_c &= y_c - y_a. \end{aligned} \tag{12.46}$$

Curves $f = (x_t + x_{t+T^*/2} - 2x_c)$, $g = (y_t + y_{t+T^*/2} - 2y_c)$ are constructed where x_t, y_t are the values of x, y at time t and $x_{t+T^*/2}$, $y_{t+T^*/2}$ are the values of x, y at time $t + T^*/2$ where T^* is the period. In view of eqn. (12.46) the zeros of f, g locate τ and $\tau + T^*/2$. Periastron is selected as being that solution for which the secondary is on the same side of the centre as the primary. (x_p, y_p), (x_a, y_a) are then known, being the values of x, y giving the zeros of f, g. The eccentricity of the true orbit can then be found from eqns. (12.43b, 12.44b, 12.45) giving

$$e = \frac{x_c}{x_c - x_p} = \frac{x_c}{x_a - x_c} = \frac{y_c}{y_c - y_p} = \frac{y_c}{y_a - y_c}. \tag{12.47}$$

The Thiele–Innes constants P, Q may also be derived from eqns. (12.43b, 12.44b, 12.45) in the form

$$\begin{aligned} P &= x_p - x_c = x_c - x_a, \\ Q &= y_p - y_c = y_c - y_a. \end{aligned} \tag{12.48}$$

In order to find the constants S, T the positive end of the minor axis of the true orbit must be located. Let the projection of this point onto the apparent orbit have coordinates (x_m, y_m). The eccentric anomaly E is 90° when the secondary is at the end of the minor axis of the true orbit. From eqns. (10.37)

$$\begin{aligned} X(E = 90°) &= \frac{r}{a}\cos v = -e, \\ Y(E = 90°) &= \frac{r}{a}\sin v = (1 - e^2)^{\frac{1}{2}}. \end{aligned} \tag{12.49}$$

Then, from eqn. (12.37a),

$$\begin{aligned} x_m &= -Pe + S(1 - e^2)^{\frac{1}{2}}, \\ y_m &= -Qe + T(1 - e^2)^{\frac{1}{2}}. \end{aligned} \tag{12.50}$$

Then, substituting for Pe and Qe from eqns. (12.45),

$$\begin{aligned} S &= (1 - e^2)^{-\frac{1}{2}}(x_m - x_c), \\ T &= (1 - e^2)^{-\frac{1}{2}}(y_m - y_c). \end{aligned} \tag{12.51}$$

The point x_m, y_m may be located through use of Kepler's Eqn. (10.31). The mean anomaly M is given by,

$$M = \left(\frac{\pi}{2} - e\right) = \frac{2\pi}{T^*}(t_m - \tau),$$

or

$$(t_m - \tau) = \frac{T^*}{2\pi}\left(\frac{\pi}{2} - e\right), \tag{12.52}$$

where t_m is the time at which the secondary lies on the minor axis of the true orbit. The values of (x_m, y_m) may be read from the curves (x, t) (y, t) once t_m is known. Hence S, T may be found.

The constants R, V may be found from the equations,

$$P^2 + Q^2 + R^2 = a^2 = S^2 + T^2 + U^2 \tag{12.53}$$

and checked using

$$PS + QT + RU = 0. \tag{12.54}$$

Eqns. (12.53, 12.54) may be obtained directly from eqns. (12.19, 12.36).

Alternatively, once $(x_c, y_c), (x_p, y_p), (x_a, y_a)$ have been located e, τ can be determined. As the period T^* has been assumed known, Kepler's equations may be solved to find E at any time and hence X, Y calculated. Eqns. (12.37a) may then be solved by least squares to give the Thiele–Innes Constants P, Q, R, S, T, U directly. Having obtained the Thiele–Innes Constants i, ω, Ω may be computed from eqns. (12.39, 12.41). a can be found from eqns. (12.38, 12.40). Eqns. (12.53, 12.54) may then be used as checks.

The method using the Thiele–Innes constants has much to recommend it. Use of trigonometric functions is reduced but not eliminated. Clearly, graphical methods can be replaced by standard numerical techniques for locating maxima and minima, determination of zeros and interpolation. Although the method (i.e. smoothing the (x, t), (y, t) curves) of obtaining an apparent orbit may seem somewhat arbitrary, the method of finding τ and e gives scope for readjustment of the (x, t), (y, t) curves so that the two equations for τ give the same value as do the four equations for e. The Thiele–Innes method at all times allows reference back to the actual observation whereas, in Kowalsky's method, once the constants A, B, F, G, H have been found, all contact with the observations is lost. Location of the centre of the apparent orbit and the projected positions of periastron, apastron and the end of the positive minor axis of the true orbit permits a rapid means of finding the elements of the orbit. Final values of the elements may be determined using corrections to the adjusted orbit and least squares analysis to find the Thiele–Innes constants.

12.4 Spectroscopic Binaries

A spectroscopic binary is detected by a periodicity in the radial velocity with respect to the Sun of one or both components of a binary. Because the motion of individual stars is observed and not the motion of the secondary about the primary as in visual binaries, the detected motion is the motion of each individual star about the centre of mass of the pair. Under these circumstances Fig. 12.3 must be redrawn. Fig. 12.4 is similar to Fig. 12.3 except that C is the centre of mass of the binary system and S represents an individual

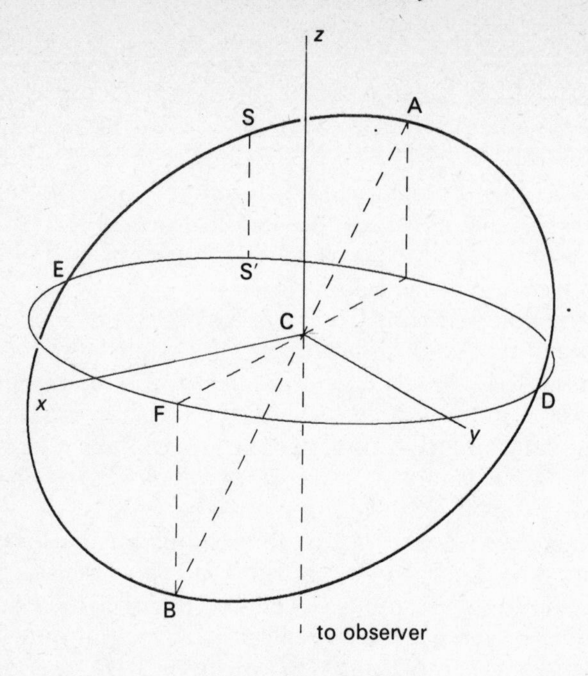

Fig. 12.4

star in motion about the centre of mass. For a spectroscopic binary the centre of mass is not observed. The radial velocity of S with respect to the Sun is determined. Clearly, since a north point of the apparent orbit cannot be obtained, the element Ω cannot be determined for a spectroscopic binary. As has been demonstrated previously neither a not i can be determined individually, but only the quantity $a \sin i$. The radial component of the orbital velocity of S is simply the rate of change of its z coordinate with time where z is as defined for visual binaries. If the centre of mass of the spectroscopic binary has radial velocity V_0 with respect to the Sun and the star S has a radial velocity dz/dt as a result of its orbital motion about the centre of mass of the system, the observed radial velocity V will be given by

$$V = V_0 + dz/dt. \qquad (12.55)$$

The z coordinate of S is given by,

$$z = r\cos v\,(\sin\omega\sin i) + r\sin v\,(\cos\omega\sin i), \qquad (12.14c)$$

as in the case of visual binaries. Then

$$\frac{dz}{dt} = \frac{dr}{dt}\sin i\,(\cos v\sin\omega + \sin v\cos\omega)$$

$$+ r\frac{dv}{dt}\sin i\,(-\sin v\sin\omega + \cos v\cos\omega). \qquad (12.56)$$

Differentiation of eqn. (10.19) gives

$$\frac{dr}{dt} = r\frac{dv}{dt}\cdot\frac{e\sin v}{1 + e\cos v} \qquad (12.57)$$

and eqn. (10.12) gives

$$r^2\frac{dv}{dt} = h = [\mu a(1 - e^2)]^{\frac{1}{2}},$$

so that

$$r\frac{dv}{dt} = na(1 - e^2)^{-\frac{1}{2}}(1 + e\cos v), \qquad (12.58)$$

where $n^2a^3 = \mu$ (eqn. 10.27). Then, use of eqn. (12.58) in eqn. (12.57) gives,

$$\frac{dr}{dt} = nae(1 - e^2)^{-\frac{1}{2}}\sin v. \qquad (12.59)$$

Use of eqns. (12.58, 12.59) in eqn. (12.56) gives,

$$\frac{dz}{dt} = K(\cos(v + \omega) + e\cos\omega), \qquad (12.60a)$$

where

$$K = na\sin i(1 - e^2)^{-\frac{1}{2}}. \qquad (12.60b)$$

Therefore the measured radial velocity V is given by

$$V = V_0 + K(\cos(v + \omega) + e\cos\omega). \qquad (12.61)$$

The elements $a\sin i$, e, τ, ω have to be derived from this equation

The period T of the binary can be determined directly from the observations. If the period is not obvious from a plot of radial velocity against time, well marked maxima and minima are sought. The interval between each pair of corresponding phases should be an integer multiple of the period. The interval is divided by integral values 1, 2, 3 . . . and the dividend common to all intervals between corresponding phases will be the period. Difficulty in establishing the period can be experienced but guidance can be found in the literature. It will be assumed that the period is known in subsequent discussion of eqn. (12.61).

12.4.1 A first approximation to the elements. An effective means of finding a first approximation to the elements of the orbit of a spectroscopic binary is to compare the observed radial velocity curve with a set of curves for known values of e and ω. The function

$$f = \cos(v + \omega) + e\cos\omega. \qquad (12.62)$$

is plotted for the complete range of values of v (from 0° to 360°) for each pair of e, ω values. The vertical and horizontal scales are chosen to be the same standard values to which the observed plot is also scaled. By visual inspection the observed curve is matched to one of the prepared plots of f. Approximate values of e, ω are then determined. Since the value of f for $v = 0$ is known from the prepared plot an approximate time τ of periastron passage can be derived for the observed star.

Again, the variation of radial velocity with time shows a maximum and a minimum. At maximum $v + \omega = 0$ and at minimum $v + \omega = 180°$. Let the measured velocities at these points be A and B respectively. Then

$$\begin{aligned} A &= V_0 + Ke\cos\omega + K, \\ B &= V_0 + Ke\cos\omega - K. \end{aligned} \qquad (12.63)$$

Clearly,

$$\begin{aligned} K &= \frac{A - B}{2}, \\ V_0 &= \frac{A + B}{2} - \frac{A - B}{2} e\cos\omega. \end{aligned} \qquad (12.64)$$

Since first estimates for e, ω are known, approximate values of K and V_0 can be determined. Since T is known, n is known and so $a \sin i$ can be determined.

This is a particularly expeditious means of finding a first approximation to the elements of a spectroscopic binary. If the same analysis were to be carried out by numerical means rather than graphical means a somewhat different procedure would need to be followed. In Fig. 12.5 a typical curve showing the variation of radial velocity with time is given.

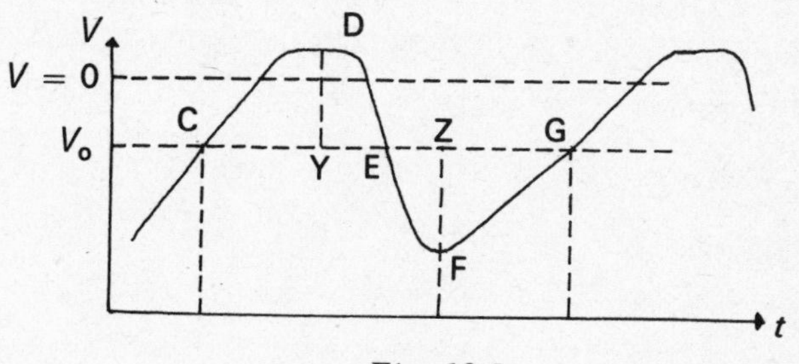

Fig. 12.5

The maximum value of V occurs at D and a minimum at F. Clearly the value of V at maximum and minimum can be interpolated numerically. Hence A and B can be found and K determined. Some method of finding V_0 must be devised. Suppose that a value for V_0 is assigned such that the line V_0, = constant, cuts the radial velocity curve in the three points C, E, G. The area bounded by the curve CDE and the line CE is given by,

$$\Delta_u = \int_{z_C}^{z_E} \frac{dz}{dt} dt = z_E - z_C \qquad (12.65a)$$

and the area bounded by the curve EFG and the line EG is given by

$$\Delta_l = \int_{z_E}^{z_G} \frac{dz}{dt} dt = z_G - z_E. \qquad (12.65b)$$

But since G is the same point as C but one period later $z_G = z_C$. Therefore, if the line defining V_0 is correctly placed,

$$\Delta_u = \Delta_l. \qquad (12.65c)$$

A set of base lines for V_0 are chosen and Δ_u, Δ_l measured for each base line, and a value of V_0 is interpolated which would make $\Delta_u = \Delta_l$. Clearly, a preliminary value of V_0 from the graphical method limits the range of choice for V_0.

Having determined a value for V_0, values for e, ω must be found. At D, $(v + \omega) = 0$ and so from eqn. (12.12), $z_D = 0$. Similarly $z_F = 0$. The area DYE, denoted by Δ_1, is then,

$$\Delta_1 = z_E - z_D = z_E = r_E \sin(v_E + \omega) \sin i \qquad (12.66a)$$

and the area FZG, denoted by Δ_2, is given by,

$$\Delta_2 = z_G - z_F = z_G = r_G \sin(v_G + \omega) \sin i. \qquad (12.66b)$$

But at E, G, $dz/dt = 0$ so that,

$$\cos(v_E + \omega) = \cos(v_G + \omega) = -e \cos\omega$$

$$= \frac{2V_0 - (A + B)}{A - B} = L. \qquad (12.67)$$

Then,

$$\sin(v_{E,G} + \omega) = \pm(1 - L^2)^{\frac{1}{2}}. \qquad (12.68)$$

In eqn. (12.68) the plus sign is associated with v_E and the minus sign with v_G. Again, taking only the magnitudes of the areas Δ_1, Δ_2,

$$\frac{\Delta_1}{\Delta_2} = \frac{r_E}{r_G} = \frac{1 + e\cos v_G}{1 + e\cos v_E} = \frac{1 + e\cos[(v_G + \omega) - \omega]}{1 + e\cos[(v_E + \omega) - \omega]},$$

$$= \frac{1 + e[L\cos\omega - (1 - L^2)^{\frac{1}{2}}\sin\omega]}{1 + e[L\cos\omega + (1 - L^2)^{\frac{1}{2}}\sin\omega]}, \qquad (12.69)$$

using eqn. (10.19). Hence,

$$e\sin\omega = \frac{\Delta_2 - \Delta_1}{\Delta_2 + \Delta_1}(1 - L^2)^{\frac{1}{2}}. \qquad (12.70)$$

Then, from eqns. (12.67, 12.70), e and ω can be obtained since Δ_1, Δ_2 can be derived from the observations.

12.4.2 Improvement of the elements. Clearly, the elements derived from a graphical comparison will contain errors, and the elements derived from numerical analysis of the raw data will also contain errors; it is desirable to be able to improve both first and second estimates. A computed form of the radial velocity variation with time can now be constructed. The difference (O − C) between the observed (O) and computed (C) radial velocities is found. Inspection of the form of (O − C) may allow a further adjustment of V_0, in which case the improved value of V_0 is adopted and a new comparison made between the observed and computed radial velocity curves. Eqn. (12.61) may be re-expressed in the form,

$$\begin{aligned} V &= f(V_0, K, e, \omega, v) \\ &= f(V_0, K, e, \omega, \tau, T), \end{aligned} \tag{12.71}$$

since v is a function of τ and T. An increment ΔV in V produces consequent increments ΔV_0 in V_0, ΔK in K, Δe in e, $\Delta\omega$ in ω, $\Delta\tau$ in τ and ΔT in T where,

$$\begin{aligned} \Delta V = {} & \frac{\partial f}{\partial V_0}\Delta V_0 + \frac{\partial f}{\partial K}\Delta K + \frac{\partial f}{\partial e}\Delta e + \frac{\partial f}{\partial \omega}\Delta\omega \\ & + \frac{\partial f}{\partial \tau}\Delta\tau + \frac{\partial f}{\partial T}\Delta T. \end{aligned} \tag{12.72}$$

Using eqn. (12.61), eqn. (12.72) may be evaluated to give

$$\begin{aligned} \Delta V = {} & \Delta V_0 + \Delta K(\cos(v + \omega) + e\cos\omega) \\ & - \Delta\omega K(\sin(v + \omega) + e\sin\omega) \\ & - \Delta v K\sin(v + \omega) + \Delta e K\cos\omega. \end{aligned} \tag{12.73}$$

But from eqns. (10.19, 10.34, 10.35, 10.36b, 10.37),

$$\begin{aligned} \frac{r}{a} &= \frac{1 - e^2}{1 + e\cos v} = 1 - e\cos E = \frac{(1 - e^2)^{\frac{1}{2}}\sin E}{\sin v} \\ &= \frac{\cos E - e}{\cos v} \end{aligned} \tag{12.74}$$

and Kepler's Equation gives,

$$E - e\sin E = n(t - \tau), \tag{10.31}$$

where $n = 2\pi/T$. The variation of v can then be found by taking derivatives with respect to e obtaining

$$\frac{\partial E}{\partial e} = \frac{1 + e\cos v}{1 - e^2}\left[\frac{\partial n}{\partial e}(t - \tau) - n\frac{\partial \tau}{\partial e} + \sin E\right], \tag{12.75}$$

from Kepler's Equation and

$$\frac{e\sin v\,(1 - e^2)}{(1 + e\cos v)^2}\frac{\partial v}{\partial e} = \frac{e(2 + e\cos v)}{(1 + e\cos v)^2} + \frac{\cos v}{(1 + e\cos v)^2} - \cos E + e\sin E\frac{\partial E}{\partial e}, \tag{12.76}$$

from eqn. (12.74). Using eqns. (12.74, 12.75) in eqn. (12.76) $\partial v/\partial e$ reduces to

$$\frac{\partial v}{\partial e} = \frac{\sin v\,(2 + e\cos v)}{1 - e^2} + \frac{(1 + e\cos v)^2}{(1 - e^2)^{\frac{3}{2}}} \times \left[\frac{\partial n}{\partial e}(t - \tau) - n\frac{\partial \tau}{\partial e}\right]. \tag{12.77}$$

Hence eqn. (12.73) becomes

$$\begin{aligned}\Delta V = \Delta V_0 &+ \Delta K\{\cos(v + \omega) + e\cos\omega\} \\ &- \Delta\omega K\{\sin(v + \omega) + e\sin\omega\} \\ &+ \Delta e K\left\{\cos\omega - \frac{\sin(v + \omega)\sin v\,(2 + e\cos v)}{1 - e^2}\right\} \\ &+ \frac{K(1 + e\cos v)^2\sin(v + \omega)}{(1 - e^2)^{\frac{3}{2}}}n\,\Delta\tau \\ &- \frac{K(1 + e\cos v)^2\sin(v + \omega)}{(1 - e^2)^{\frac{3}{2}}}(t - \tau)\,\Delta n.\end{aligned} \tag{12.78}$$

ΔV is (O − C) and v is evaluated using the preliminary elements using Kepler's Equation and eqn. (10.36b). For each observed point an estimate of (O − C) can be made and a set of equations of the form (12.78) obtained. Values of ΔV_0, ΔK, etc., can then be found by solution of this set of equations giving corrections to the elements. Since eqn. (12.72) is a first order Taylor approximation, it is necessary to find good estimates of the preliminary elements. The final elements can be checked as follows. With the "final" elements again compute the radial velocity curve and compute the "final" residuals. Subtract the preliminary residuals from the "final" residuals and add to eqn. (12.78) evaluated with respect to the final elements, using the known values for ΔV_0, ΔK, etc. The result should be nearly zero at every point. It is usually required that each sum shall not exceed 0·1 km s^{-1}. $a \sin i$ can be derived, once final values for e, K, n are available, from eqns. (12.60b, 12.64).

12.5 Eclipsing Binaries

Eclipsing binaries cannot be treated in the same way as visual or spectroscopic binaries. Accordingly, only a brief treatment will be given here. The information available is the observed light curve giving the variation of brightness of the binary with time. In principle the light curve of an eclipsing binary has two minima corresponding to the passage of the smaller star behind the larger and the passage of the smaller star across the larger. It frequently happens that one of the minima is undetectable. The interpretation of the light curve depends upon the nature of the light output by the stars of the pair. Limb darkening has to be taken into account as well as mutual gravitational distortion of the stars in a close pair, heating of the atmosphere of one star by the other and so on. In consequence the treatment of eclipsing binaries is not a simple geometrical problem and a realistic solution requires additional astrophysical assumptions. The treatment here will be restricted to a particularly simple case in order to illustrate the geometrical principles.

In Fig. 12.6 the variation of the light output of an eclipsing binary is illustrated. The deeper minimum is called the primary and corresponds to the eclipse of the star of greatest surface brightness. This star will be denoted by a subscript 1. The shallower minimum is called the secondary and corresponds to the eclipse of the star of lesser surface brightness. This star will be denoted by a subscript 2. The eclipse of the smaller by the larger star is called an occultation. The eclipse of the larger by the smaller star is called a transit. Eclipses can be total, partial or annular.

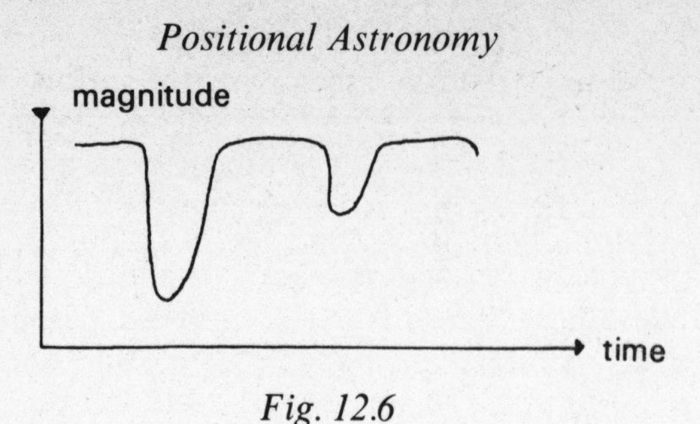

Fig. 12.6

12.5.1 A simple model of an eclipsing binary. A simple model of an eclipsing binary can be constructed. The smaller star moves around the larger in a circular orbit at a uniform rate. Both stars are assumed to be observed as circular uniformly illuminated discs (with no limit darkening). It will be assumed that a total eclipse of the smaller star occurs. These assumptions lead to a particularly straightforward model which allows a discussion of the geometrical situation without astrophysical complication. The orbital plane of the binary pair is assumed to be inclined at angle i to the line of sight. In Fig. 12.7 the plane of the circular orbit of the smaller star S about the larger star P is inclined at angle i to the line of sight. θ is measured uniformly in the direction of increasing θ from the point I on the orbit nearest the observer (inferior conjunction). Denoting the separation PS by D, the position of S in its orbit is then fixed by D and θ.

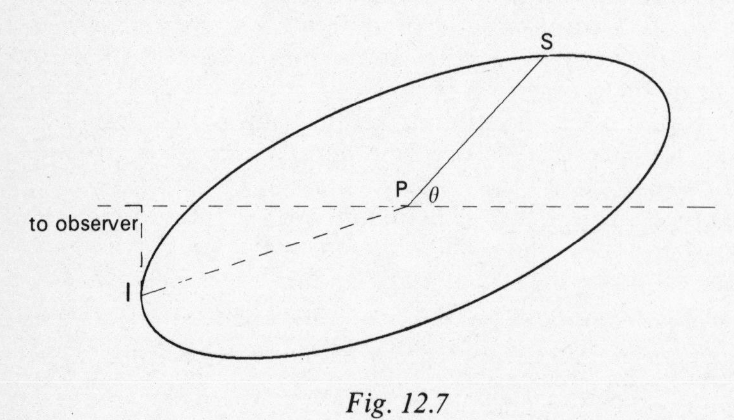

Fig. 12.7

Let the radius of the star P be r_1 and the radius of the star S be r_2 where $r_1 > r_2$. Let L_1 be the luminosity of the larger star and L_2 be the luminosity of the smaller star. L_1 and L_2 are assumed to be normalised such that

$$L_1 + L_2 = 1. \tag{12.79}$$

Eqn. (12.79) gives the luminosity of the pair when neither component is eclipsed. Fig. 12.8 represents an occultation of the smaller star by the larger. Let α be the fractional area of the smaller star which is occulted. The luminosity of the system is then

$$l_s = 1 - \alpha L_2. \tag{12.80}$$

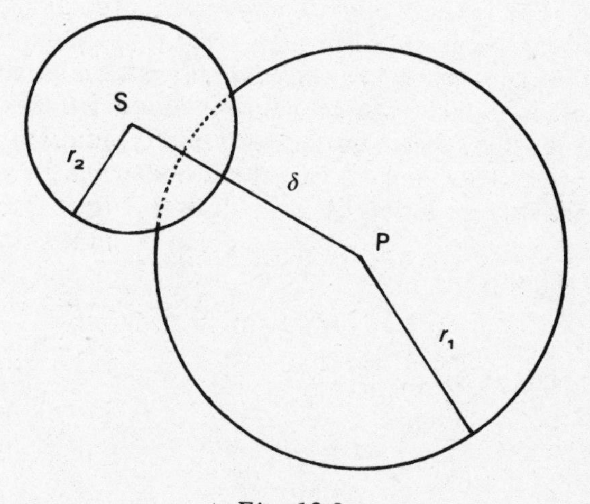

Fig. 12.8

Exactly one half period later the same area of the larger star will be obscured. The fractional area will be in proportion to the square of the ratio of the radii k so that the luminosity of the system will be

$$l_p = 1 - k^2 \alpha L_1 \tag{12.81}$$

where $k = r_2/r_1$.

If an eclipse is total the minima exhibit a constant value of the luminosity. If the eclipse is partial such a flattening in the minima is not observed. If the smaller star is totally hidden when occulted the subsequent transit will be annular. Let λ_s be the combined luminosity during the total occultation of the secondary and λ_p be the combined luminosity during transit then

$$\begin{aligned} \lambda_s &= 1 - L_2, \\ \lambda_p &= 1 - k^2 L_1, \end{aligned} \tag{12.82}$$

so that,

$$(1 - \lambda_s) + \frac{(1 - \lambda_p)}{k^2} = L_2 + L_1 = 1,$$

or

$$k^2 = \frac{1 - \lambda_p}{\lambda_s}. \tag{12.83}$$

If both minima are observed and each shows a flat bottom then the values of λ_p and λ_s can be determined and hence k can be found. If, on the other hand, the secondary minimum is not available but the primary is and the primary minimum has a flat bottom corresponding to a luminosity λ, then

$$\lambda \begin{cases} = 1 - L_2 & \text{for a total eclipse,} \\ = 1 - k^2 L_1 & \text{for an annular eclipse.} \end{cases} \tag{12.84}$$

Hence, from eqns. (12.80, 12.81)

$$\alpha = \frac{1 - l}{1 - \lambda}, \tag{12.85}$$

in either case where l is the luminosity of the binary at any stage during the eclipse. The corresponding value of α may be calculated from eqn. (12.85) knowing only l and λ. If there is no flattening of the minimum the value of α has to be treated as an unknown. However, in the present case it will be assumed that a flat part of a minimum is available. The time, corresponding to the line about which the primary minimum has symmetry, increased or decreased by half the known period, gives the instant of inferior conjunction. Having found the instant of inferior conjunction the value of θ at any instant within the period may be determined.

12.5.2 The determination of i and r_1. The position of the smaller star S with respect to the larger P is determined by θ, its position angle from inferior conjunction. Let δ be the separation of the centres of the two stars measured by a terrestrial observer were he able to resolve the binary into its component stars. Taking the orbit of the smaller star about the larger to have unit radius (i.e. $D \equiv 1$)

$$\delta^2 = D^2 \sin^2 \theta + D^2 \cos^2 \theta \cos^2 i = \cos^2 i + \sin^2 i \sin^2 \theta, \tag{12.86}$$

where i is the inclination of the plane of the orbit to the line of sight (displacements perpendicular to the line of sight are not altered but all distances parallel to the line of sight are foreshortened by a $\cos i$ factor). δ may also be written in the form

$$\delta = r_1 + pr_2, \tag{12.87}$$

where p is the fraction of r_2 which must be added to r_1 to equal δ, and r_1, r_2 are in units of D. Hence

$$\frac{\delta}{r_1} = 1 + p\frac{r_2}{r_1} = 1 + kp. \tag{12.88}$$

p is a function of α and k. Hence,

$$\cos^2 i + \sin^2 i \sin^2 \theta = r_1^2(1 + kp)^2 = r_1^2\phi^2(k, \alpha). \tag{12.89}$$

Values of p and ϕ can be determined for specified geometrical situations. Suppose values θ_1, θ_2, θ_3 corresponding to three values α_1, α_2, α_3 are selected from the light curve. Then,

$$\begin{aligned}
\cos^2 i + \sin^2 i \sin^2 \theta_1 &= r_1^2\phi^2(k, \alpha_1), \\
\cos^2 i + \sin^2 i \sin^2 \theta_2 &= r_1^2\phi^2(k, \alpha_2), \\
\cos^2 i + \sin^2 i \sin^2 \theta_3 &= r_1^2\phi^2(k, \alpha_3).
\end{aligned} \tag{12.90}$$

Elimination of $\cos^2 i$, $\sin^2 i$ using eqns. (12.90) gives

$$\frac{\sin^2 \theta_1 - \sin^2 \theta_2}{\sin^2 \theta_2 - \sin^2 \theta_3} = \frac{\phi^2(k, \alpha_1) - \phi^2(k, \alpha_2)}{\phi^2(k, \alpha_2) - \phi^2(k, \alpha_3)} = \psi(k, \alpha_1, \alpha_2, \alpha_3), \tag{12.91}$$

where ψ is a function of k, α_1, α_2, α_3. Since the instant of inferior conjunction can be determined the time scale for the light curve may be replaced by a θ scale. θ may then be assumed to be a known quantity. If A, B are defined by

$$A = \sin^2 \theta_2, \qquad B = \sin^2 \theta_2 - \sin^2 \theta_3 \qquad (12.92)$$

eqn. (12.91) may be written in the form

$$\sin^2 \theta = A + B\psi(k, \alpha), \qquad (12.93)$$

where the subscript 1 is omitted. If two values θ_2, θ_3 of θ are selected A, B may be determined from eqn. (12.92). The corresponding values of α, namely α_2, α_3 may then be determined from eqn. (12.85). For any other value of θ and the corresponding value of α, $\psi(k, \alpha)$ can be determined from eqn. (12.93). Since ψ is determined from ϕ which, in turn, is constructed from k, p, a tabulation of ψ as a function of k and α (since α is determined by p) can be calculated from a series of assumed values of k (in the range (0, 1)) and p (in the range $(-1, 1)$). If ψ and α are known from the observation, k can be determined by inverse interpolation among the tabular values of ψ.

At the instant when the eclipse begins $\theta = \theta'$, $\alpha = 0$, $\delta = r_1 + r_2$ and $p = 1$. At this instant

$$r_1^2(1 + k)^2 = \cos^2 i + \sin^2 i \sin^2 \theta', \qquad (12.94a)$$

or,

$$\sin^2 \theta' = \frac{r_1^2}{\sin^2 i}(1 + k)^2 - \cot^2 i. \qquad (12.94b)$$

When totality is just beginning $\theta = \theta''$, $\alpha = 1$, $\delta = r_1 - r_2$ and $p = -1$. At this instant

$$r_1^2(1 - k)^2 = \cos^2 i + \sin^2 i \sin^2 \theta'', \qquad (12.94c)$$

or,

$$\sin^2 \theta'' = \frac{r_1^2}{\sin^2 i}(1 - k)^2 - \cot^2 i. \qquad (12.94d)$$

Elimination of $\cot^2 i$ between eqns. (12.94a, c) gives

$$4kr_1^2 = \sin^2 i\{\sin^2 \theta' - \sin^2 \theta''\}, \qquad (12.95a)$$

and elimination of $r_1^2/\sin^2 i$ from eqns. (12.94b, d) gives

$$4k \cot^2 i = (1 - k)^2 \sin^2 \theta' - (1 + k)^2 \sin^2 \theta''. \quad (12.95b)$$

Use of eqn. (12.93) in eqns. (12.95a, b) gives

$$4k \cot^2 i = -4kA + B\{(1 - k)^2\psi(k, 0) - (1 + k)^2\psi(k, 1)\}, \quad (12.96a)$$

$$4kr_1^2 \operatorname{cosec}^2 i = B\{\psi(k, 0) - \psi(k, 1)\}. \quad (12.96b)$$

Since k is known, $\psi(k, 0)$, $\psi(k, 1)$ may be determined from the tabulation of ψ for the same values α_2, α_3 used to determine A, B (eqns. (12.92)). Eqn. (12.96a) is used to determine $\cot i$ and r_1 is then derived from eqn. (12.96b).

No guidance has been given on how a choice should be made of points on the light curve from which to determine A, B. Standard treatises on eclipsing binaries should be consulted on procedures for making the choice.

It is clear that an analysis of the light curve of an eclipsing binary gives i, k, r_1, on the basis of a simple geometrical model. No information on masses can be obtained though a more extensive analysis can give important astrophysical data. Studies of eclipsing binaries principally give direct information on relative stellar radii. It is therefore evident that a binary system which is both an eclipsing and spectroscopic binary could in principle give information simultaneously on both the masses and radii of the component stars.

Appendices

A Relations between Celestial Coordinate Systems

Equatorial coordinates are denoted by α, δ. Ecliptic coordinates are denoted by (λ, β). Horizon coordinates are denoted by (a, A). Hour angle is denoted by H, zenith distance by $z = 90° - a$ and sidereal time by S_t. ϕ is the astronomical latitude of the observer. Galactic coordinates of system II are denoted by l, b. Equations are given the same numbers as in the main text to facilitate reference to derivations.

1. Equatorial and horizon coordinates

$$\begin{aligned} \sin\delta &= \sin\phi\cos z + \cos\phi\sin z\cos A, \\ \cos\delta\cos H &= \cos\phi\cos z - \sin\phi\sin z\cos A, \\ \cos\delta\sin H &= \sin z\sin A, \\ \sin z\cos A &= \cos\phi\sin\delta - \sin\phi\cos\delta\cos H, \\ \cos z &= \sin\phi\sin\delta + \cos\phi\cos\delta\cos H. \end{aligned} \tag{3.10}$$

Sidereal time S_t is related to hour angle H and right ascension α by

$$S_t = H + \alpha \tag{3.3}$$

Convention: $0° \leqslant A \leqslant 180°$ measured East or West from the North Point of the Horizon.

2. Equatorial and ecliptic coordinates

$$
\begin{aligned}
\sin\delta &= \cos\varepsilon\sin\beta + \sin\varepsilon\cos\beta\sin\lambda,\\
\cos\delta\sin\alpha &= -\sin\varepsilon\sin\beta + \cos\varepsilon\cos\beta\sin\lambda,\\
\cos\delta\cos\alpha &= \cos\beta\cos\lambda,\\
\cos\beta\sin\lambda &= \sin\varepsilon\sin\delta + \cos\varepsilon\cos\delta\sin\alpha,\\
\sin\beta &= \cos\varepsilon\sin\delta - \sin\varepsilon\cos\delta\sin\alpha.
\end{aligned}
\tag{3.8}
$$

3. Equatorial and galactic coordinates

$$
\begin{aligned}
\sin\delta &= \sin b\cos I + \cos b\sin I\sin(l - 33°),\\
\cos\delta\sin(\alpha - 282°{\cdot}25) &= \cos b\cos I\sin(l - 33°)\\
&\quad - \sin b\sin I,\\
\cos\delta\cos(\alpha - 282°{\cdot}25) &= \cos b\cos(l - 33°),\\
\cos b\sin(l - 33°) &= \cos\delta\cos I\sin(\alpha - 282°{\cdot}25)\\
&\quad + \sin\delta\sin I,\\
\sin b &= \sin\delta\cos I - \cos\delta\sin I\\
&\quad \times \sin(\alpha - 282°{\cdot}25).
\end{aligned}
\tag{4.5}
$$

In reducing to or from Galactic Coordinates (α, δ) will refer to epoch 1950·0.

B Summary of Formula for Displacements of Position

The formulae are expressed either in rectangular equatorial coordinates (x, y, z) or equatorial coordinates (α, δ). S_t denotes sidereal time and H denotes hour angle. Where appropriate undisplaced coordinates are denoted by a subscript zero and displaced coordinates are denoted by a subscript one. Alternatively only the corrections are given. The equations are given the same numbers as appear in the relevant chapter.

1. Precession

$$\begin{aligned} x_1 &= x_0X_x + y_0Y_x + z_0Z_x, \\ y_1 &= x_0X_y + y_0Y_y + z_0Z_y, \\ z_1 &= x_0X_z + y_0Y_z + z_0Z_z, \end{aligned} \tag{7.14}$$

where,

$$\begin{aligned} X_x &= \cos\zeta_0\cos\theta\cos z - \sin\zeta_0\sin z, \\ Y_x &= -\sin\zeta_0\cos\theta\cos z - \cos\zeta_0\sin z, \\ Z_x &= -\sin\theta\cos z, \\ X_y &= \cos\zeta_0\cos\theta\sin z + \sin\zeta_0\cos z, \\ Y_y &= -\sin\zeta_0\cos\theta\sin z + \cos\zeta_0\cos z, \\ Z_y &= -\sin\theta\sin z, \\ X_z &= \cos\zeta_0\sin\theta, \\ Y_z &= -\sin\zeta_0\sin\theta, \\ Z_z &= \cos\theta, \end{aligned} \tag{7.15}$$

and

$$\begin{aligned} \zeta_0 &= (2304''{\cdot}250 + 1''{\cdot}396T_0)T + 0''{\cdot}302T^2 \\ &\quad + 0''{\cdot}018T^3, \\ z &= \zeta_0 + 0''{\cdot}791T^2, \\ \theta &= (2004''{\cdot}682 - 0''{\cdot}853T_0)T - 0''{\cdot}426T^2 \\ &\quad - 0''{\cdot}042T^3. \end{aligned} \tag{7.10}$$

In eqn. (7.10) the initial epoch is $1900{\cdot}0 + T_0$ and the final epoch is $1900{\cdot}0 + T_0 + T$.

2. Nutation

$$\begin{aligned} \mathrm{d}x &= -\Delta\psi(y\cos\varepsilon + z\sin\varepsilon), \\ \mathrm{d}y &= \Delta\psi x\cos\varepsilon - \Delta\varepsilon z, \\ \mathrm{d}z &= \Delta\psi x\sin\varepsilon + \Delta\varepsilon y, \end{aligned} \tag{7.23}$$

where $\Delta\psi$ is the nutation in longitude and $\Delta\varepsilon$ is the nutation in obliquity and ε is the obliquity of the ecliptic. The correction for nutation should be used to modify the expression for precession by matrix multiplication.

$$\begin{aligned} d\alpha &= \Delta\psi\{\cos\varepsilon + \sin\varepsilon \sin\alpha \tan\delta\} - \Delta\varepsilon \cos\alpha \tan\delta, \\ d\delta &= \Delta\psi \sin\varepsilon \cos\alpha + \Delta\varepsilon \sin\alpha. \end{aligned} \tag{7.38}$$

3. Approximate formulae for precession

$$\begin{aligned} dx &= -y_0(\psi' \cos\varepsilon_0 - \lambda') - z_0(\psi' \sin\varepsilon_0) \\ &= -my_0 - nz_0, \\ dy &= x_0(\psi' \cos\varepsilon_0 - \lambda') = mx_0, \\ dz &= x_0(\psi' \sin\varepsilon_0), = nx_0, \end{aligned} \tag{7.33}$$

where ψ' is the luni-solar precession in longitude, λ' is the planetary precession.

$$\begin{aligned} m &= \text{annual change in } (\zeta_0 + z) = \psi' \cos\varepsilon_0 - \lambda', \\ n &= \text{annual change in } \theta = \psi' \sin\varepsilon_0, \end{aligned} \tag{7.34}$$

where m is the annual general precession in right ascension and n is the annual general precession in declination.

$$\begin{aligned} d\alpha &= m + n \sin\alpha_0 \tan\delta_0, \\ d\delta &= n \cos\alpha_0, \end{aligned} \tag{7.29}$$

where $d\alpha$, $d\delta$ are interpreted as annual changes.

4. Aberration

$$dx = \frac{1}{c}\{-X'(1 - x^2) + Y'(xy) + Z'(xz)\}, \tag{6.96}$$

$$dy = \frac{1}{c}\{X'(xy) - Y'(1 - y^2) + Z'(yz)\}, \tag{6.97}$$

$$dz = \frac{1}{c}\{X'(xz) + Y'(yz) - Z'(1 - z^2)\}, \tag{6.98}$$

where c is the velocity of light, $-X'$, $-Y'$, $-Z'$ are the components of velocity of the Earth.

$$\mathrm{d}\alpha = \frac{\sec\delta}{c}(X'\sin\alpha - Y'\cos\alpha),$$
$$\mathrm{d}\delta = \frac{1}{c}(X'\cos\alpha\sin\delta + Y'\sin\alpha\sin\delta - Z'\cos\delta). \tag{6.99}$$

5. Day numbers

Day Numbers are used to give the combined correction for precession, nutation and aberration.

Bessels' day numbers

$$A = n\tau + n\frac{\Delta\psi}{\psi'}, \qquad B = -\Delta\varepsilon, \qquad C = -\frac{Y'}{c_*},$$
$$D = \frac{X'}{c_*}, \qquad E = \frac{\lambda'\Delta\psi}{\psi'}, \tag{7.43}$$

where τ is the fraction of the year elapsed since its beginning or the time until the start of the next year, c_* is the velocity of light. The displacements in terms of Besselian Day Numbers take the form:

$$\alpha = \alpha_0 + A\left(\frac{m}{n} + \sin\alpha_0\tan\delta_0\right) + B\cos\alpha_0\tan\delta_0$$
$$+ C\cos\alpha_0\sec\delta_0 + D\sin\alpha_0\sec\delta_0 + E, \tag{7.39, 7.41, 7.44}$$

$$\delta = \delta_0 + A\cos\alpha_0 - B\sin\alpha_0 + C(\tan\varepsilon_0\cos\delta_0$$
$$- \sin\alpha_0\sin\delta_0) + D\cos\alpha_0\sin\delta_0. \tag{7.40, 7.41, 7.44}$$

The multipliers of the Day Numbers are normally expressed as star constants.

Independent day numbers.

$$f = \frac{m}{n}A + E, \qquad g\sin G = B, \qquad g\cos G = A,$$
$$h\sin H = C, \qquad h\cos H = D, \qquad i = C\tan\varepsilon. \tag{7.45}$$

B Formulae for Displacements of Position

The displacement in terms of the Independent Day Numbers takes the form,

$$\begin{aligned} \alpha &= \alpha_0 + f + g \sin(G + \alpha_0) \tan \delta_0 \\ &\quad + h \sin(H + \alpha_0) \sec \delta_0, \\ \delta &= \delta_0 + g \cos(G + \alpha_0) + h \cos \\ &\quad \times (H + \alpha_0) \sin \delta_0 + i \cos \delta_0. \end{aligned} \tag{7.46}$$

6. Proper motion

$$\begin{aligned} x_1 &= x_0 \cos\{\mu(t_1 - t_0)\} \\ &\quad - \sin\{\mu(t_1 - t_0)\}(y_0 \sin \phi_0 + x_0 z_0 \cos \phi_0)/(1 - z_0^2)^{\frac{1}{2}}, \\ &\simeq x_0 + \mu_x(t_1 - t_0) - \tfrac{1}{2}\{\mu(t_1 - t_0)\}^2 x_0, \\ y_1 &= y_0 \cos\{\mu(t_1 - t_0)\} \\ &\quad + \sin\{\mu(t_1 - t_0)\}(x_0 \sin \phi_0 - y_0 z_0 \cos \phi_0)/(1 - {z_0}^2)^{\frac{1}{2}}, \\ &\simeq y_0 + \mu_y(t_1 - t_0) - \tfrac{1}{2}\{\mu(t_1 - t_0)\}^2 y_0, \\ z_1 &= z_0 \cos\{\mu(t_1 - t_0)\} \\ &\quad + \sin\{\mu(t_1 - t_0)\}(1 - z_0^2)^{\frac{1}{2}} \cos \phi_0, \\ &\simeq z_0 + \mu_z(t_1 - t_0) - \tfrac{1}{2}\{\mu(t_1 - t_0)\}^2 z_0, \end{aligned} \tag{7.53, 7.57}$$

where ϕ_0 is the angle between the great circle defining the direction of proper motion and the meridian through the star,

$$\begin{aligned} \mathrm{d}\alpha &= \mu_\alpha = \mu \sec \delta_0 \sin \phi_0, \\ \mathrm{d}\delta &= \mu_\delta = \mu \cos \phi_0. \end{aligned} \tag{7.56}$$

Proper motion at different epochs referred to same equator and equinox.

$$\begin{aligned} \mu'_\alpha &= \mu_\alpha + \left(2\mu_\alpha\mu_\delta \tan \delta_0 - \frac{2\mu_\alpha V_R}{r}\right)(t_1 - t_0), \\ \mu'_\delta &= \mu_\delta - \left(\mu_\alpha^2 \sin \delta_0 \cos \delta_0 + \frac{2\mu_\delta V_R}{r}\right)(t_1 - t_0), \end{aligned} \tag{7.64}$$

where V_R, r, are the radial velocity and distance respectively of the star.

Proper motion at the same instant referred to different equators and equinoxes.

$$\mu_\alpha'' = \mu_\alpha + n(t_1 - t_0)(\mu_\alpha \cos \alpha_0 \tan \delta_0 + \mu_\delta \sin \alpha_0 \sec^2 \delta_0), \tag{7.68}$$

$$\mu_\delta'' = \mu_\delta - n(t_1 - t_0)(\mu_\alpha \sin \alpha_0).$$

7. Stellar parallax

$$dx = \pi\{X(1 - x^2) - Y(xy) - Z(xz)\}, \tag{6.90}$$

$$dy = \pi\{-X(xy) + Y(1 - y^2) - Z(yz)\}, \tag{6.91}$$

$$dz = \pi\{-X(xz) - Y(yz) + Z(1 - z^2)\}, \tag{6.92}$$

where X, Y, Z are the geocentric rectangular equatorial co-ordinates of the Sun in units of the semi-major axis of the Earth's orbit.

$$d\alpha = \pi \sec \delta(-X \sin \alpha + Y \cos \alpha),$$

$$d\delta = \pi(-X \cos \alpha \sin \delta - Y \sin \alpha \sin \delta + Z \cos \delta). \tag{6.93}$$

8. Planetary parallax

These formulae apply except in the case of the Moon.

$$dx = \frac{\rho}{r_0}[xz \sin \phi' + \cos \phi'\{-(1 - x^2) \cos S_t + xy \sin S_t\}], \tag{6.121}$$

$$dy = \frac{\rho}{r_0}[yz \sin \phi' + \cos \phi'\{xy \cos S_t - (1 - y^2) \sin S_t\}], \tag{6.122}$$

$$dz = \frac{\rho}{r_0}[-(1 - z^2) \sin \phi' + \cos \phi'\{xz \cos S_t + yz \sin S_t\}], \tag{6.123}$$

where ρ is the geocentric distance of the observer, r_0 is the geocentric distance of the planet and ϕ' is the geocentric latitude of the observer.

B Formulae for Displacements of Position

$$\begin{aligned} d\alpha &= -\frac{\rho}{r_0}\cos\phi' \sec\delta \sin H, \\ d\delta &= -\frac{\rho}{r_0}(\sin\phi'\cos\delta - \cos\phi'\sin\delta\cos H). \end{aligned} \tag{6.124}$$

9. Refraction

$$dx = -k\left[\frac{xz\sin\phi + \cos\phi\{xy\sin S_t - (1-x^2)\cos S_t\}}{z\sin\phi + (x\cos S_t + y\sin S_t)\cos\phi}\right], \tag{6.84}$$

$$dy = +k\left[\frac{-yz\sin\phi + \cos\phi\{(1-y^2)\times\sin S_t - xy\cos S_t\}}{z\sin\phi + (x\cos S_t + y\sin S_t)\cos\phi}\right], \tag{6.85}$$

$$dz = +k\left[\frac{(1-z^2)\sin\phi - z(x\cos S_t + y\sin S_t)\cos\phi}{z\sin\phi + (x\cos S_t + y\sin S_t)\cos\phi}\right], \tag{6.86}$$

where k is the constant of refraction ($= 60''\!\cdot\!4$ at S.T.P.) and ϕ is the latitude of the observer. In eqns. (6.84, 6.85, 6.86) k should be expressed in radians.

$$\begin{aligned} d\alpha &= k\tan\zeta\sec\delta\sin\eta, \\ d\delta &= k\tan\zeta\cos\eta, \end{aligned} \tag{6.87}$$

where η is the parallactic angle. The formulae above should only be used for zenith distance $\zeta < 75°$.

10. Diurnal aberration

$$dx = -\frac{V_3}{c}\{xy\cos S_t + (1-x^2)\sin S_t\}, \tag{6.109}$$

$$dy = \frac{V_3}{c}\{(1-y^2)\cos S_t + xy\sin S_t\}, \tag{6.110}$$

$$dz = \frac{V_3}{c}z\{-y\cos S_t + x\sin S_t\}, \tag{6.111}$$

where c is the velocity of light and V_3 is the velocity of the observer as a result of the Earth's rotation.

$$\left.\begin{aligned} d\alpha &= \frac{V_3}{c}\cos H \sec \delta = 0''\cdot 32 \cos \phi \cos H \sec \delta, \\ d\delta &= \frac{V_3}{c}\sin H \sin \delta = 0''\cdot 32 \cos \phi \sin H \sin \delta. \end{aligned}\right. \qquad (6.112)$$

11. Motion under gravity

Notation:

m, M_*—the masses of gravitationally interacting mass points; M_* is at the focus of the elliptical orbit taken by m.

r—separation of the mass points.

v—true anomaly.

T—period required by m to make one complete transit of its elliptical orbit.

e—eccentricity.

τ—time at which m makes its nearest approach to M_*.

$M = \frac{2\pi}{T}(t - \tau)$—mean anomaly ($t$ is the time).

E—eccentric anomaly.

W—speed of mass point m in its orbit.

Formulae:

$$r^2\dot{v} = h \qquad (10.12)$$

$$\left(\frac{2\pi}{T}\right)^2 a^3 = \mu = G(M_* + m) \qquad (10.27)$$

$$r = \frac{h^2/\mu}{1 + e\cos v} = \frac{a(1 - e^2)}{1 + e\cos v} \qquad (10.19)$$

$$r \sin v = a(1 - e^2)^{\frac{1}{2}} \sin E \qquad (10.34, 10.37)$$

$$r \cos v = a(\cos E - e) \qquad (10.35)$$

$$\cos v = \frac{\cos E - e}{1 - e\cos E} \qquad (10.36b)$$

$$W^2 = \mu\left(\frac{2}{r} - \frac{1}{a}\right) \tag{10.23}$$

$$M = E - e\sin E - \text{Kepler's Equation} \tag{10.31}$$

C Useful Data

These data have been compiled either from the Explanatory Supplement to the Astronomical Ephemeris or from *Astrophysical Quantities* (3rd Ed.).

1. Miscellaneous

Gravitational constant	G =	$6{\cdot}670 \,.\, 10^8$ dyn cm^2 g^{-2}.
Velocity of light	c =	$2{\cdot}997\,925 \,.\, 10^{10}$ cm s^{-1}.
Astronomical unit	1 A.U. =	$1{\cdot}495\,979 \,.\, 10^{13}$ cm.
Parsec	1 pc =	206 264·8 A.U.
	=	$3{\cdot}0857 \,.\, 10^{18}$ cm,
	=	3·2616 ly.
Light year	1 l.y. =	$9{\cdot}4605 \,.\, 10^{17}$ cm
	=	$6{\cdot}324 \,.\, 10^4$ A.U.
Constant of aberration	κ =	20″·47 (adopted value).

2. The Sun

Solar mass	$M_\odot$ =	$1{\cdot}989 \,.\, 10^{33}$ g.
Solar radius	$R_\odot$ =	$6{\cdot}960 \,.\, 10^{10}$ cm.
Solar luminosity	$L_\odot$ =	$3{\cdot}826 \,.\, 10^{33}$ erg s^{-1}.
Semi diameter at mean distance	$S_\odot$ =	15′59″ . 63 = 959″ . 63
Inclination of Solar Equation to ecliptic		7°15′
Period of sidereal rotation (for heliographic coordinates)		25·38 days.
Longitude of ascending node	Ω =	73°40′ + 50″·25t, t in years from 1850·0.
Solar parallax		8″·80 (adopted value).

3. The Earth

Earth mass	$M_\oplus$ =	$5{\cdot}976 \,.\, 10^{27}$ g
Polar radius		6356·912 km (adopted value).

Equatorial radius		6378·388 km (adopted value).
Flattening	f	= 1/297 (adopted value).

4. *The Moon*

Lunar mass	$M_{☾}$	$= 7{\cdot}350 \,.\, 10^{25}$ g.
Lunar radius	$R_{☾}$	= 1738·2 km.
Semi diameter at mean distance		15′32″·6 (geocentric).
Mean equatorial horizontal parallax		3422″·54.
Mean distance from Earth		384 400 km (range 356 400 to 406 700 km).
Inclination of orbit to ecliptic		5°8′43″.

5. *Time*

24^{h} Mean Solar Time = $24^{h}03^{m}56^{s}{\cdot}56$ Mean Sidereal Time.

24^{h} Mean Sidereal Time = $23^{h}56^{m}04^{s}{\cdot}09$ Mean Solar Time.

1 mean solar second ≡ 1 ephemeris second (to one part in 10^{8}).

1 ephemeris second = 1/31 556 925·975 of tropical year for 1900·0.

6. *Length of year*

	days
Julian	365·25
Tropical (equinox to equinox)	365·2422
Sidereal (fixed star to fixed star)	365·2564
Anomalistic (perihelion to perihelion)	365·2596
Eclipse (lunar node to lunar node)	346·6201

7. *Length of the month*

Synodic (new moon to new moon)	29·5306
Tropical (equinox to equinox)	27·3216
Sidereal (fixed star to fixed star)	27·3217
Anomalistic (perigee to perigee)	27·5546
Draconic (node to node)	27·2122

8. *Besselian Year (B.Y.)*

B.Y.	J.D.
1900·0	2 415 020·313
1950·0	2 433 282·423
1970·0	2 440 587·267
1975·0	2 442 413·478
1980·0	2 444 239·689

J.D. = 2 433 282·423 + 365·2422 (B.Y. = 1950·0)

9. *Planetary Data*

Planet	Semi-diameter at Unit distance (″)	Semi-major axis of orbit (AU)	Semi-major axis of orbit (10^6 km)	Sidereal Period (tropical years)	Mean Daily Motion $n(°)$
Mercury	3·37	0·3871	57·9	0·240 85	4·092 339
Venus	8·46	0·7233	108·2	0·615 21	1·602 131
Earth	8·80	1·0000	149·6	1·000 04	0·985 609
Mars	4·68	1·5237	227·9	1·880 89	0·524 033
Jupiter	98·37	5·2028	778·3	11·862 23	0·083 091
Saturn	82·8	9·5388	1427·0	29·457 7	0·033 460
Uranus	32·9	19·1819	2869·6	84·0139	0·011 732
Neptune	31·1	30·0578	4496·6	164·793	0·005 981
Pluto	4·1	39·44	5900	247·7	0·003 979

Planet	Inclination i	Eccentricity e
Mercury	7°00′10″·6 + 6″·3T	0·205 615 + 0·000 020T
Venus	3°23′37″·1 + 3″·6T	0·006 820 − 0·000 048T
Earth		0·016 750 − 0·000 042T
Mars	1°51′01″·1 − 2″·3T	0·093 312 + 0·000 094T
Jupiter	1°18′31″·4 − 20″·5T	0·048 332 + 0·000 164T
Saturn	2°29′33·1″ − 14″T	0·055 890 − 0·000 345T
Uranus	0°46′21″ + 3″T	0·0471
Neptune	1°46′45″ − 34″T	0·0085
Pluto	17°10′	0·2494

The epoch for i, e, Ω, ω is 1900·0 and T is the time in centuries after 1900·0.

	Mean Longitude		
	of node	of Perihelion	of Planet at 1950 Jan. 0·5
Planet	Ω	ω	
Mercury	47°08′43″ + 4266″T	75°53′54″ + 5596″T	33°10′06″
Venus	75°46′50″ + 3240″T	130°09′51″ + 5063″T	81°34′19″
Earth		101°13′11″ + 6180″T	99°35′18″
Mars	48°47′12″ + 2786″T	334°13′6″ + 6626″T	144°20′07″
Jupiter	99°26′20″ + 3640″T	12°43′ + 5800″T	316°09′34″
Saturn	112°47′10″ + 3140″T	91°05′ + 7050″T	158°18′13″
Uranus	73°28′50″ + 1800″T	169°03′ + 5800″T	98°18′31″
Neptune	130″40′50″ + 3950″T	43°52′ + 2400″T	194°57′08″
Pluto	109°00′ + 5000″T	224°	165°36′09″

10. *Galactic coordinates*

Equatorial Coordinates of Galactic Pole:

$$\alpha_G = 12^h 49^m{\cdot}0, \quad \delta_G = +27°24'. \tag{4.2}$$

Equatorial Coordinates of Galactic Centre:

$$\alpha_M = 17^h 42^m{\cdot}4, \quad \delta_M = -28°55'. \tag{4.1}$$

Inclination of Galactic Plane to Equator: $I = 90° - \delta_G = 62°36'$.

The above values refer to parameters defining Galactic Coordinates of System II.

The equatorial coordinates refer to the equator and epoch of 1950·0.

11. *Precessional constants*

Reference should be made to Table 7.1, eqn. (7.10) of Chapter VII.

12. *I.A.U. systems of constants*

The I.A.U. has given (Trans. IAU, XIIB, 594, 1966) a list of fundamental constants which differ from some of the values listed above. This reflects recent attempts to improve fundamental values and to achieve self consistency. The I.A.U. alternatives to values given above are the following.

Astronomical unit	$1{\cdot}49600.10^{13}$ cm.
Constant of aberration	$20''{\cdot}496$
Solar parallax	$8''{\cdot}794$
Equatorial radius of Earth	6378·160 km
Flattening (Earth)	1/298·25.

Index

Page numbers in *italic* indicate major derivation, section or definition.

Index

Index